Applied Industrial Energy and Environmental Management

Applied Industrial Energy and Environmental Management

Zoran K. Morvay

UNDP, Croatia

Dušan D. Gvozdenac

Faculty of Technical Sciences, University of Novi Sad, Serbia

A John Wiley and Sons, Ltd, Publication

This edition first published 2008
© 2008 John Wiley & Sons Ltd

Registered office
John Wiley & Sons Ltd, The Atrium, Southern Gate, Chichester, West Sussex,
PO19 8SQ, United Kingdom

For details of our global editorial offices, for customer services and for information
about how to apply for permission to reuse the copyright material in this book please see our website
at www.wiley.com.

Wiley also publishes its books in a variety of electronic formats. Some content that appears in print
may not be available in electronic books.

Designations used by companies to distinguish their products are often claimed as trademarks. All brand names
and product names used in this book are trade names, service marks, trademarks or registered trademarks of
their respective owners. The publisher is not associated with any product or vendor mentioned in this book.
This publication is designed to provide accurate and authoritative information in regard to the subject matter
covered. It is sold on the understanding that the publisher is not engaged in rendering professional services.
If professional advice or other expert assistance is required, the services of a competent professional should
be sought.

Library of Congress Cataloging-in-Publication Data

Morvay, Zoran.
  Applied industrial energy and environmental management / Zoran Morvay,
  Dušan D. Gvozdenac.
    p. cm.
  Includes bibliographical references and index.
  ISBN 978-0-470-69742-9 (cloth)
  1. Factories—Energy conservation.   2. Industries—Energy conservation.
  3. Environmental protection.   I. Gvozdenac, Dušan D.   II. Title.
  TJ163.5.F3.M67 2008
  658.2′6—dc22

                                           2008032217

A catalogue record for this book is available from the British Library.

ISBN: 978-0-470-69742-9 (H/B)

Set in 9/11pt Times by Integra Software Services Pvt. Ltd, Pondicherry, India
Printed in Great Britain by CPI Antony Rowe, Chippenham, Wiltshire

To Jehona, Boran and Ljiljana
They have let me borrow the time to prepare this book!
Z. Morvay

To Nevena, Nenad and Branka,
For their understanding
D. Gvozdenac

Contents

About the Authors

Dr. Z.K. Morvay, holds a PhD in electrical engineering and has more than 20 years of teaching experience at universities in Croatia, the UK and Thailand. During the early 1990s he worked for a UK energy and environmental consultancy firm, managing projects for international development banks and aid organizations, technical assistance projects for governments and consultancy projects and feasibility studies for industry. He moved to Thailand to manage the local office just before the economic crisis in South East Asia in 1997. He developed, designed and managed the implementation of a number of energy and environmental management projects for large industrial international companies from Japan, the USA, the UK and Thailand, which often evolved into cost management projects because the theme of the day at that time was cost cutting. In addition, he has practical experience of change management, training needs assessment and support for organizational learning. He is the author of a number of professional papers and of a handbook and manuals on energy and environmental subjects. Currently, he manages a national energy efficiency project in Croatia.

Dr. D.D. Gvozdenac is Professor and Chairman of Thermal Engineering at the University of Novi Sad, Serbia. He holds a PhD in mechanical engineering and has more than 20 years of teaching experience in Serbia and Thailand. This is complemented by having spent the last 15 years as an international consultant working on energy projects in Europe, Africa, South East Asia and China. He was also the director of a national energy efficiency agency and a regional energy efficiency centre. He has managed a number of energy audits in industry, solving the problems of energy performance improvement and designing and managing the implementation of solutions. He has also designed and delivered training courses for the operators of industrial energy and utility systems, and has developed best practice and preventive maintenance manuals for industrial equipment and machines. He is the author of a number of professional papers, editor of conference proceedings and author of training manuals. Currently, he is working on the development of a national cogeneration project in Thailand.

Preface

Energy and environmental management starts with people!

Owing to concern with regard to globalization and climate change, the rules of business are changing fast and frequently. Each day seems to bring new technology, a new partner or competitor, or a new way of working. Against such a dynamic and complex backdrop, some things remain constant: a need for the developed and developing world to manage their energy resources and impacts on the environment and to improve energy efficiency continuously and to reduce harmful emissions. Companies operating in either world are required to achieve continual improvement of their energy and environmental performance.

The motivation for this book was initiated by years of experience, firstly of teaching and training energy and environmental management subjects and then by working internationally as consultants for development banks, international aid organizations, large multinational companies, as well as for small local businesses. This experience has evolved over the last two decades through work on energy efficiency and environmental projects in Europe and Asia. The last decade was marked by economic crises and market meltdown in South East Asia, the liberalization of energy markets, industrial globalization, a surging appetite for energy in the developing world, increasing and constantly high energy prices, energy security concerns, a growing awareness of climate change, the introduction of emissions trading, green financing and voluntary and compulsory arrangements designed to combat climate change.

The second half of the 1990s was a particularly instructive period when we were working in South East Asia at a time of well-publicized economical upheaval. This has given us inside experience of the impact of a sudden economical downturn on business performance in general and energy performance in particular.

When businesses slow down, cost cutting becomes imperative and energy costs and energy and environmental performance suddenly attract greater interest from management. Although energy efficiency and environmental management were supposed to be familiar terms at that time, the implementation of a successful energy and environmental management program in a real industrial environment still proved to be a difficult task for any team assigned to it. Its complexity arose from the need to bring together *people*, *procedures* and *technologies* in order to achieve consistent and lasting performance improvement.

The context of energy and environmental management is changing fast, but the basic principles and knowledge that have been essential for years, are still relevant today. The underlying know-how remains stable and valid and provides a solid grounding that needs to be refined from time to time in order to reflect the impact of the changes that occur. This generates a need for continuous learning and creation of new knowledge in order to cope with the challenges of an ever changing business environment.

One of the most useful lessons that we have learned from the years spent working with people is about the range and breadth of issues of which they are not aware, do not understand or are not confident about, which in turn hinders them in implementing energy and environmental management programs. These issues range from queries about simple technical details or how to do some calculations, on up to missing the entire concept or context for energy and environmental management. But the questions are always about the practical aspects of evaluation of current performance or implementation of some improvement measure, rather than about the theoretical foundations.

At the top management level, energy performance is often considered to be a more technical than management topic and the cross-functional character of energy and environmental management has not been sufficiently appreciated. Consequently, there is no management structure to deal with energy performance continuously and consistently. Energy and environmental management has of course a strong technical component, but if the focus is on technical aspects only, the results will be limited. Energy and environmental management needs to *focus on people* because technical expertise and sophisticated equipment will fail to produce results unless people are committed and receptive to the changes needed for performance improvement.

Normally, people involved in energy and environmental management are from various educational backgrounds and from all levels of management. What they have in common is that they are assigned responsibilities to operate and manage an *existing plant* efficiently!

Working with people with diverse backgrounds united by common goal has always presented the challenge of finding the most effective way of helping them to understand and implement Energy and Environmental Management System (EEMS) procedures and techniques. We have been searching for an important element of knowledge in the vast amount of technical and managerial know-how that will have the so called 'Ah-hah . . .' effect, i.e., an element that will trigger understanding and that will result in practicable knowledge. In other words, it has been a search for the appropriate level and content of knowledge relevant to a given application context.

Figure 1 illustrates the basic idea that the required knowledge in an industrial environment will be less detailed but more conceptual or general as we move from the technical to the management level. The opposite works for an academic environment where the higher up the pyramid one moves, the more specialized is the knowledge that is required. Of course, there are always links and feedback between

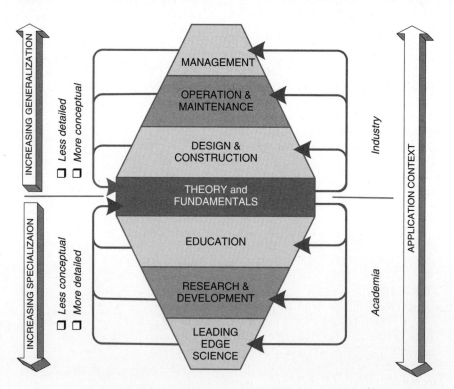

Figure 1   Knowledge levels for a given application context

the different levels that hold the body of knowledge together and stimulate its growth in both the applied and scientific areas. In industrial energy and environmental management, the focus is not on theory and fundamentals, not even on design considerations, but on *managing performance of an existing plant*. As one moves up the managerial ladder, the required knowledge will shift from the more detailed to the more conceptual.

Many environmental management issues in industry are closely related to energy use and therefore are also covered in the book from the perspective of performance management. This approach is based on the fact that the environmental impacts of industrial operations are the consequences of energy use and the processing of raw materials. Hence, improving the efficiency of energy and raw material use, followed by waste avoidance and minimization, are the logical first steps and most cost effective means of environmental management.

Successful EEMS must provide a framework not only for managing companies' energy and environmental affairs, but also for enabling and stimulating continuous learning, creation and management of knowledge, which is the key asset in maintaining a company's competitiveness. Companies that can learn and manage their knowledge effectively are the foundations of the knowledge economy.

The focus on people and performance management of existing plants is the guiding principle for establishing the scope, depth and style of presentation in this book. Therefore, in its scope, the book will cover the management aspects of energy and environmental performance improvement in Part I and the technical aspects in Part II.

Regarding the depth of considerations, the book does not develop detailed principles and theories, but rather stresses the application of these theories to specific performance improvement solutions. If we consider the required knowledge level related to an application context (Fig. 1), then the book aims at providing support to practitioners in the field involved with developing and implementing solutions for industrial energy and environmental management problems. However, anybody who has completed a basic management or engineering course will also find this book to be a useful guide in resolving practical operational efficiency improvement problems.

We have resisted the temptation to make the book an energy management encyclopedia. Therefore, a number of relevant subjects (lighting, heating, degree days, industrial water systems, drying process, environmental abatement technologies) are deliberately omitted, but references are given to other books where these subjects are well covered. Our aim is to focus on the management aspects and technical issues which are more specific to an industrial environment.

In its structure, the book acknowledges the fact that energy and environmental management are highly interdisciplinary subjects; hence few people will attempt to read it from cover to cover with the same level of interest and attention. Therefore, it consists of three parts and some modules that can be studied independently.

The book starts with the introduction of a framework for energy and environmental management in industry. It is followed by Part I which explains the context and concept of energy and environmental management, focusing on practical methods for performance assessment, evaluation and improvement at industrial plants. It also recognizes the need for an adequate understanding of the production processes which are the main users of resources in a company. It follows a systematic approach to identify the significant internal and external interactions and key factors affecting performance within a company. It emphasizes the importance of and gives guidance for dealing with people, i.e., management aspects, as a prerequisite for lasting energy and environmental performance improvement and it proceeds with practical guidelines on how to introduce energy and environmental management systems in an industrial environment. Part I is concluded by a proposal for an integrated performance management approach that can be easily built onto an EEMS when established and operational.

Part II deals with the technical aspects of industrial energy and environmental performance management, starting with a definition of industrial energy systems. This is followed by the analysis of five industrial energy systems: (1) Industrial Steam System, (2) Electrical Energy System, (3) Compressed Air System, (4) Refrigeration System and (5) Industrial Cogeneration. Industrial cogeneration as a system differs from the classical understanding of Industrial Energy Systems. Today, the trend of using cogeneration in industry, as a means of improving the overall performance of energy use, is very significant and for that reason it is considered separately. Each chapter contains a section on the environmental impacts of the considered industrial energy system.

Part III is a special feature of this book. It includes a 'TOOLBOX' and a set of software to be found on the accompanying website at www.wiley.com/go/morvayindustrial. It contains many useful 'tools' for the implementation of energy and environmental management programs. These 'tools' are: data on materials' and fuels' properties, analytical methods, definitions, procedures, questionnaires and software that support the practical application of the methods elaborated in the first two parts of the book. The software we have developed and used during our consulting practice facilitates the calculations and problem solving for specific solutions for energy and environmental performance improvement, financial evaluation and performance monitoring. It will help readers to understand the relevant topics and assess their own industrial energy systems. The context of why, when and how to use the tools and software is provided consistently throughout the book but, again, a knowledgeable reader should be able to use the tools directly without referring to the book. The TOOLBOX also contains definitions of terms or basic explanations of the subjects that are used in the book but which could divert the reader's attention if explained in the main text.

This book embodies our philosophies and consulting experience in developing and applying energy and environmental management solutions for performance improvement in industrial plants. As with our endeavors in training and implementing specific solutions, where we have also learned from the very people we have trained or assisted, we welcome comments and suggestions, in this case from our readers.

In closing, we express our appreciation to Luka Lugaric and Aleksandar Vukojević who have edited the text and illustrations and have made the software more user friendly.

Introductory Chapter

Framework for Energy and Environmental Management in Industry

1 Introduction

The consumption of energy is increasing at a fast pace while available resources remain limited. The global need for energy is increasing by more than 2 % a year. This may not sound like very much, but it represents almost a doubling of energy consumption over a period of 30 years.

Out of the total amount of primary energy (Box 1), around 80 % comes from fossil fuels. The current consumption of fossil fuels, particularly oil, is not sustainable in the long term. Over 3.5 billion tons of crude oil are consumed every year. A major part of this is consumed by the transport sector, with the industrial sector as the second largest oil consumer.

Energy consumption has a significant impact on our natural environment. Recently, a consensus has arisen that there is clear evidence that climate change is caused by human activity, mostly related to the use of energy.

Box 1: Types of Energy

> **Primary Energy** *refers to all types of energy extracted or captured directly from natural resources. Primary energy can be further divided into two distinctive groups:*
>
> - *Renewable (solar, wind, geothermal, tidal, biomass);*
> - *Non-renewable (fossil fuels: crude oil, coal, natural gas, oil shale, etc.)*
>
> *The primary energy content is generally referred to as **toe** (ton of oil equivalent). The primary energy content of all fuels can be converted to **toe** on the basis of the conversion factors:*
>
> $$1\,toe = 11\,630\,kWh = 41\,870\,MJ.$$
>
> **Intermediate energy** *refers to forms of energy that are produced by converting primary energy into other forms.*
> **Final energy** *refers to the form of energy that consumers buy or receive in order to perform desired activities. Industry uses final energy for various **services** and **utilities** such as electrical drives, cooling, transport, lighting, production of compressed air, production of heat for own use or for sale, etc. (Fig. 1).*

The recent and sustained increase in oil prices during 2007 has reminded us again that all but renewable energy sources are depletable, hence energy prices will continue to grow as energy reserves shrink.

Applied Industrial Energy and Environmental Management   Zoran K. Morvay and Dušan D. Gvozdenac
© 2008 John Wiley & Sons, Ltd

It is common knowledge that a more efficient use of energy and resources is a relatively quick and painless way to reduce a company's costs and lessen the harmful impact on the environment. However, until recently, relatively few companies paid sufficient attention to these issues, and particularly at the board level, energy and environmental management were not given a great deal of prominence.

Governments around the world have therefore responded by introducing various voluntary and mandatory energy and environmental policies that are changing the business climate and the regulatory environment in which companies operate. There is also growing public pressure and an expectation that businesses should be carried out in a more environmentally responsible manner, and that resources need to be used more efficiently.

Hence, energy and environmental management has become a strategic issue that has to be dealt with at the highest level within an organization. An increasing number of companies understand that they have social responsibilities to the communities in which their businesses operate, and they are trying to respond with appropriate energy and environmental corporate strategies.

Formulating a strategy is a fairly easy task, but implementing it and achieving results is difficult. Normally, energy and environmental strategies are formulated at the top of an organization, but implementation involves everyone because energy is used in every corner of a business. The need to involve various internal functional units in strategy implementation makes the job complex enough, which is then aggravated further by the external compliance requirements that the organization is subject to. Therefore, it comes as no surprise when companies embark on ambitious energy and environmental performance improvement programs the results are often disappointing.

The success depends on understanding organizations' manufacturing operations, the internal and external context for strategy implementation, how energy is used and where, and what the sources of significant environmental impacts are. Based on such an understanding, the critical part is capturing the specific knowledge necessary for converting a strategy from a raw concept into a detailed plan, and ultimately into results.

The rest of this chapter will describe briefly the general characteristics of energy use in industry, the environmental impacts of industrial operations, and highlight the most important aspects of external energy and environmental policy and the regulatory framework relevant to energy and environmental management in industry.

2 Energy Use by Industrial Operations

The energy and manufacturing sectors consume by far the bulk of primary energy in the industrial sector. Figure 1 shows the various conversion steps, from extracting primary energy to transforming it into useful final energy services.

A variety of fuels is used in industry. Oil and natural gas by far exceed energy consumption from any other source in the industrial sector. Industrial operations use the energy in three major ways:

- to produce heat and generate electricity;
- to produce utilities like compressed air, chilled water, hot water, etc.
- as raw material input to the production process.

The extent, to which a process, a company, or even a country consumes energy, is described by its energy intensity. The energy intensity is statistical concept which is defined as energy consumption per unit of output at different levels of aggregation. At the national level, the energy intensity is measured as a ratio between total final consumption and gross domestic product:

$$EI = \frac{FC}{GDP} \tag{1}$$

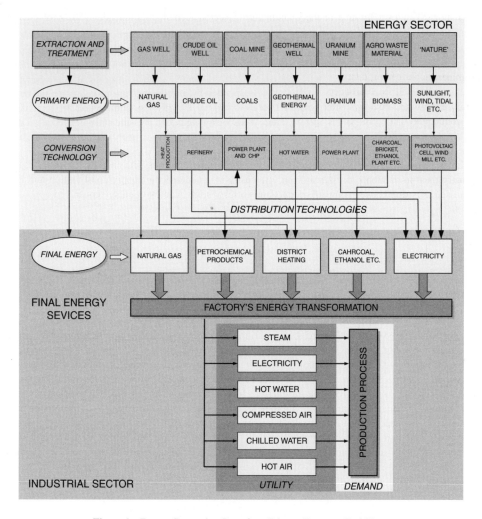

Figure 1   Energy Conversion Steps from Primary Energy to End-Users

where:
EI   = Energy intensity, national level
FC   = Total final consumption, national level
GDP = Gross domestic product

The Standard Industrial Classification (SIC) groups industrial establishments according to their primary economic activity. Each major industrial group is assigned a two-digit SIC code. The SIC system, which serves as a framework for the collection of energy consumption and output data, divides manufacturing into 20 major industry groups and non-manufacturing into 12 major industry groups. The intensity of energy end-use in industry varies according to requirements specific to the type of industries such as manufacturing, agriculture, forestry, fishing, construction and mining operations. Six of the major 20 industry groups in the manufacturing sector account for more than 80 % of energy consumption:

(1) Food and kindred products,
(2) Paper and allied products,
(3) Chemical and allied products,
(4) Petroleum and coal products,
(5) Stone, clay, and glass products,
(6) Primary metals.

Table 1 summarizes the key characteristics of energy-using industries with an overview of each. For industry or a production process, the energy intensity is called specific energy consumption and it is expressed as a ratio between units of product and corresponding amount of energy consumed.

$$SEC = \frac{E}{N_{PO}} \qquad (2)$$

where:
SEC = Specific energy consumption
E   = Energy consumed to produce N_{PO}
N_{PO} = Units of product

Table 2 shows specific energy consumptions for different industries. These data are sector aggregated and therefore can be used only for the rough benchmarking of energy consumption among different industries from the same sector.

Energy use relative to manufacturing output has been falling more or less continuously in most developed countries since the 1970s. This is caused by a structural shift away from energy intensive products, and also by improvements in energy efficiency and changes in individual energy intensities within each manufacturing subsector. There are various databases on energy consumption that contain the data necessary to undertake comparisons of energy use in industry. For instance, the ODYSSEE database provides comprehensive data on energy consumption by end-use sectors, as well as energy efficiency and CO_2 related indicators. The ODYSSEE indicators are macro-indicators, defined at the level of the economy as a whole, the level of a sector, and the level of an end-user. Seven types of indicators are considered in order to monitor energy efficiency trends or to compare energy performances:

- energy intensities relating energy consumption to macro-economic variable;
- unit consumption or specific consumption relating energy consumption to physical indicator of activity;
- 'bottom-up' energy efficiency index to provide the synthesis of energy efficiency trends assessed at disaggregated level;
- adjusted indicators to make cross-country comparisons that attempt, as far as possible, to adjust structural differences between countries (climatic, economic or technical);
- diffusion indicators for monitoring the diffusion of energy efficient equipment and practices;
- target indicators to set up for each country a target or benchmark in comparison to countries with better performances;
- CO_2 indicators complement energy efficiency indicators.

The Energy Information Administration of the USA Department of Energy (DOE) and the International Energy Agency (IEA) also maintain very comprehensive databases on industrial energy consumption.

Table 1 General Characteristics of Industrial Energy Consumption. Reproduced from: Energy Information Administration, Office of Energy Markets and End Use, Manufacturing Consumption of Energy 1991, DOE/EIA-0512(91)

	SIC*	Major Industry Group	Description
High-Energy Consumers	20	Food and kindred products	*This group converts raw materials into finished goods primarily by chemical (not physical) means. Heat is essential for their production, and steam provides much of the heat. Natural gas, fuel oil, by-product and waste fuels are the largest sources of energy in this group. All, except food and kindred products are the most energy-intensive industries.*
	26	Paper and allied products	
	28	Chemicals and allied products	
	29	Petroleum and coal products	
	32	Stone, slay and glass products	
	33	Primary metal industries	
High Value-Added Consumers	34	Fabricated metal products	*This group produces high value-added transportation vehicles industrial machinery, electrical equipment, instruments, and miscellaneous equipment. The primary end-users are motor- driven physical conversion of materials (cutting, forming, and assembly) and heat treating, drying, and bonding. Natural gas is the principal energy source.*
	35	Industrial machinery and equipment	
	36	Electronic and other electric equipment	
	37	Transportation equipment	
	38	Instruments and related products	
	39	Miscellaneous manufacturing industries	
Low-Energy Consumers	21	Tobacco manufacturers	*This group is the low energy-consuming sector and represents a combination of end-use requirements. Electric motor drives are one of the key end-users.*
	22	Textile mill products	
	23	Apparel and other textile products	
	24	Furniture and fixtures	
	25	Lumber and wood	
	27	Printing and publishing	
	30	Rubber and miscellaneous plastics	
	31	Leather and leather products	

* SIC – Standard Industrial Code
Source: Energy Information Administration, Office of Energy Markets and End Use, Manufacturing Consumption of Energy 1991, DOE/EIA-0512(91).

Table 2   Specific Energy Consumption in Industry[1]. Reproduced from: ODYSSEE Energy Efficiency indicators in Europe database

INDUSTRY	Thermal	Electricity[2]		Thermal to total	Primary energy[1] Total
		End-user consumption			
	toe/t	MWh/t	toe/t	%	toe/t
Brewing	0.050	0.10	0.0086	85 %	0.0720
Brick	0.075	0.05	0.0043	95 %	0.0860
Cement	0.080	0.11	0.0095	89 %	0.1043
Chocolate	0.200	2.00	0.1720	54 %	0.6409
Dairy	0.150	0.50	0.0430	78 %	0.2602
Engineering	0.300	2.75	0.2365	56 %	0.9063
Flour products (pasta, etc.)	0.040	0.16	0.0138	74 %	0.0753
Foundry	0.300	0.90	0.0774	79 %	0.4984
Ham, sausage	0.100	0.35	0.0301	77 %	0.1772
Hot pressing	0.250	0.75	0.0645	79 %	0.4153
Ice cream and cake	0.100	0.75	0.0645	61 %	0.2653
Milk processing	0.020	0.10	0.0086	70 %	0.0420
Non-ferrous metal	0.070	0.30	0.0258	73 %	0.1361
Paper and pulp	0.150	0.43	0.0370	80 %	0.2448
Plastic	0.050	0.55	0.0473	51 %	0.1713
Rubber	0.100	5.00	0.4299	19 %	1.2023
Textile (dyeing)	0.750	0.75	0.0645	92 %	0.9153
Textile (spinning)	0.450	7.50	0.6449	41 %	2.1035
Wood	0.020	0.06	0.0052	79 %	0.0332

Notes:
[1] Electrical energy consumption is multiplied by 1/0.37 to get **primary energy for its generation**. This assumes an average efficiency of electricity generation of approximately 37 %.
[2] $1\ toe = 11.63\ MWh$.

3 Environmental Impacts of Industrial Operations

The biggest environmental impacts in industry come from the use of fossil fuels (oil and coal). The fossil fuels are converted into useful energy through a combustion process. The combustion process results in significant impact on the natural environment and generate emissions into the air, water and soil (Table 3). The main emissions into the air from the combustion of fossil fuels are SO_2, NOx, particulate matters, heavy metals, and greenhouse gasses such as CO_2.

Table 3 Environmental Impacts by Source, Type and Substances

SOURCE RELEASE — AIR (A) WATER (W) LAND (L)	SUBSTANCES												
	Particulate Matter	Oxides of Sulfur	Oxides of Nitrogen	Oxides of Carbon	Organic Compounds	Acids, Alkalis, Salts, etc.	Hydrogen Chloride/Fluoride	Volatile Organic Compounds	Metals and their Salts	Chlorine (as Hypochlorite)	Mercury and/or Cadmium	PAHs	Dioxins
Fuel Storage and Handling	W				W			A					
Water Treatment	W								W		W		
Exhaust Gas	A	A	A	A	A		A	A	A			A	A
Exhaust Gas Treatment	W				W				W,L		W		
Site Drainage Including Rainwater	W				W								
Waste Water Treatment	W				W	W							
Cooling Water Blow-down	W				W				W	W	W		
Cooling Tower Exhaust								A					

Table 4 Sources of Emissions Released by Combustion Plants Related to Total Emissions. Reproduced from: EMEP/CORINAIR Emission Inventory Guidebook – 2006, http://reports.eea.europa.eu/EMEPCORINAIR4/en/page002.html

Source Category	Contribution to total emissions, [%]							
	SO$_2$	NOx	NMVOC	CH$_4$	CO	CO$_2$	N$_2$O	NH$_3$
Combustion plants over 300 MW, including: Public power plants District heating plants Industrial combustion plants	85.6	81.4	10.2	5.5	16.8	79.0	35.7	2.4
Combustion plants from 50-300 MW, including: Public power plants District heating plants Commercial and institutional boilers Industrial combustion plants	6.4	5.4	1.1	0.6	3.1	6.5	1.9	0.2
Combustion plants below 50 MW, including: Public power plants District heating plants Commercial and institutional boilers Industrial combustion plants	0.2	0.3	0.1	0.05	0.1	0.2	0.01	0 (N1)
Gas turbines used in: Public power plants District heating plants Commercial and institutional installations Industry	0	0.39	0.07	0.06	0.05	0.35	0.02	n.a.
Stationary engines used in: Public power plants District heating plants Commercial and institutional installations Industry	0.04	0.10	0.04	0 (N1)	0.01	0.02	0 (N1)	n.a.

Notes:
N1   Emissions are reported, but the precise number is below rounding limit
n.a.   Data not available

An example of the contributions of the main sources of emissions released by combustion plants related to total emissions is given in Table 4.

Even if fuels are not converted into final energy in an industrial plant itself, the electricity and heat that the plant uses have to be produced elsewhere, therefore the plant is responsible for indirect environmental impacts, the extent of which will depend on the power plant technologies and fuel sources used to generate it in a particular country. In Europe, a simplified approach has been introduced to derive emission factors in order to account for the environmental effects of electricity and heat. It is called the 'European Electricity and Heat Mix', and it provides multiplication factors for calculating the impact of the use of electricity by a business. To create 1 GJ (277.8 kWh) of electrical energy, the average primary energy of fossil fuels is 2.57 GJ (efficiency of electricity generation is 38.9 %). This production assumes consumption of 9.01 kg

Table 5 Resources Used and Emissions Caused by Use 1000 kWh of Electricity, http://www.epa.gov/tri/

Resource Used		Emissions	
Oil (kg)	32.43		
Gas (m3)	24.91		
Coal (kg)	0.47		
Brown Coal (kg)	124.69		
		SO_2 (kg)	0.36
		CO_2 (kg)	420.12
		NO_2 (kg)	0.58

of fuel oil, $6.92\,nm^3$ of natural gas, 0.13 kg of coal and 34.64 kg of brown coal. The emissions for such electricity generation are 0.1 kg SO_2, 116.71 kg CO_2 and 0.16 NO_2. For example, the use of 1000 kWh of electricity will create environmental impacts as shown in Table 5.

Apart from environmental impacts arising from energy use and conversion of fossil fuels, the processing of raw materials through a string of manufacturing operations is another source of major impacts on the natural environment. Most countries nowadays have national legislation in place that sets the limits for allowed emission values from various pollution sources, and prescribes procedures for emissions monitoring and compliance assessment. The types of impact are as diverse as the types of raw materials used and technologies applied for their processing. Particularly hazardous substances are always covered by legislation and environmental permits. Some examples will be provided latter in the text (see **7.1**).

4 End Use Energy Efficiency

Energy efficiency can be defined in various ways (Box 2) but essentially, it always refers to a ratio between useful energy or final energy services and input energy.

Box 2: Definition of Energy Efficiency

> *To define the **energy efficiency** of an installation or a system, it is necessary to determine **system boundaries** and to define precisely all mass and energy flows that pass through the boundary. Then 'input' energy, 'useful' energy or system output and 'losses' need to be defined. The table below gives examples of steam turbine, heat exchanger, boiler and refrigeration systems where system boundaries are determined, input energy and useful energy are identified, and energy efficiency is defined in different ways.*

	Schematic	**Energy Efficiency Definition**
Steam Turbine		*Isentropic Efficiency:* $$\eta_i = \frac{H_1 - H_2}{H_1 - H_2}$$ *Where:* $H = m \cdot h = Enthalpy, kJ$ $m = Mass\ flow\ rate, kg/s$ $h = Specific\ enthalpy, kJ/kg$ $h_{2'} = Enthalpy\ of\ steam\ after\ isentropic\ expansion\ from\ point\ 1\ to\ the$ $pressure\ equal\ to\ real\ process.$

Box 2: Continued

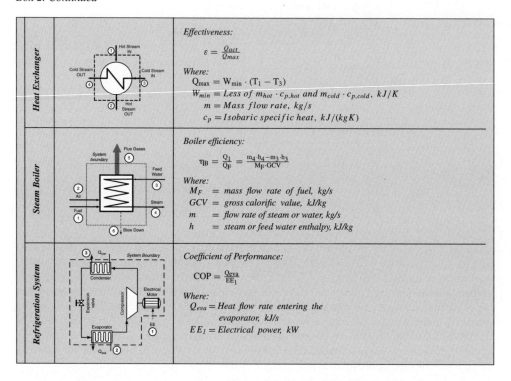

| | | Effectiveness:

$$\varepsilon = \frac{Q_{act}}{Q_{max}}$$

Where:
$Q_{max} = W_{min} \cdot (T_1 - T_3)$
$W_{min} = Less\ of\ m_{hot} \cdot c_{p,hot}\ and\ m_{cold} \cdot c_{p,cold},\ kJ/K$
$m = Mass\ flow\ rate,\ kg/s$
$c_p = Isobaric\ specific\ heat,\ kJ/(kgK)$ |

Improving energy efficiency, i.e. obtaining more final energy services from less energy – is the surest and most direct way of increasing sustainability of the use of energy resources and decreasing the negative aspects – environmental pollution and financial costs – associated with using energy and producing goods. The economic potential of even more efficient energy use will continue to grow with new technologies and with cost reduction resulting from the economy of scale.

Energy efficiency is beneficial to everyone. It lengthens the time for which fossil fuels will be available to meet the world's growing energy needs, for consumers – energy efficiency saves money, and for everyone – improved energy efficiency reduces environmental hazards and greenhouse gas emissions.

5 Efficiency of Using Raw Materials

Closely related to, and of equal importance to energy efficiency is the efficiency of using raw materials, their waste minimization and recycling. Changes in energy and material use efficiency are driven by higher energy and raw material prices, technical improvements, energy conservation and waste minimization programs that are promoted by energy and environmental policies.

A considerable proportion of manufacturing costs is directly related to the raw materials necessary for production. The use of raw materials is both an input element, in terms of the amount of raw materials needed for production and an output element, in terms of the amount of raw materials that is ultimately wasted. Materials accounting or, in other words, measurement and tracking materials' usage within a process, helps improving the efficiency of raw materials usage. Materials balance calculations can be used to verify measurements/estimates of waste quantities, based on the amount of input materials and product yield.

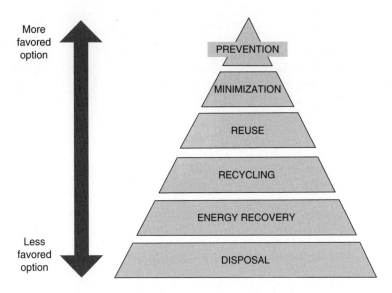

Figure 2   Order of Preference for Waste Management Options

In industry, waste is generally reduced firstly by improved quality in manufacturing operations, i.e. minimization of rejects and rework, and then by using more efficient manufacturing processes and better materials. The application of waste minimization approaches has led to the development of innovative and commercially successful replacement products. Waste minimization has proven benefits for industry and the wider environment:

- It reduces raw material costs.
- It reduces the cost of transport and processing raw materials and the finished product.
- It reduces the waste disposal cost to other parties (including collection, transport, processing and disposal).

The strategies for efficiency improvement of raw material use are based on so-called '3 Rs' approach – *Reduce, Reuse and Recycle* – which classifies waste management strategies according to their desirability. Waste management has evolved over the past decade, but the basic concept has remained the cornerstone of most waste minimization strategies (Fig. 2). The aim of waste management is to extract maximum practical benefits from certain raw material and to generate minimum amount of waste.

Effective market forces and good information can accelerate performance improvements, but market failures and barriers can inhibit efficiency gains. In such cases, the government's energy and environmental policy interventions are useful in focusing market interest on resources use efficiency. These include legislation, codes, standards, voluntary agreements, and special financing arrangements.

6 Global Energy Policy Framework

The international community shares the concern for secure energy supply (mostly gas and oil) and for the environmental impact of energy use. In particular, the concern for the consequences of climate change and depletion of natural resources brought the countries together at the Rio Earth Summit back in 1992, where the United Nations Framework Convention on Climate Change (UNFCCC) was adopted, calling for stabilization of greenhouse gases (GHG) concentrations in the atmosphere.

Climate change is one of the most serious challenges facing humanity in the 21st century. It is a truly global problem that will trigger the fundamental transformation of fossil fuel powered industrial economies. The first steps of this global process have already begun with the Kyoto Protocol and other policy instruments being implemented worldwide. Regulatory limits on GHG emissions materially impact on businesses that are producers or large-scale consumers of fossil fuels. They will accelerate the development of renewable energy and energy efficiency markets, and they will fundamentally impact on strategies and profits in industries such as power, cement, steel and oil refining.

The UNFCCC sets an overall framework for intergovernmental efforts to tackle the challenge posed by climate change. It recognizes that the climate system is a shared resource whose stability can be affected by industrial and other emissions of carbon dioxide and other heat-trapping gases. The ultimate objective of this Convention is the stabilization of greenhouse gas concentrations in the atmosphere at a level that will prevent dangerous anthropogenic interference with the climate system.

The main goal of energy policies in the context of climate change prevention and sustainable development (Box 3) is to explore ways of reducing the amount of energy used to produce a product, a service or a unit of economic output, and indirectly to reduce related emissions into the natural environment.

The concept of sustainability originated in the forestry industry and basically means only cutting down the amount of wood that you subsequently replenish with new growth. This idea was developed further in 1987 by the UN World Commission on Environment and Development, paving the way for a new policy of long-term sustainable growth that is meant to 'meet the needs of the present without compromising the ability of future generations to meet their own needs'. The term '*sustainability*' was introduced into the world of politics and business at the Earth Summit held in Rio de Janeiro in 1992.

Box 3: Sustainable Development

The most common definition of sustainable development says that it is 'development that meets the needs of the present without compromising the ability of future generations to meet their own needs'.

The concept of 'sustainable development', which made its first appearance in the 1970s, did not really enter the mainstream until the World Commission on Environment and Development (the Brundtland Commission) of 1987 provided a working definition that has largely been accepted ever since.

It is possible to trace a line of conceptual development from the Commission's work from the 1990 Intergovernmental Panel on Climate Change (IPCC) meeting in Geneva, to the historic Rio Earth Summit of 1992. Agenda 21, the Rio Declaration on Environment and Development, and the Statement of Principles for the Sustainable Management of Forests was adopted by more than 178 Governments at the United Nations Conference on Environment and Development (UNCED) held in Rio de Janeiro, Brazil in June 1992.

Agenda 21 is a comprehensive plan of action to be taken globally, nationally and locally by organizations of the United Nations System, Governments, and major groups in every area in which there is human impact on the environment. The Agenda has affirmed the views that the integration of environmental and development concerns and paying greater attention to them will lead to the fulfillment of basic needs, improved living standards for all, better protected and managed ecosystems and a safer, more prosperous future. Hence, the concept of eco-efficiency has come into being, based on the idea that improved efficiency benefiting business processes will also automatically benefit the environment.

In March 1995, politicians held a climate summit in Berlin where they agreed that legally binding CO_2 reduction targets should be ready for signing by industrialized countries. In December 1995 in Rome, scientists completed the second IPCC report, agreeing for the first time that human activities are altering the climate.

Energy, in its different forms, is required as continuous input to all industrial processes. The total energy consumption of the industrial sectors of developed countries contributes to around 30–40 % of total energy demand. In China, for instance, the share of industrial energy demand goes up to 70 % of total energy use. The consequences of burning fossil fuels are releases of greenhouse gasses (Box 4), which are responsible for perceived climate changes and global warming.

Box 4: Greenhouse Gasses and Climate Change

The earth's climate is dependent on the energy balance, which is affected by the sunlight reaching the earth, the amount reflected and the amount absorbed by the earth's surface. About 30 % of the incoming sunlight is reflected back into space, while the remaining 70 % warms up the earth and its atmosphere before being radiated back into space as infrared light.

The greenhouse effect, in which the atmospheric concentration of water vapour and CO_2 traps infrared radiation, keeps the planet sufficiently warm to support life. 'Greenhouse gases' (GHG are those gaseous constituents of the atmosphere, both natural and anthropogenic, that absorb and re-emit infrared radiation. Naturally occurring greenhouse gases include water vapour, carbon dioxide, methane, nitrous oxide, and ozone.

Certain human activities, however, add to the levels of most of these naturally occurring gases:

Carbon dioxide (CO_2): *released into the atmosphere when solid waste, fossil fuels (oil, natural gas, and coal), and wood or wood products are burned.*

Methane (CH_4): *emitted during the production and transport of fossil fuels (coal, natural gas, and oil) from the decomposition of organic wastes in municipal solid waste landfills, and the raising of livestock.*

Nitrous oxide (N_2O): *emitted during agricultural and industrial activities, as well as during combustion of solid waste and fossil fuels.*

Other greenhouse gases *that are not naturally occurring include: hydrofluorocarbons (**HFCs**), perfluorocarbons (**PFCs**), and sulfur hexafluoride (**SF$_6$**), which are generated in a variety of industrial processes.*

Increasing emissions and accumulation of anthropogenic GHG gasses, which are mainly carbon dioxide (CO_2), methane (CH_4), and nitrous oxide (N_2O); over the last two centuries, the amount of trapped infrared radiation has increased leading to the global warming and climate change.

'Climate change' means a change of climate which is attributed directly or indirectly to human activity that alters the composition of the global atmosphere and which is in addition to natural climate variability observed over comparable time periods.

The international community has responded to these threats by introducing a number of conventions and protocols, the most famous being the Kyoto Protocol (Box 5).

Box 5: The Kyoto Protocol

The Kyoto Protocol to the UNFCCC strengthens the international response to climate change. *Adopted by consensus at the third session of the Conference of the Parties (COP3) in December 1997, it contains legally binding emissions targets for Annex I (developed) countries for the post-2000 period.*

By arresting and reversing the upward trend in greenhouse gas emissions that started in these countries 150 years ago, the Protocol promises to move the international community one step closer to achieving the Convention's ultimate objective of preventing 'dangerous anthropogenic (man-made) interference with the climate system'.

The developed countries commit themselves to reducing their collective emissions of six key greenhouse gases by at least 5 %. *The six gases are to be combined in a 'basket', with reductions in individual gases translated into 'CO_2 equivalents' that are then added up to produce a single figure.*

Countries will pursue emissions cuts in a wide range of economic sectors. *The Protocol encourages governments to cooperate with one another, improve energy efficiency, reform energy and transportation sectors, promote renewable forms of energy, phase out inappropriate fiscal measures and market imperfections, limit methane emissions from waste management and energy systems, and protect forests and other carbon 'sinks'.*

The EU and its Member States ratified the Kyoto Protocol in late May 2002, *and Russia in 2004, enabling the Kyoto Protocol to finally come into force in February 2005.*

It has become clear that no nation can achieve ultimate results on its own but only in a global partnership for sustainable development. Governments have recognized the need to redirect international and national plans and policies in order to ensure that all economic decisions fully take into account any environmental impact. And the message is producing results, making eco-efficiency a guiding principle for business and governments alike.

7 Energy and Environmental Policies

Although the Kyoto Protocol is still not universally supported, there is no doubt in anybody's mind that issues regarding limited reserves of fossil fuels and the environmental impact of their use must be taken seriously. National governments are responding to climate change challenges by introducing variety of energy policies, technical standards and environmental legislation aimed at increasing efficiency of energy consumption and reducing the environmental impact of energy use.

These policies and instruments for their implementation create a framework against which businesses must assess their own energy policies and implement energy and environmental management programs. These policies usually have mandatory and voluntary components, which are often combined in an integrated approach that brings together a number of policies and programs aimed at creating strong overall policy framework that addresses a variety of needs related to promoting the improvement of efficiency in energy use and in environmental performance.

Some of the policy instruments will be briefly described in the text that follows.

7.1 Integrated Pollution Prevention and Control (IPPC)

The EU has a set of common rules on environmental permissions for industrial installations, which extend also to cover energy use at facilities. These rules are set out in the so-called IPPC Directive of 1996, http://ec.europa.eu/environment/ippc/index.htm. The IPPC stands for Integrated Pollution Prevention and Control. In essence, the IPPC Directive is about minimizing pollution from various sources throughout the European Union. All concerned companies are required to obtain an authorization (permit) from the authorities in the EU countries. Unless they have a permit, they are not allowed to operate. The permits must be based on the concept of Best Available Techniques (or BAT).

The same Directive has established the European Pollutant Emission Register – the first Europe-wide register of industrial emissions into the air and water. A comparable instrument in the USA is the Toxics Release Inventory (TRI). The number of facilities submitting reports to the USA TRI is about 20 000 and it is comparable with the total number of facilities involved in the European Union.

7.2 Energy Markets Deregulation and Liberalization

The operation of electric power systems in the 1970s was a vertically integrated activity with a strongly monopolistic character and with a public supply service obligation in their location area. Their activities were under the scrutiny of a regulatory authority accountable, directly or indirectly, to the government.

The 1980s brought changes to this landscape. Achievements in technology, a political environment favorable to 'free markets' and opposed to monopolies, of the state in particular, as well as modifications to the ownership structure by placing distribution and production assets in the hands of private ownership have changed the perception of energy sector development completely. An important part of the sector, that related to network services (transmission, distribution), still remains a natural monopoly which needs monitoring by the state by regulating access to the network in a non-discriminatory way, as well as by providing protection mechanisms for underprivileged consumers. Privatization in Chile and the UK marked the beginning of this process.

7.3 Consumers' Choice in the Liberalized Energy Market

This liberalization is aimed at introducing supply competition in the electricity and gas markets. Competition is supposed to benefit consumers through lower prices and by broadening the choice of available services (Fig. 3).

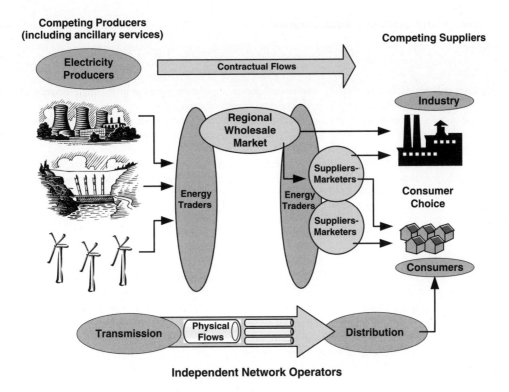

Figure 3   Liberalized Electricity Market

In addition to affording choice to consumers, liberalization has also introduced complexity into the wholesale and retail energy markets. A consequence of the complexity of the wholesale electricity market is price volatility at times of peak demand and supply shortages. Final consumers (large and small) are at the receiving end of these complexities. They can buy from retailers or directly from the generator (if their consumption is large enough). This competition creates the potential for lower prices and enhanced customer service, but it also creates complexity arising from the range of products that are offered.

Industrial energy buyers need to be well informed in order to make best decisions regarding the purchase of their fuels and electricity. They need to consider how to avoid price spikes during extremely hot or cold seasons, how to mitigate the risk of fluctuating market prices for electricity and how to forecast more accurately and manage energy expenses in each quarter.

That is where an energy management system, if in place, can help by providing accurate data on current and reliable forecasting of future energy consumption profile and amounts, as a basis for negotiating optimal energy supply contracts with an energy trader.

7.4 Emissions Trading

The aim of emission trading is to achieve emissions reductions at the lowest economic cost. Emission trading occurs when a factory reduces its emissions and then transfers ownership of the emission reduction to another party. Emission allowances are typically given by regulators to large sources of pollution, and allow those sources to release a prescribed amount of a pollutant. Surplus allowances can be sold, traded, or banked for future use.

As an instrument for reducing carbon dioxide emissions, the European Union Emission Trading Scheme (EU ETS) became effective in January 2005. It was developed not only to support the reduction of CO_2 emissions, but also to offer companies an incentive for developing and using new technologies. Under the ETS, the EU imposes CO_2 quotas on more than 15 000 company-owned power plants and factories. Businesses that exceed their limits must either buy allowances from companies that emit less, or pay a penalty.

7.5 Compulsory Energy Efficiency Programs

There are only a few compulsory energy efficiency programs aimed at industrial facilities. It is only recently that the EU introduced a Directive aimed at the energy performance of buildings that makes it mandatory to carry out energy audits of large commercial buildings at regular intervals. In that respect, the energy conservation program (ENCON) in Thailand is unique, because it began in the early 1990s and it is still running as a compulsory program for improving energy efficiency in industry and large buildings.

7.5.1 ENCON Program in Thailand

Energy consumption in Thailand showed a rapid increase during the 1980s and 1990s, and reached the highest recorded rate of 15 % in 1997. Such an increasing trend had to be slowed down in order to prevent massive government spending so as to increase the capacity of energy supply plants in Thailand, as well as meeting the need to improve the nation's environmental performance.

Therefore, in order to introduce the efficient use of energy, the Energy Conservation program made it mandatory for large industrial users and commercial buildings to appoint a person responsible for energy, carry out energy audits by registered consultants, report results to the Government for approval, prepare targets and plans for energy efficiency improvements, implement the plans and report on the results of implementation. Recently, the scheme has been updated, but designated factories and buildings are still obliged to follow a basic four-step procedure, as described in Box 6.

Box 6: Steps of the ENCON program in Thailand

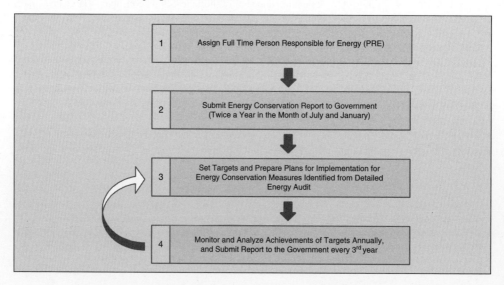

7.6 *Voluntary Programs*

Voluntary programs are part of energy polices based on targets and incentives that industry may subscribe to or apply for. There is a number of such schemes around the world, but we will point out just one of them that addresses energy and environmental issues.

7.6.1 Eco-Management and Audit Scheme (EMAS)

The Eco-Management and Audit Scheme (EMAS) is the EU voluntary scheme for organizations willing to commit themselves to evaluate, improve and report on their environmental performance. The EMAS is established to evaluate and improve the environmental performance of organizations and to provide relevant information to the public and other interested parties. The objective of the EMAS is to promote continual improvement in the environmental performance of organizations by:

(1) establishing and implementing environmental management systems;
(2) systematic, objective and periodic evaluation of the performance of such systems;
(3) providing information on environmental performance and open dialogue with the public and other interested parties;
(4) active involvement of employees in the organization and appropriate initial and advanced training that make active participation in the above-mentioned tasks possible.

In its requirements, the EMAS is closely related to ISO 14000 environmental management standards.

8 Industries' Self-Motivation for Effective Energy and Environmental Performance

As energy prices increase, industries are more and more aware that energy is an expensive commodity that has to be used efficiently in order to reduce the costs that it entails. When companies realized that improving energy efficiency was not only about energy performance, but also about the business performance as a whole, i.e. about improving profits, the motivation for energy management became evident.

However, the improvement of environmental performance is associated with costly measures and interventions without obvious and immediate financial benefit to a company; hence for years improvement of environmental performance has not been high on the agenda. Therefore, the introduction of environmental legislation is necessary in order to enforce minimum standards of environmental performance.

Increasingly, over the last decade, companies themselves are recognizing that care for the environment is part of good corporate behavior, which besides having obvious benefits for the global environment that we all share, also brings with it the 'goodwill' benefit of environmentally conscious customers that prefer to buy goods and service from companies that employ so-called 'green' business practices.

Major international corporations are subscribing voluntarily nowadays to schemes like the Global Reporting Initiative (Box 7), Business Roundtable Principles of Corporate Governance (Box 8), Corporate Sustainability (Box 9) and the US Climate Action Partnership (Box 10).

Box 7: Global Reporting Initiative

The Global Reporting Initiative (GRI) is a multi-stakeholder process and independent institution whose mission is to develop and disseminate globally applicable Sustainability Reporting Guidelines. These Guidelines are for voluntary use by organizations for reporting on the economic, environmental, and social dimensions of their activities, products and services. Established in 1997, GRI became independent in 2002, and is an official

Box 7: Continued

collaborating centre of the United Nations Environment Program. The GRI provides a reporting framework that outlines a core content broadly relevant to all organizations regardless of size, sector, or location. The GRI content Index covers:

- *Vision and strategy;*
- *Governance structure and management Systems;*
- *Economic performance indicators;*
- *Environmental performance indicators, consisting of:*
  - *Materials;*
  - *Energy;*
  - *Water;*
  - *Biodiversity;*
  - *Emissions, effluents and waste;*
  - *Suppliers;*
  - *Products and services;*
  - *Compliance;*
  - *Transport;*
- *Social performance indicators.*

Most of the indicators in the GRI guidelines are covered by technical protocols providing detailed definitions, procedures, formulae and references so as to ensure consistency and comparability in reported information. The relevant example is the GRI Energy Protocol which is intended to clarify the measurement expectations for individual energy performance indicators.

Box 8: Business Roundtable Principles of Corporate Governance

The Business Roundtable is recognized as an authoritative voice on matters affecting American business corporations. It supports the following guiding principles of corporate governance:
    The paramount duty of the board of directors of a public corporation is to select a Chief Executive Officer and to oversee the CEO and other senior management in the competent and ethical operation of the corporation on a day-to-day basis.
    It is responsibility of management to operate the corporation in an effective and ethical manner in order to produce value for stockholders.
    Although the Principles do not spell out the environmental concerns explicitly, they are covered by the requirement to operate corporations in an 'ethical manner'. Also the Business Roundtable trusts that the most effective way to enhance corporate governance is through conscientious and forward-looking action by a business community that focuses on generating long-term stockholder value with the highest degree of integrity.

Box 9: Corporate Sustainability

***Corporate Sustainability** is a business approach that creates long-term shareholder value by embracing opportunities and managing risks deriving from economic, environmental and social developments. Corporate sustainability leaders achieve long-term shareholder value by gearing their strategies and management to harness the market's potential for sustainability products and services while at the same time reducing and avoiding sustainability costs and risks.*
    The quality of a company's strategy and management and its performance in dealing with opportunities and risks deriving from economic, environmental and social developments can be quantified and used to identify and select leading companies for investment purposes.

Corporate sustainability performance is and investable concept. **The Dow Jones Sustainability Index** defines corporate sustainability assessment criteria along these three dimensions:

ECONOMIC ↔ ENVIRONMENTAL ↔ SOCIAL,

each having an **average weighting of 50, 20 and 30 % respectively**.

Leading sustainability companies display high levels of competence in addressing global and industry challenges in variety of areas:

Strategy: Integrating long-term economic, environmental and social aspects in their business strategies while maintaining global competitiveness and brand reputation.

Financial: Meeting shareholders' demands for sound financial returns, long-term economic growth, open communication and transparent financial accounting.

Environmental: Managing environmental performance of the business according to highest standards, and continuously improving efficiency of using energy, water and raw materials.

Customers and products: Fostering loyalty by investing in customer relationship management and product and service innovation that focuses on technologies and systems, which use financial, natural and social resources in an efficient, effective and economic manner over the long-term.

Governance and stakeholders: Setting the highest standards of governance and stakeholder engagement, including corporate codes of conduct and public reporting.

Human: Managing human resources to maintain workforce capabilities and employee satisfaction through best-in-class organizational learning and knowledge management practices and remuneration and benefit programs.

Box 10: The US Climate Action Partnership

Ten industry giants – with business operations spanning the utilities, manufacturing, chemicals and financial-services sector – joined forces with four environmental groups to pressure for setting mandatory limits on CO_2 emissions. The group calls for a market based emission trading program. Under a 'cap and trade' system, the Government gives or sells permits to business, allowing them certain levels of emissions. Companies that emit less than allowed get credits they can then sell to companies that need them to meet the standard because they are emitting more than their designated amounts.

As of now, the US has a voluntary emission–trading system through the Chicago Climate Exchange (CCX). CCX is the first US pilot program for voluntary trading of GHG emissions which, apart from emissions trading among emitters, also provides for offset projects in North America and limited offset projects in Brazil. Starting from an already existing regional SO_2 trading scheme, the project aims at extending the latter by GHG emissions trading.

The coalition is aiming to reduce GHG emissions by 10 % to 30 % over next 15 years.

This coalition presents a strong sign that climate change debate has become less of a debate and more of a consensus. Expanding the GHG market into a global market is believed among industrialists, will ensure that the climate change is being addressed effectively and efficiently and will stimulate innovation in addressing the climate change.

9 Environmentally Responsible Investing

Socially responsible corporate governance is further reinforced by the emergence of environmentally conscious investors that prefer to buy shares in 'green' companies. The underlying principle of sustainable investment is that the economy does not exist in a vacuum, but it is tied to its environmental and social milieu. Society and the environment have a significant impact on the economy, and vice versa.

For instance, if raw materials are in short supply, companies that are prudent in their use of finite resources have systematic cost and revenue advantages over companies that are more reckless in their consumption of resources. Further, fair conduct towards employees and other stakeholders pays off because contented employees are harder working and more motivated, happy suppliers are more reliable and satisfied customers are more loyal.

More than 10 years ago the banks realized that there was a positive correlation between environmental and social compatibility, on the one hand, and commercial success, on the other. Hence, the influence of environment and social themes is growing steadily in the asset management business. From environmental viewpoint, the main focus is on *eco-efficiency*. This means that companies which keep environmental pollution per dollar of value added as low as possible will receive a favorable rating. Some 10 years ago, the first investment fund appeared based on this principle. That fund invested not only in environmental technologies, as was common practice at that time, but also in any other sector where the above-mentioned environmental criteria were applied.

The recent surge in natural resource prices and higher and sustained energy prices are also pushing clean technologies and other green investments into the foreground. Besides the impact of higher energy prices, an equally compelling market driver is acceptance that environmental responsibility is raising global issues for both companies and governments.

Creating *green financial products* is an outcome of the convergence of capital markets and environmental and social issues. Investors have moved beyond socially responsible investing (SRI) and are demanding greater disclosure by companies of their climate change risks and overall 'greenness'. We are starting to see the emergence of environmental score cards for both GHG and clean technology exposure. Being 'green' is now good for business and is something that more and more shareholders expect from and look for in companies. Energy and environmental issues have thus become closely intertwined and have emerged as board level issues.

Heightened awareness of environmental and energy issues also creates greater investment opportunities for clean technologies that are more energy efficient and more environmentally benign. As a result, more and more investment opportunities and vehicles through which to make those investments in the nexus of energy and the environment will emerge. There will be more hedge funds, equity funds and venture funds, as well as more investable financial products targeting investors. This increases the demand for companies to report on sustainability issues, and to build on their reporting capacity, moving towards greater coverage of environmental and social issues, transparency and consistency in the way that companies report upon their impact on the environment and in the delivery of improvements.

But more importantly, it also increases demand for companies to manage effectively their energy and environmental performance, alongside their overall business performance, in order to achieve the improvements that can then be reported upon.

The text that follows will describe a management approach which yields continuous energy and environmental performance improvements, reduces overall business costs and provides a framework for effective reporting on business performance to internal and external stakeholders.

10 Where to Look for Energy and Environmental Performance Improvements

Improvements in performance may come from more effective energy use in production, increased efficiency of energy supply and waste minimization. Generally, improvements can be found both in the way that people operate equipment and in the efficiency of equipment and technologies involved in a production process (Fig. 4). In order to achieve lasting reductions in environmental impacts and energy requirements for production of a particular product, a combination of several factors is required:

- improved operational and maintenance procedures;
- improved controls (involves both people and technology);

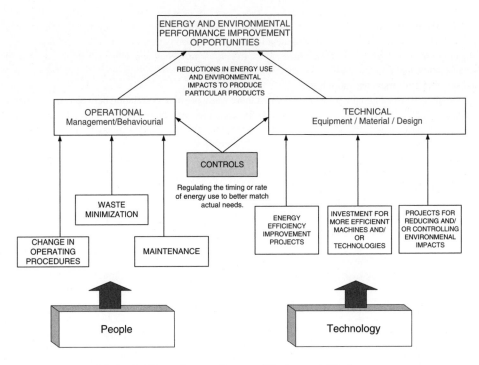

Figure 4 Ways to Improve Energy and Environmental Performance

- waste avoidance and minimization;
- efficient equipment, and above all
- skilled and dedicated people.

It needs to be emphasized that energy management and pollution control starts with increasing the *efficiency of existing operations* and by minimizing waste. Whenever some raw materials are processed, energy is required. Excessive waste in material processing will also cause excessive energy consumption. Variations in efficiency of materials processing will be reflected in variations of energy use. *Impacts on the environment* are a *consequence of processing raw materials and energy use* in that process. Therefore, we stress that *energy management is a driver for environmental management and overall operations performance management.*

From these perspectives, we will continue to elaborate in Part I of the book, focusing on the managerial, organizational and human aspects of energy and environmental performance improvement.

Part II of the book will cover the technical opportunities and measures for the energy efficiency improvement of industrial energy and utility systems and the resulting environmental performance improvements. Environmental technologies considerations are not covered, but useful references will be provided.

11 Bibliography

V. Bukarica, Z. Morvay, Z. Tomšić (2007) Evaluation of energy efficiency policy instruments effectiveness – case study Croatia, IASTED International Conference on Power and Energy Systems (EuroPES), Spain (English).
http://www.eia.doe.gov/emeu/efficiency.
http://ec.europa.eu/environment/ippc/index.htm.
http://www.iea.org.

IETA, www.ieat.org.
http://www.ebrd.com/.
http://eippcb.jrc.es/pages/Directive.htm.
http://www.emas.org.uk/.
http://www.eper.cec.eu.int/eper/.
http://ec.europa.eu/environment/water/water-framework/index_en.html.
http://www.iso.org/iso/en/iso9000-14000/index.html.
http://www.odyssee-indicators.org.
http://www.unece.org/env/lrtap/.
http://unfccc.int/.
http://www.worldbank.org/.
Z. Morvay (2002) 'Options for restructuring of a vertically integrated power utility', invited presentation for EGAT's internal seminar (Electricity Generation and Transmission Company of Thailand), Chang Mai, Thailand.
Z. Morvay, Kosir, M., Limari, S., Hamiti A. (2003) Restructuring of a vertically integrated electric power utility for a competitive regional electricity market, International Conference on Energy and Environment, Pukhet, Thailand, February (English).

Part I

Energy and Environmental Management System in Industry (EEMS)

1

Introducing the Energy and Environmental Management System

Energy and environmental management is about people and machinery performance management.

1.1 Introduction

Industrial plants use large amounts of fossil fuels and other raw materials taken from the earth's natural resources and convert them into products and useful energy. The use and conversion of primary energy always results in waste, emissions and effluents that have an impact on all environmental media. ***Environmental pollution*** through industrial operations is a highly regulated area where operating in ***compliance*** with regulations and standards carries additional requirements and costs.

> *The prudent management of energy and the efficient use of natural resources are two major preconditions for environmental management. It is important to recognize from the start that impact on the environment and pollution are the consequence of energy use and the processing of material resources. If the use of energy and materials is optimized, the resulting environmentals impact will be minimized! Where no energy or materials are used, there is no impact on the environment.*

Efficiency is not only important as an indicator of the careful treatment of natural resources, it is also an indicator of the emissions released in order to produce a unit of production or energy. Good environmental practice is simply good management. Environmental problems are often the symptoms of inefficiency and waste of resources. Achieving best operational practices through the implementation of energy management is the first logical step to introduce systematic environmental management. When unnecessary energy consumption is eliminated and waste minimized, the impact on the environment will already be reduced to a minimum and the remaining level of environmental releases and discharges should then be subjected to appropriate disposal and abatement techniques.

Otherwise, if a company decides before considering anything else to install a flue gas treatment facility so as to reduce SO_2 and NOx emissions, and subsequently embarks on energy efficiency improvement projects which will result in a significant reduction of energy consumption, the company will find itself

Applied Industrial Energy and Environmental Management   Zoran K. Morvay and Dušan D. Gvozdenac
© 2008 John Wiley & Sons, Ltd

with an oversized treatment facility that costs much more to procure and to operate than was actually required.

Energy management is a prerequisite for environmental management, but these two performance management systems are complementary to each other. The scope of energy management can be easily extended to include environmental concerns, at least with regard to monitoring of performance. When both performance areas – energy and environment – are addressed at the same time, we can refer to it as *integrated energy and environmental management*.

Improving environmental performance sometimes creates considerable cost but ensures compliance and demonstrates social responsibility. From the experience of numerous companies worldwide, improving energy performance often proves to be the most profitable investment a company can make, as we will demonstrate through selected case studies. Energy performance improvement can be a driver for the improvement of environmental performance and the financial benefits from energy savings can offset some of the costs of environmental improvements.

Energy and environmental management in industry is about controlling energy and environmental performance throughout industrial operations with the objective of achieving the company's goals by lowering energy consumption and minimizing the impact on the environment due to the use and conversion of energy, water and any input materials. The foundations for both energy and environmental management are similar if not the same, therefore they will be elaborated upon concurrently in this chapter.

1.2 Definition of terms

We now explain what the term 'Energy and Environmental Management System' (EEMS) stands for in this book. This is particularly important because EEMS contains the words 'system' and 'management', which are so ubiquitous nowadays. However, it seems that wherever we encounter them, their meaning is different.

Some of the ambiguity is derived from the fact that Management Systems are not tangible and easily identifiable assets within a company. Some comes from the lenient use of the same terminology in different areas that come together in industrial operations. Often, we have discussions with people in industry who are striving to understand what this environmental 'Management System' that they have struggled so hard to develop in order to get their ISO 9000 or 14 000 certification actually is. They understand quality and environmental concerns but the 'Management System' part is often hard to grasp. On the other hand, various manufacturers claim that a versatile electricity meter is an 'energy management device', or that an automatic or computerized control system is the 'Energy Management System'. This is not quite the case, as we will explain here.

Usually, the term 'Environment' stands for the surroundings of a specified system. In the context of this book, *'Environment'* stands for a natural surroundings of a company or an organization which may be influenced or polluted by the company's or organization's activities. The effects of industrial activities are referred to as *environmental impacts* upon the major environmental media, which are air, water and land.

Assuming that the term *'Energy'* stands for all forms of heat, power and utilities – electricity, fuel oil, solid fuels, gas (NG, LPG), water (chilled, hot, treated, industrial), steam, air (compressed, cold, hot) – and does not need much explanation at this point, we can focus on the 'Management System' part. We will start with the term 'System'.

1.2.1 System

Formally, a system can be viewed as a collection of functional components interacting in order to achieve an objective or to perform a task within defined boundaries. Every system has some inputs and outputs and is interfaced with its surrounding environment (see Fig. 1.1). From this viewpoint, almost anything can qualify to be called 'a system' – that is why the use of the term is so common.

If a system is to be controlled, the essential feature is feedback – information about output deviations from target values, brought back to the input side (see Fig. 1.1). This information must be based on measurements. Inputs can be regulated based on the feedback, in order to control the system so that satisfactory performance can be maintained.

Companies can also be viewed as man-made systems of coordinated processes and interrelated parts working in conjunction with each other, operated by people, aimed at accomplishing the core objectives of the business and a number of other goals (organizational and individual).

Such a system – a company or a business system – is always in dynamic interaction with its surroundings, determined by a constant flow of resources into the system, and flow of system outputs, intended and incidental, out into the environment.

Very broadly, the company resources fall into four categories:

1	**Natural resources**	Materials, energy, land . . .
2	**Man-made resources**	Technology, machines, buildings . . .
3	**Human resources**	People.
4	**Capabilities**	Organizational routines, procedures and **knowledge** determining a company's ability to perform, i.e. provide services or deliver products.

To regulate such a system is clearly outside the scope of an ordinary automatic control system because increased complexity and the uncertainties caused by impacts from surrounding and human factors must be taken into account.

This problem brings us to the concept of 'Management'.

1.2.2 Management

The founder of management philosophy Henry Fayol specified the main management tasks or processes that are still around today (Table 1.1).

On the operational level, management is concerned with optimizing and controlling the use of company resources in order to achieve specified objectives. The most important resources are knowledgeable,

Table 1.1   Main Management Tasks

1	**Planning**	Deciding on how to use resources in order to achieve given targets.
2	**Coordination**	Communication between the company's functional units.
3	**Organization**	Organizing people to get the best out of their potential.
4	**Staffing**	Hiring, motivating and developing people as the most valuable company resource.
5	**Controlling**	Supervising, supporting, communicating, motivating and guiding people in order to achieve required performance.
6	**Budgeting**	Planning and securing the financial means for company operation.
7	**Reporting**	Enabling the flow of information and control of policy implementation.

experienced and resourceful people and a company's capabilities and awareness of how to perform – the 'know-how'!

Therefore, management process is not only task oriented but also, and even more importantly – people oriented! Effective management process must be based on data analysis and information from a business system. Management by facts is a management concept that should prevail rather then a concept of management by opinion.

The key objective of any management process is to manage the various aspects of a company's *performance*. Input for a management process is *information* and the outputs are **decisions**. This concept is illustrated in Figure 1.1.

1.2.3 Performance

A critical issue for any business function or operation is its *performance*.

Companies perform by achieving their goals and satisfying customers. Common questions that managers are facing in their quest to improve business performance are:

- How well do we perform?
- What is the efficiency and effectiveness of our operations in meeting business objectives and satisfying customers?
- How do we compare with our competitors?
- What can we do to improve the performance of our business?
- How can we monitor progress and recognize the areas of strong or poor performance?

However, before we proceed further, let us examine what the word 'performance' stands for.

Performance can be defined as an ability to complete a task or operation according to a *specified standard*. The standards may be defined as measures, yardsticks or benchmarks for assessing the deviations of actual performance as compared to preset requirements, as a basis for *managerial control*.

Overall business performance depends on the effective allocation and use of resources in order to produce some level of output with the *least cost* and with the desired quality. Traditionally, business performance is measured by money and expressed by numerous financial ratios but the essential business performance indicator is PROFITABILITY.

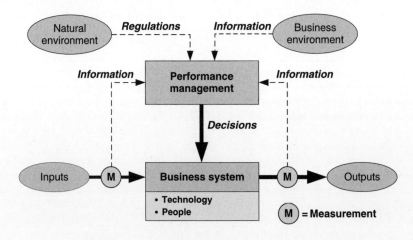

Figure 1.1   Overview of Performance Management Process

More generally, performance measurement means the systematic collection of data on business activities at all levels and on all functions of a company:

- business planning;
- financial planning;
- sales/marketing planning;
- distribution requirements;
- capacity planning;
- master production schedule;
- material requirements planning;
- procurement;
- inventory management;
- bills for material management;
- manufacturing control;
- work floor scheduling;
- compliance and permits;
- product costing, etc.

Company profitability is derived from the company's willingness and ability to improve, and from the profit-ability of employees! Therefore, people at every level must have *information* that will enable them to see beyond their local functions and to understand interactively how their individual actions increase or decrease profitability.

1.2.4 Information

It is essential to have appropriate information upon which management decisions are based. Managers need to know about employee performance and about how the business will perform if they control and improve operations adequately.

Information should not be confused with data. Data are often gathered in an *ad hoc* fashion without regard to user needs. A list of numbers does not convey much meaning without explanation. To become useful, data must be made into information by analyzing and presenting them in a form which is appropriate to a particular type of user.

Data gathering should be coordinated. Data from one source should be shared by all who are interested in that aspect of operation, instead of the same data being collected separately by multiple users. Managers should be able to delegate the evaluation of data to other employees who will decide what is important for a given purpose and who will then present this information in a way easily understandable by others.

Information management is about developing a means to collect, handle and distribute the right information to the right people. The right information is relevant, accurate, timely and presented in an appropriate form. Information technology (IT) must provide an efficient infrastructure for information management according to the performance management needs. However, making technology match business needs is not the way that IT has traditionally worked; it has usually been the other way around. Until the end of 1990s, business consumers of information technology were in a buy-it-at-all-cost mode, ending up with complex IT systems, with numerous applications not working together, or unable to share the same data. Nowadays, IT should not be regarded as something special but should simply be treated as a normal performance enhancing element of a company conforming to its actual needs. Companies need to go from using a bunch of standalone 'islands' of software to using applications that can work with one another. *Performance enhancement is a people-driven process supported by IT!*

Web-based solutions provide an IT infrastructure that helps to disseminate communication and information in a clear, consistent and intended manner to a large number of users no matter where they are.

Web services allow for the sending and receiving of data over the Internet from any application in a system and at the required time. No special hardware and no extra communication lines are required and it works at fraction of the cost of the old way. Since the content is constantly available, the audience can review it and update and interact with it when and where they want. It is an effective solution for large or distributed organizations, enabling them to monitor consistently and uniformly the performance of all business units, wherever they may be, communicate strategic messages clearly, reduce costs and support organizational learning.

1.2.5 Performance Indicators

The role of a performance indicator (PI) is to make complex systems understandable or easily comprehensible. An effective indicator or a set of indicators helps in determining the current position against established objectives.

Performance indicators quantify information on a trend or phenomenon in a manner that promotes understanding of a performance problem by both plant operators and decision makers. Performance indicators enable the measuring and tracking of progress towards goals in a comparable, consistent and easily comprehensible manner.

The process of selecting PI must necessarily start from a precise understanding of the process being addressed and of monitoring objectives. The goal of PI is to enable the monitoring and evaluation of applied energy and environmental performance improvement measures.

But information on PI alone does not solve the problem. What is required is *knowledge* on how to interpret PIs, and how to relate causes to consequences, and which are instances of either good or bad performance.

1.2.6 Knowledge

Knowledge is nowadays widely recognized as the only sustainable source of competitive advantage.

Like the other related concepts: truth, belief, and wisdom, there is no single definition of knowledge on which scholars agree. We may propose a definition, which says that *knowledge is a developed understanding* about a subject, based on facts and gained through education and experience.

The acquisition of knowledge involves complex cognitive processes: perception, learning, communication, association, emotions and reasoning. The term 'knowledge' is also used to describe the confident understanding of a subject, potentially with the ability to use it for a specific purpose. Knowledge is the goal of education, and the product of experience. The part of knowledge that is more easily definable involves the accumulation and assimilation of multiple pieces of information, providing a structure for it in the form of relationships between the information, and internalizing, or personalizing that knowledge by bringing it from the outside 'in' to the mind.

An important part of knowing is *learning*, i.e., the process of knowledge development and creation.

Knowledge is gained through many different learning 'modes', and the basic division is on

- *kinesthetic learning* – based on hands-on work and engaging in activities.
- *visual learning* – based on observation and seeing what is being learned.
- *auditory learning* – based on listening to instructions/information.

Knowledge is the basis for management by facts as opposed to 'management by opinion'. Facts are *unknown* until they are established through a deliberate process of the analysis of relevant information, which allows informed decisions to be made and significantly reduces the risk of opinionated decisions. The most powerful tool that managers can use for facts discovery – is a *question!* Managers have to ask themselves and their subordinates '*Why*?' Why such an event or such data pattern appeared? What does it mean? What has caused it? Managers also have to encourage workers to ask themselves 'Why?'

The workers should seek actively to make sense of data and information in order to develop their own understanding of underlying events in the process and combination of influencing parameters that have resulted in an observed data pattern or in an unexpected operational event.

Decisions and actions should be based on an analysis of factual and measured data and information. The analysis is based on knowledge and understanding of underlying business processes, systems and technologies. Informed management decisions, when implemented, improve business results, performance of processes and systems, communication and accountability.

1.2.7 Management System

Putting 'management', 'performance', information', 'knowledge' and 'system' together with 'people', we are arriving at the '*Management System*'. Every company, department or person has a system that it uses to *manage* events and other people even when it is a 'no-system' system. The purpose of a system, formal or informal, is to establish strategies that consistently enable achievement of the business objectives.

A management system is an organizational structure supported by a *body of knowledge* that integrates *people*, *procedures and technology* (information, communication and measuring equipment, i.e. infrastructure for data collection and processing) in order to control and optimize the use of resources and company performance (Fig. 1.2).

People play a major role in management and manufacturing processes and their own performance is critical for the company's overall performance. Managers are people, and they manage other people – staff and workforce, that in turn operate machines and provide services.

Of course, a management system can have various objectives: *to manage manufacturing, marketing, finance, quality, environment, energy, etc.* A proficient management system is *a tool* that supports the improvement of skills, rewards correct actions, builds integrity of competent workforce and creates and captures knowledge!

1.2.8 Environmental Impacts

Environmental impacts can be defined as any change to the environment, whether adverse or beneficial, wholly or partially resulting from the activities, products or services of a company.

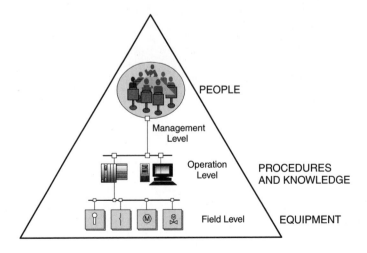

Figure 1.2   Concept of Management System

The environmental impacts from industrial operations can be categorized as follows:

- emissions to the atmosphere including GHG (greenhouse gases);
- discharges to water;
- solid and other wastes;
- contamination of land and underground water;
- use of water, raw materials, energy and other natural resources
- noise, odor, dust, vibration and visual impact;
- effects on specific parts of the environment and ecosystems.

This should include effects arising from normal and abnormal conditions, incidents, accidents and emergencies. Significant environmental impacts, including GHG and resources consumed, need to be listed and quantified thus forming an *inventory of resources use, emissions, effluents and wastes*, as a basis for environmental management.

1.2.9 Energy Performance

The *energy performance* of a technical device that converts energy from one type into another is defined as *efficiency* and is expressed by a dimensionless *output/input ratio*.

The energy performance of an activity is expressed by the ratio of energy units and the quantified results or aspects of this activity.

From a business perspective, energy is a cost item that can sometimes amount to up to 25 % (or more, depending on the industry) of the overall production costs. Energy costs are usually the largest production cost where significant scope for improvement still exists. The basic bottom line equation, where energy is figured out through the costs it incurs, brings business and energy performance together:

Sales Revenue – Production Costs = Gross Profit

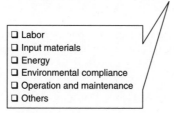

If we can reduce any of the cost items, *profit* – the key measure of overall business performance – will improve.

1.2.10 Environmental Performance

Environmental performance can be defined in a number of ways. It can be expressed by:

- quantified annual releases to the air and water;
- annual amount of waste disposed;
- amount of waste recycled;
- number of spills or accidents;
- notices of violations or fines;

- costs of environmental compliance;
- number of environmental awards;
- participation in voluntary environmental programs, etc.

Environmental performance objectives can be just

- *compliance oriented* – to satisfy mandatory external regulatory requirements; but, it should also be
- *improvement oriented* – to better manage operational performance, rather then just measure its environmental impact, in order to assess risks and address root causes of environmental problems, and to reduce costs of environmental compliance.

1.2.11 Performance of Materials Use

One of the most neglected areas in performance management is *material productivity*, or performance of material use. Productivity is the most important economic indicator for understanding and predicting economic prosperity and business competitiveness. Productivity improvement increases competitive strength of companies and economies in the international trade (see Box 1.1).

Box 1.1: On Productivity

Generally, productivity is the measure of efficiency of using resources to produce goods and services for the market. It is measured by computing the ratio of output index to input index. Labor productivity denotes productive efficiency of labor, while total productivity denotes efficiency of using all factors of production, such as capital, labor, raw materials, and energy.

General factors contributing to changes in productivity are:

- *technological progress;*
- *education and training;*
- *economy of scale;*
- *improved resource allocation;*
- *legal-human environment.*

The internal factors affecting productivity variations are:

- *output: mix, variety, quality;*
- *production factors:*

  - *technology intensity, age of plant and machinery;*
  - *scale of operations;*
  - *design for manufacturing;*
  - *labor skills and motivation;*
  - *raw materials and parts;*

- *capacity utilization;*
- *organization of functions and tasks;*
- *performance management practice.*

The external factors include demand, relative input prices, competition rules, government ownership, unionism, laws and regulations, etc.

The common questions that arise on material productivity are:

- How much of total expenses are spent on input material?
- How efficiently are resources used?
- What is the potential for:
  - reduction of raw material losses;
  - reduction of raw material use;
  - reduction of packaging material use and waste;
  - recycling of waste.

- What are the annual costs of treatment and disposal of waste materials?

The answers to all these questions are seldom readily available. Raw materials can amount to 30 %–80 % of manufacturing costs, and losses of 5 %–25 % are not uncommon. The cost of waste is often more than four times greater than the facility realizes. Actual costs can be established only by the continuous measurement and evaluation of the performance of the use of materials.

1.2.12 Environmental Management

Environmental management should consider the relevant legislation, regulation and standards and should aim at achieving environmental compliance, controlling and reducing pollution and environmental impacts from resource use in industrial operations. The evolution of environmental management shifts the focus from compliance issues to a more risk-based focus on potential liabilities associated with process changes and day-to-day operations. The business value of environmental management comes from fewer impacts on the environment, reduced risk of accidents and reduced costs of compliance.

It is surprising to see how many companies, having analyzed their *environmental costs*, discover that these costs are at an order of magnitude higher than the company initially supposed. These costs arise from

- capital and operational costs for pollution control equipment;
- waste disposal fees;
- environmental training;
- monitoring, record keeping and reporting, etc.

Environmental management establishes and implements an environmental policy, sets its goals and expectations, establishes a system for monitoring the environmental performance of industrial facilities and implements procedures for continuous environmental performance improvements and reduction of present and avoidance of future environmental compliance costs.

As this definition indicates, environmental abatement technologies and engineering solutions for assuring environmental compliance are not included within the scope of considerations in this book but useful references are provided at the end of this chapter.

1.2.13 Energy Management

Energy management is concerned with the efficient use of energy, water and other material resources, waste minimization in manufacturing operations and continuous improvement of performance of resources use in a company. Energy management specifically links and relates energy use to production output, aimed at achieving the required level of output with the minimum use of energy and other resources. *Energy management* implements an energy policy, sets its goals and expectations, establishes a system for monitoring energy performance and implements procedures for continuous energy performance improvements.

Improvement in energy performance will be reflected directly as the increase in profits of a business. A simple example illustrates its potential:

IF	energy costs are 20 % of all production costs
AND IF	gross profit is at 15 % of all production costs
AND IF	10 % improvement of energy performance is achieved
THEN	the result is 2 % decrease in production costs **which is equivalent to 13.33 % increase in profit!** (all other factors assumed to remain constant)

This profit increase will happen with no increase in sales volume, which is a more painful way of increasing profit! Worth exploring, isn't it? Knowing that when you focus on energy performance improvement very hard, as a side benefit you will achieve performance improvements in other cost categories as well!

Energy equals cost – and energy management equals cost management. Thus, energy performance improvement means business *profit improvement*! That is why *top managers* also need to be interested and involved in an energy management program. This is emphasized here and it will be repeated later, because experience shows that top management commitment is essential for a successful energy management program. In order to make energy management sufficiently important as a topic for top managers, comprehensive arguments for energy management, as a relevant and positive contribution to overall business performance, should be used.

1.3 Energy and Environmental Management System

With energy management and environmental management defined separately, we can proceed to introducing the term 'Energy and Environmental Management' (EEM), and to propose a definition of 'Energy and Environmental Management Systems' (EEMS).

EEM is an integrated management approach concerned with the energy and environmental performance of a plant. Consequently, when we refer to 'Energy and Environmental Management' (EEM), it always relates to the energy and environmental performance of a process that uses energy and other resources, and the performance of people who operate the process. In fact, any energy and environmental management has to focus on people and the way in which they operate and maintain machines and processes as the key factor for optimization of energy and environmental performance. For large companies, manufacturing activities should be broken down into areas with specific process operations or tasks with recognizable inputs, activities, and outputs. Performance improvement will come as a result of implementing optimized operational procedures at the work floor level.

When manufacturing processes are complex, a high level of knowledge and adequate process information will be required for performance evaluation and management with continuous monitoring, control and follow-up information, supported by accurate measurement of data on energy, production and environmental impacts.

Therefore, applying the EEM requires a performance measurement system and involving people with advanced evaluation skills and the ability to make decisions supported by the necessary information. These key components – people, performance evaluation procedures and performance measurement and IT equipment – together with an underlying knowledge are the cornerstones of the EEMS concept (Fig. 1.3).

We can now outline the concept of the Energy and Environmental Management System (EEMS) in the way it will be elaborated upon in this book, as an energy and environmental performance improvement implementation framework that ties together:

- people with skills and assigned responsibilities;
- performance measurement system;
- performance indicators;
- performance evaluation based on direct measurements;
- performance monitoring procedures.

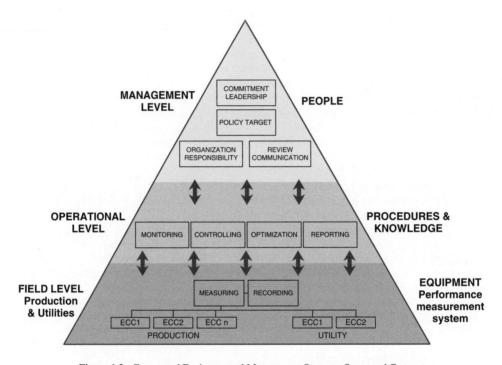

Figure 1.3   Energy and Environmental Management System – Scope and Concept

Energy and environmental management (EEM) focuses on clearly specified performance targets and objectives related both to people's performance and underlying process performance. An energy and environmental management system (EEMS) is a tool for achieving these targets and objectives through a system of metering, monitoring and evaluation of energy and environmental performance. Finally, we can offer the definition of Energy and Environmental Management System as:

a specialized body of knowledge with an underlying organizational and implementation structure which integrates interrelated elements, such as

- people with skills and assigned responsibilities;
- declared policies with clear objectives and targets;
- defined procedures and practices for implementation;
- established metering system for performance monitoring;
- action plan for continuous improvements;
- reporting system for checking on progress and communicating results

with the goal of achieving continuous energy and environmental performance improvements.

In their intent, the terms EEM and EEMS are similar inasmuch they both focus on continuous performance improvement, hence they will be used intermittently throughout the book. The main difference between the terms is that EEM refers mostly to a management concept, a program or an approach, while EEMS is a tool or a specific implementation project that realizes the EEM objectives.

1.4 Objectives of Energy and Environmental Management

Energy in all its forms – electricity, fuel oil, solid fuels, gas (natural, LPG), water (chilled, hot, treated, industrial), steam, air (compressed, cold, hot) – is a medium that crosses departmental and functional borders within a company, because it is used everywhere! It is used by **people** and by machines that are operated by people. Correspondingly, performance of energy use concerns everybody in a company. Likewise, the activities of all the people in a company result in more or less significant impacts on the environment, depending on their place of work and type of manufacturing operation.

Energy management should be concerned with both the *cost effectiveness* of energy distribution to and use by a manufacturing process, and with *efficiency* of energy transformation, generation, conversion and distribution within an *existing plant*. Environmental management is concerned with regulatory compliance, minimization of environmental impacts from manufacturing operations and reduction of emission, effluents and waste. Both energy and environmental management are concerned with the efficiency of raw material use and waste minimization, because any unnecessary wastage means at the same time increased energy use and increased environmental pollution.

The objective of an energy and environmental management system is *continuous improvement of energy and environmental performance* throughout the plant with the main goal of reducing the operational costs, minimizing waste and reducing the environmental impact of the company's operations.

To achieve these objectives, EEM must have good technical foundation, but equally, if not more importantly, a strong managerial component! Energy is used everywhere in the company and human behavior affects energy and environmental performance, for that reason *dealing with people* is of key importance for long-lasting performance improvement!

1.5 Dynamics of Energy and Environmental Management

An existing facility has been designed on the basis of certain parameters which almost inevitably change by the time it is in operation. Over the course of their lifetime, the performance of the machines will deteriorate due to wear and tear or inadequate maintenance. Then there are always variations in performance because of the essentially dynamic nature of industrial operations (Fig. 1.4). The volume and type of orders are changing, production output varies, the quality of raw material differs, and people's behavior is a source of constant uncertainty. Even the work of the most experienced and dedicated operators can fluctuate, operational problems may occur, changes in the environment affecting performance can happen.

The challenges are to operate the facility in the best possible manner against the background of a changing environment. The critical factors for success will be the availability of reliable and relevant data.

The key variable for performance assessment is *production output*. We cannot speak about energy and environmental performance as an absolute category without regard to the production output. Different levels of production output, or various product types, require differing amounts of energy and should result in different amounts of emission, effluents and waste.

Therefore, energy and environmental performance data will need to be gathered continually, alongside production output data for particular products and production operations. Aggregated monthly data on energy, production and environmental impacts are usually available, but this data are not sufficient for an informed performance evaluation because the impact of the human factor, i.e. how people operate the machinery, cannot be derived from such data.

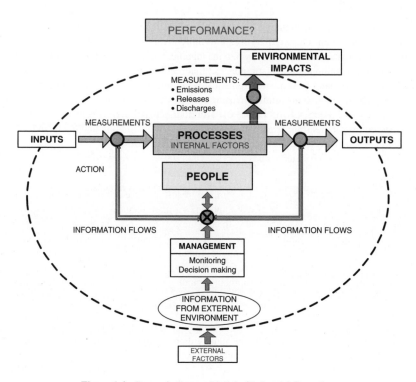

Figure 1.4   Dynamic System Model of Industrial Operations

An elaborate performance measurement system that will provide the necessary disaggregated daily data on important operational aspects is an essential tool for performance monitoring and evaluation, of both people and machinery at the place of work and decision making on improvement measures.

This also clearly suggests a necessity to assign *responsibility* for energy and environmental performance, decentralized down to the places of work where points of energy use are, and *focused on people* performance in each business segment that consumes energy.

Unless responsibility for energy and environmental performance is assigned explicitly to a specific person, it is illusory to expect performance to improve. This fact brings us back to people as the key factor for energy and environmental performance improvement.

1.6 Human Aspects of Energy and Environmental Management

> *Introduction of energy and environmental*
> *management*
> *is introduction of CHANGE!*

The main objective of EEM is energy and environmental performance improvement. To achieve performance improvements, people will need to *change* their *attitude* about energy use and improve operational practices and routines. This is the truly difficult part – *to change people's behavior*!

Employees must share the ownership of energy and environmental performance problems – no amount of monitoring, measurement or investment in technology can do the job without the aid and touch of a human hand. When considering energy and environmental management, most people tend to think of information first (data, monitoring, measurement, analysis and decision making), followed by procedures and equipment. Unfortunately, even with sophisticated equipment and exact information, unless people are persuaded to change their entire attitude to using energy, vast areas of day-to-day inefficiency will still remain. Machinery and processes will be operated without concern about energy efficiency, optimizers will be by-passed, meters read incorrectly, lights left on, taps left running and doors and windows left open and, as a consequence, environmental performance will deteriorate.

Information remains inert data unless it is processed and acted upon by people. Investment in machinery remains lifeless hardware unless people operate it intelligently and carefully. This is why, for an effective energy and environmental management, we have to include the third dimension – *human involvement* – where the main challenge is to change the people's attitude to energy use. This is the same for any other objective of performance management. Usually, this is the most difficult task for the managers because they also need to change themselves along the way. Managers need practical guidance in order to help them steer the process of implementing EEMS from introduction to successful implementation. This is what we will offer here, using as little professional jargon and theory as possible.

In the following sections of this chapter, we will emphasize the relevant aspects of change while dealing with people when introducing the EEMS project.

1.6.1 Change

> *Change means a new way of doing things!*

In this context, *change* means a qualitative new way of treating energy and environment. As such, it always means partial restructuring of established habits and routines. EEM requires new patterns of behavior to evolve – both at the level of individual actors and at the general organizational level. Accordingly, change is understood as an innovation of social, communicative and organizational features which foster taking up EEM techniques.

Implementation of EEMS can be viewed as implementing a project of *change*. To achieve transformation, i.e. to improve energy and environmental performance, people must be willing and capable of changes. *Willingness* has three dimensions:

- awareness,
- motivation; and
- skills.

Awareness means that people, from top management down to the workshops, know what energy forms are, what may be the environmental impact of their activities, what the energy costs are, and where the potential for energy savings is.

Motivation is necessary in order to trigger people to change their attitude to energy and use of materials in order to cut energy costs through more efficient and responsible operation of machines and other energy uses, providing that they have adequate skills and competencies.

These prerequisites of change – *Skills-Awareness-Motivation* – can be shown together in an *S-A-M* diagram (Fig. 1.5). If any of them is missing or is inadequately developed, a training program will need to be devised and implemented. There is one more vital ingredient for the success of change – *leadership*! Without leadership any change project, EEMS included, will lose its purpose and eventually fail. Leaders

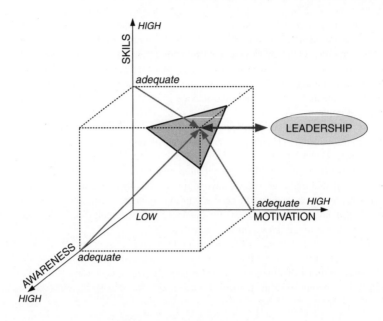

Figure 1.5 Prerequisites for Change

at all levels of an organization must ensure that plans are put rapidly into action, rather then striving for perfection and precise analyses. Leaders should inspire and motivate people and provide them with the tools and knowledge for achieving the targeted energy and environmental performance improvement goals. In order to do that, leaders will often need to change themselves first in order to get ready and lead the staff to performance improvement.

When we are close to the indicated area in the S-A-M box (Fig. 1.5), i.e. where skills, awareness and motivation are high, with leaders at the helm of the EEMS ship, the change process will be on its way, and energy and environmental performance may start to improve. Nevertheless, there is a long way to go before arriving there.

1.6.2 Aspects of Change

Our experience has shown that there were a variety of impulses, coincidental events, personal contacts and fostering factors, which created different framework conditions for change at every company we have worked with. These factors range from the culture of the country where companies operate, organizational culture, social relationships within a company, to management and decision-making styles, level of skills and competencies and, of course, willingness and motivation for change.

At every such instance, our role has been clear: we are an external partner and external manager of change, supporting the internal stakeholders in achieving change objectives. Each group or individual stakeholder has had their own motivation and perceived difficulties and barriers for change, which need to be properly identified and understood in order to be able to move forward.

This is not to say that an external partner is a must because, surely, these are companies able to carry out the change process by themselves. However, sometimes external partners may provide the first shove for energy and environmental (EE) improvement projects, or help to legitimize and encourage the proposed EE performance improvement projects inside the company, especially in cases where key internal actors do not belong to top management.

The most important internal stakeholders are always top managers. They are the decision makers, and without their involvement and the decisions they need to make, the change process cannot even be triggered, let alone completed successfully. Their motivation comes from reduction of costs, saving money, innovation, continuous development and, of course, the improved profitability of the company. External factors such as regulatory framework, green image, social responsibility or simply orders from headquarters play a role as an initiation event, grabbing the attention of the top management for energy and environmental issues.

A critical role in the process of change is always that of the so-called '*change manager*', i.e. the person(s) identified by others as the key promoter and supporter of change. Such a person is usually of a technical background with a good understanding of the production, energy and environmental aspects of the business and with an interest in identifying areas where the company can save time and money.

The next important role is that of '*change agents*', persons made responsible for energy and/or environmental issues (in this context responsible for change) in each functional unit of the business. Working with energy/environment is a way for them to achieve economic benefit for the company and to gain some personal recognition.

The success of the change process depends on the dedication and commitment of internal change managers and change agents. Usually, their technical knowledge and abilities are not matched by those of decision-making powers. Sometimes, it is difficult for technical staff to gain access to decision makers. If this is the case, then external partners may be invited to provide support by reinforcing their messages and actions at the top management level.

What also proved to be important was the early achievement of some visible results, i.e. verifiable energy savings, waste or emission reductions, etc. Such a success will boost the confidence of change agents, increase their recognition by management and, usually, provide stronger support and greater good will from the management.

The performance improvements have to be well illustrated and documented in order to be understandable for the company's non-experts and the general workforce. This helps in removing the usual barriers when addressing some aspects of change, as described in the following text.

1.6.3 Barriers to Overcome

There are many perceived barriers to improving energy and environmental performance. Some of these barriers can be attributed to:

- ignorance;
- fear of change;
- vagueness of the subject;
- desire to preserve the status quo.

The other most common barriers can be summarized as shown in Figure 1.6. There may be other barriers not addressed here, but in practice, if the reality of skills-awareness-motivation circumstances is assessed and treated adequately through a training program, with leaders taking charge, the barriers will gradually melt away.

1.6.4 Unawareness of Opportunities

Another barrier to energy and environmental management is expressed as: 'We don't know what we can do'. This may be true in the short term, but there is much that can be done to identify the real opportunities.

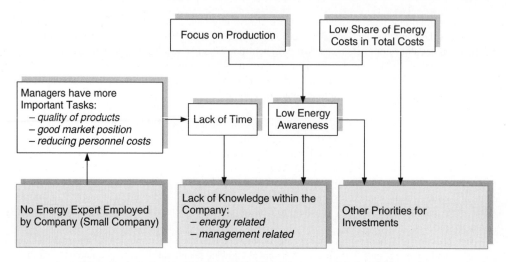

Figure 1.6 Some Barriers for Energy and Environmental Performance Improvement

One way is to hire an expert to identify and quantify opportunities by conducting an *energy and environmental audit* (see Part I Chapter 4). Although this will obviously incur some costs, it may reveal a number of opportunities for energy saving and environmental performance improvements.

Alternatively, for no cost you can ask: 'Where, why and how energy is used?'; 'How much could we save?'; 'What would we need to change and save even more?'; 'Where are the sources of environmental impacts?'; 'Can we reduce the amount of waste that we generate?' and so on. The answers to questions such as these are always very revealing and invariably uncover opportunities in the no-cost and low-cost categories of improvements, capable of rapid implementation.

1.6.5 De-motivating Aspects

Although the costs of resources, including energy may occupy the minds of managers, they are certainly not the main worry of employees. There are a few reasons for this:

- Most people tend to concentrate on the job at hand rather then on the means or facilities used. Energy is a particularly 'invisible' resource and it is taken for granted.
- Most people may occasionally think about energy and environmental issues but daily operations are repetitive, subconscious and rather boring: a monotonous routine.
- Most people do not view energy costs in the same way at the work place as they do in their own homes because the costs are not paid directly by them.
- People tend not to see direct links between their actions and environmental performance.

These factors provide real challenges for management. How do you overcome disinterest and lack of motivation and at the same time build on growing energy and environmental performance concerns?

1.6.6 Motivating People

If the above reasons can be classified as 'demotivating' factors, nowadays there is a relatively new 'motivating' factor: global *climate change concern*. There is an increasing general awareness of energy as a diminishing resource, costing more and more money, and greater knowledge on climate change resulting from energy generation and raw materials use.

Motivation and thus willingness to cooperate enthusiastically with management tends to increase

- if people are given the opportunity of personal involvement in making decisions about actions that will affect them and their performance;
- if people are properly informed about the real situation, problems and reasons for decisions;
- if people are given the authority to decide on the most effective way in which to carry out their own work;
- if people are given recognition for their personal contribution;
- if people believe that their managers are genuinely interested in them as individuals;
- if people are given incentives and rewards (in addition to their regular remuneration) for exceptional efforts – not only of material but also emotional value to them;
- if people also understand the consequences of failure for them personally.

When actually implementing an EEMS project, the greatest challenges can be expected to be found in manufacturing departments where energy and environmental performance is not usually one of the standard reporting items, as opposed to utility departments where energy is the main subject of reporting.

When a proper energy and environmental performance measurement system is installed and operational, it can reveal some unflattering information about the energy and environmental performance that can make people feel defensive. Consequently, an EEMS project will likely require more changes and therefore more motivational activity in the production areas. Dealing with employees in these departments in a sensible way and having enthusiastic leadership in this location within your organization is of high importance.

This brings us back to the issue of *leadership* that we have already introduced earlier as the key component of successful EEMS project.

1.6.7 Providing Leadership

We all know that good management is much more than just planning and administering. It also requires leadership and *effective interaction with human beings.*

In order to succeed, an EEMS project must have champion(s) or leader(s) who will relentlessly pursue opportunities for energy and environmental performance improvements. These individuals must be true believers in the needs and possibilities for performance improvement and must have sufficient personal and professional authority to be respected, listened to and followed as leaders. But they must also be aware of what leadership is about as a role and what is or should be their management style in order to successfully mobilize people to follow them. Box 1.2 gives a quick overview of *leaders' roles and responsibilities.*

Box 1.2: On Leadership – Leaders' Roles and Responsibilities

I. ROLES

I.1 Building and maintaining the team

People expect their leader to help them achieve their tasks and to build the synergy of teamwork.

I.2 Developing the individual

They also expect the leader to be receptive and responsive to their individual needs and help them to develop and improve themselves.

I.3 Achieving the task

The leader has clear responsibility to achieve the task with the team assigned and, if the task is not achieved, only the leader is to be blamed.

Box 1.2: Continued

II. RESPONSIBILITIES

II.1 Defining the task

Leaders have to assign tasks to the people that are

- *clear;*
- *specific;*
- *time-limited;*
- *realistic;*
- *capable of evaluation.*

The task must convey a clear sense of direction to a person, i.e. why it is given and what is required in order to carry it out.

II.2 Planning

Planning means building a mental bridge from where we are now to where we want to be when we achieve the given task or objective. In short, planning gives the answer to the question 'who does what, when and how?'

II.3 Briefing

A function of communicating objectives and plans to the team, usually in a face-to-face manner. The sister function of communicating is listening!! It is equally, if not more, important.

II.4 Controlling

Controlling is a function ensuring that all team efforts and energy will result in achieving the objectives.
    Sometime teams are inefficient, like old boilers, but the leader needs to tune them up and get maximum efficiency out of them.

II.5 Evaluating

Whenever we review the progress or lack of it, we evaluate our people. One thing for sure: no team or person is perfect.
    Therefore, evaluation must happen in an environment where:

- *objectives are clear and realistic;*
- *there is an atmosphere of openness – two-way communication;*
- *there is an ability to handle failures.*

II.6 Motivating

There is always a lot of talk on motivation. But firstly, leaders must be motivated themselves, then they should give clear targets to the staff, give them feedback on progress, if any, provide fair rewards, and give recognition.
    People have to see leaders as an example and they need to be inspired by leaders.

II.6 Organizing

Making workable plans and structuring people to facilitate two-way communication and teamwork. It also involves a measure of control and supervision.
    It requires active and continuous attention by leaders. Because people are like equipment (if we do not maintain and take care of it, equipment will soon run down). Similarly, if we do not organize, motivate and supervise team work, people will lose their sense of direction and motivation.

II.7 Providing an example

'Leadership by example!'

If managers want to do more than just follow the principles of motivation and really want to get the best out of people, they will have to become more aware of their management style. Managers, and particularly energy and environmental managers, need to be able to create true awareness of resource values and of the need to improve energy and environmental performance and avoid unnecessary waste. This does not just mean issuing unilateral instructions. It means *interactive dialogue*, i.e. *two-way communication*.

Two-way communication does not only:

Demand   ⇔ *it also asks people*;
Impose   ⇔ *it also obtains agreement on targets*;
Measure data ⇔ *it also provides feedback*;
Order   ⇔ *it also motivates by incentives*.

Good leaders also know how to attract good people and how to establish strong and diverse leadership teams with clearly defined roles and commitment to common goals. They build cultures that inspire and motivate, and they provide the tools and learning to help their people succeed.

Leaders must be decisive. They must ensure that their strategies are directionally correct and put them into action promptly, rather then striving for perfection or precise analyses. They need to mobilize quickly and with flexibility in order to evolve their strategies over time.

An effective implementation of EEMS is based upon work that is performed across functional boundaries; hence it requires the creation of cross-functional teams and leaders that can enable different talents to work together. It must start at the top of the organization. Therefore, to convince the workforce of the importance of EEMS project and to provide the right leadership, the whole management structure must be involved into the project as follows:

1. TOP MANAGEMENT TEAM:

   - Factory Manager
   - Financial Manager
   - Human Resource Manager
   - Production Managers
   - Utility Managers
   - EEMS Manager(s)
   - A&M Coordinator

2. OPERATION TEAM:

   - EEMS Manager(s)
   - QA manager
   - Team Leaders – Line Managers, Supervisors (Departmental Teams)
   - Meter Readers and/or PC Operators
   - A&M Coordinator

3. PERFORMANCE IMPROVEMENT TEAMS:

   - Team Leaders
   - Shift Foremen
   - Machine Operators

People at every organizational level are change agents who, when properly empowered, inspired and guided, come alive with vitality, intelligence, and exceptional profitability potential that can make or break an EEMS project.

Let us consider how we can gradually move forward, starting first with top management.

1.6.8 Top Management's Visible Commitment to EEM

Every change process is painful enough even if all the stakeholders are firmly on board. For a subject such as energy and environmental management, which does not immediately strike one as the highest priority, this is even more important. The first task is to make energy and environmental management important and to establish it as a topic for top management.

If the *commitment of top management* cannot be achieved during the project start up, the activity is destined to fail. In order to achieve top management support, suitable arguments and adequate presentation have to be prepared by internal or external key actors. The argument for EEM as a relevant and positive contribution to business performance has to combine pure economic considerations with additional benefits, focusing on the company's multi-dimensional aspects of competitiveness, such as cost reduction, environmental compliance, improved quality, 'green' image, etc.

Few people would be immediately enthusiastic about any proposed change. It is particularly so when change will bring an additional responsibility, like that for energy and environmental performance. In such cases, one is likely to encounter a 'tree-in-the-wind' effect, as illustrated in Figure 1.7. In this metaphor, the tree is a symbol for the organizational hierarchy. When the wind of change blows, the shaking will be strongest at the top, but as we go down to the bottom of the tree, the lower branches (lower levels of the organization) are calm, protected and unaffected by the big noise up at the top.

As we have said earlier, energy is used everywhere and by everyone in a company and every individual's behavior affects the environmental performance, so an EEMS project must involve everyone in order to give the best results. Consequently, all employees must be mobilized to assume adequate responsibility for energy and environmental performance based on their position within the organization. This can be achieved only if employees are convinced that EEMS is really the matter of high importance to the company.

That is another reason why top management's *visible commitment* to EEM is important. If there is no visible commitment at the top, it is highly unlikely that there will be any thought of change at the lower levels of the company's hierarchy.

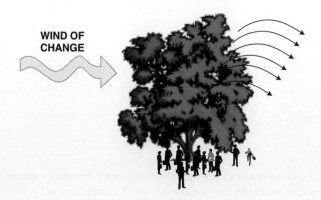

Figure 1.7 A Tree in a 'Wind of Change'

1.6.9 Decisions to be Made

Further, a very pragmatic reason for top management's active involvement is a number of decisions that have to be made at the outset of EEMS implementation, such as:

- adopting an energy and environmental policy;
- approving an energy and environmental management action plan;
- assigning responsibilities;
- allocating a budget;
- deciding on rewards and incentives;

An energy and environmental policy should stimulate and support the introduction of energy and environmental teams and social circles and other forms of internal coordination structures such as meetings, working groups, etc. Formal responsibilities have to be supported by corresponding changes in company culture, leeway to act, management support and sufficient resources. Attempts to enforce *organizational changes* without hierarchical backing incorporate the risk that only superficial adaptation will take place.

New rules of conduct with regard to energy and environmental issues, responsibilities, broad involvement of staff, effective communication structures, as well as internal rewarding of suggestions schemes contribute to an institutionalization and strengthening of EEM activities within the company.

Decisions on *rewards and incentives* are particularly sensitive. Although it may be argued that continuous improvement of energy and environmental performance is in everyone's job description, some additional incentives will surely help to overcome people's natural resistance to change. Some companies have tried offering financial rewards to the people or groups that have achieved energy cost reduction but others are reluctant to offer financial incentives due to the possible abuse of the scheme or the negative effect on people who feel excluded from it. The alternatives to direct financial incentives include:

- recognition;
- certificates;
- energy/environmental personal identification, i.e. employee of the month;
- bonus for recreational activities;
- educational fund for staff's kids;
- contribution to a charity fund of choice, etc.

Measures on rewards and incentives can be supplemented by a mix of motivational and awareness activities such as:

- energy and environmental exhibition;
- energy and environmental day;
- energy and environmental motto contest;
- case study visit.

How people are inspired, empowered and rewarded for their ability to adapt to changing conditions is the means by which objectives are achieved. Motivating and mobilizing people to become enthusiastic about an EEM program depends strongly on *explicit* and *specific communication* about the scope of change, i.e. the *reasons, objectives and targets*. The important thing is to keep the employees regularly informed and updated on the progress of the energy and environmental management program. Therefore, performance changes and improvements should be reported on periodically to employees. So, another part of the energy and environmental management activities must be regular communication and public relations, making people aware of what is happening, why it is happening and what they can do to help. The success of the project has to be documented and communicated in order to provide a sense of success to all the actors involved. Feedback of positive developments contributes to an increased motivation to continue.

Internal communication is another area that requires top management decisions on the ways, means and frequency of channeling information about an EEM program's activities, objectives, targets and results; *top-down and bottom-up*. It includes issues like:

- *What needs to be communicated*:
    1. Company energy and environmental policy and action plan
    2. Monthly energy consumption charts
    3. Quantified environmental impacts, related targets and improvements
    4. Energy and environmental tips and knowledge
    5. Reward/incentives

    Possible Communication Channels:

    1. *Top – Down*
        - formal announcements
        - monthly message from management
        - internal staff meetings
        - training courses
        - notice-boards
        - newsletter, posters, stickers, leaflets
        - lunch time video show
        - energy and environmental portable library
        - internal radio.
    2. *Bottom–Up*
        - employee suggestion scheme
        - weekly/monthly report
        - opportunities for performance improvements (OPI).

Organizational profitability is derived from the *profit-ability* of employees. If employees do not have the concrete means to change, align and improve their habits – the same means that are understood and utilized by management for the improvement of its own strategies – there will be breakdowns in communications.

Large companies may benefit if a dedicated person, an *A&M Coordinator*, is assigned on a permanent basis to take care of internal communication on energy and environmental awareness and performance. Performance improves and profit increases when the workforce can truly see and map the cause and effect of each individual's action.

If the results are communicated outside the company and gain some recognition there, it will foster the motivation and provide confidence for the key actors that they are on the right track. They can also find some inspiration through external contacts and networking within energy, environment and business associations.

1.6.10 Visibility

Holding people accountable for specified results is a critical point on the path to success. Defining objectives, setting performance targets and measuring achievements over time are essential steps in converting performance improvement strategy from a raw concept into a detailed plan and ultimately into results. This would not happen without the persistent and visible commitment of the top management.

What does '*visible commitment*' mean? It means that top management is directly and clearly associated with the EEMS project by doing all of the following (and possible more):

- Not only defining, but also declaring EEMS project objectives.
- Facilitating and monitoring the progress in achieving the targets.
- Providing feedback on the staff's energy and environmental performance reports.
- Encouraging the staff to make proposals for performance improvements.
- Providing recognition for achievements.
- Having energy and environmental performance regularly on the agenda of management review meetings.

All of these measures must be carried out continuously in order to ensure the endurance of any performance improvement. How does one do that? That is what the Energy and Environmental Management System is all about, and we will explore it step by step throughout the rest of the book.

1.6.11 Working with People Towards a Successful EEMS Project

1.6.11.1 Managers

The manufacturing processes are places where additional value is created. The value creation is carried out by people grouped together according to the organizational arrangements needed for a process to work. *Managers* are the most important group within a company because they cement people within the groups into teams connected by common goals. The people within groups and groups among themselves establish relationships. Therefore, organizations are relational systems and can achieve high performance and continuous improvement only by empowering and motivating their people to do so.

The realization of EEMS cannot be seen as a single and purely market driven economic decision but requires the initiation and management of related social processes in the firm involving various internal and external actors. The success of companies working with EEMS is dependent on different internal factors such as company culture, technical capability or social integration of energy and environmental issues. However, since companies as social units interact with their settings, external conditions, including policy programs, directly or indirectly influence the process of implementing energy efficient (EE) and environmentally friendly technologies or establishing EE behavior.

Managers need to integrate a number of factors in order to enhance organizational ability to adopt changes and perform. Their focus must be on the common goal – performance improvement – which requires the commitment of the people and agreement upon the course of action.

Companies differ in characteristics and requirements – so there is no 'one-size-for-all' approach that can secure a successful EEMS project. This is especially so when it comes to people because success in working with people is highly dependent on cultural and social aspects, which are 'soft' issues. At the beginning of the process, the company is exposed to key events, which stimulate the key actor (change manager) to think about an EEMS project and initiate planning, decision and implementation processes. This is influenced by the existing context of personal motivation, experience, energy and environmental know-how and organizational conditions.

Companies represent social systems which are influenced by networks of internal and external inter-relations and interactions. Key actors within these networks – *managers providing leadership* – are instruments for stimulation, realization and diffusion of energy efficiency and environmental improvement activities. To be effective as a manager requires one to be good at managing people and relationships within and across functional groups.

The workforce is a living organism. If leadership is to effect dramatic changes in a company, the workforce needs to know how to follow and how to respond correctly to changing events. Managers inspire constancy and unity of purpose for continual improvement of organizational profitability. All stakeholders are willing to provide or accept leadership when needed and will follow directives when goals, means and deadlines are clear, and when they feel competent and able to do so.

1.6.11.2 Employees

Employees, with their roles, interactions, skills and (de)motivational factors are key dimensions of the realization of EEMS. Employees are organized into functional units or teams but a unit is not just a mere organizational box! *Units consist of people – the unit is people!* Where a single individual may despair of accomplishing a seemingly complex task, units or teams nurture, support and inspire each other. The challenge for team leaders is to form a cohesive unit from people of varying capabilities, personalities and work styles. The team must understand how performance standards and targets are established. They need to accept them as a guide for a common purpose and direction.

Each team member has to be assigned to a right role based on his or her skills. If the skills are not adequate, training is required. Regular team progress meetings are essential in order to build team spirit, to develop better understanding of the tasks ahead and to get the members working together effectively toward the targets. Any initial hesitation and insecurity has to be overcome quickly and a 'We can do it!' attitude has to prevail. The teams may need some external support in order to reach that point and some positive reinforcement of their performance, but once they are there, we are on the right track!

In our experience, to build and achieve people's '*confidence*' is the key ingredient for success.

When teams work in a spirit of cooperation and share confidence and mutual respect, this enables them to make decisions effectively and with a minimum of internal disagreement, which will lead them towards the achievement of the established targets. A competent and motivated workforce will be able to manage day-to-day uncertainties proficiently, always improving its ability to do so, measuring the effect of its actions by feedback in order to adjust the next action, thus improving skills and learning by doing.

1.7 Initiating Training, Awareness and Motivation Programs

> *Training comes first!*

It would seldom be the case that a company is in a position to embark on a major performance improvement program without a need for additional *skills development*. More often, an internal *training program* would be required in order to raise awareness, improve individual qualifications and provide the necessary skills for the workforce. Training is a process of teaching and developing skills. Training can be carried out in a formal (training courses) or informal (meetings and workshops) way for different target groups and on various subjects. An important training objective should be to develop a positive 'it can be done' attitude among employees. Such confidence will improve their '*self-efficacy*', i.e. ability to adopt and implement the process of change. Self-efficacy for further action is increased by a better knowledge of the energy and environmental aspects of production and new norms and ways of treating energy and environmental issues gained during implementation.

Energy and environmental related qualification and training increases internal know-how and contributes to the higher motivation and confidence of employees. These individual training and learning processes represent a crucial element for initial success and continued action.

1.7.1 Management

No group is more important to EEM than the *management team* which must allocate resources and provide support to the EEM program. In addition, it often provides input and assistance with the process itself. Specific ways to work with the management team should be carefully planned and executed. One approach is to conduct a series of preparatory and progress review workshops in order to get the program going. The 'workshop' is a structured meeting with clear objectives, agenda, some presentations and discussions.

An introductory workshop titled 'Managers' Roles and Responsibilities for EEMS' can launch the process with the aim of clarifying roles and responsibilities, shaping critical skills and changing perception, as well as enhancing support for EEMS (Box 1.3). The objective is to have managers leave the workshop with an improved perception of the impacts of EEMS on business, the relevance of their area of responsibility and a clearer understanding of the role of EEMS. If the workshop is successful, managers will have a renewed commitment to making EEMS more effective in their departments.

Box 1.3: Management Roles and Responsibility for EEMS

MANAGEMENT TEAM

1. *Monthly review of energy and environmental performance reports from operation team.*
2. *Making sure that all action plans are as scheduled.*
3. *Making decisions and preparing budget for all activities concerning EEMS.*

   - *Rewards/incentive*
   - *Awareness & Motivation(A&M) activities*
   - *EEMS Instrumentation*

OPERATION TEAM

1. *Weekly review of energy and environmental performance reports from departmental teams.*
2. *Providing feedback to performance improvement teams.*
3. *Providing technical support to performance improvement teams.*
4. *Communication – horizontal and vertical.*

EEMS MANAGER

1. *Organize and manage EEMS operation:*

   - *Meters – installation, maintenance and calibration*
   - *Meters – reading verification and check-ups*
   - *Supervising performance data key-in and processing*
   - *Report to management team*
   - *Giving feedback to operation teams on their performance directly*
   - *Giving technical support to operation teams, if required*

TEAM LEADERS

1. *Responsible for energy and environmental performance in their areas of work.*
2. *Organize A&M training for their staff.*
3. *Organize and motivate their staff in order to improve energy and environmental performance following general A&M training.*
4. *Communication.*
5. *Be the leader in their area of responsibility.*
6. *Actively identify opportunities to reduce energy and raw materials consumption while maintaining production quality and quantity.*
7. *Take corrective action.*

A&M COORDINATOR

1. *Responsible for coordination and implementation of all A&M activities as approved by the management, on an on-going basis.*

How many preparatory workshops will be required depends on the extent to which energy and environmental management has been developed in a company prior to the decision to embark on a major energy and environmental performance improvement program. To assess the extent of current energy and environmental management practice, a simple questionnaire may prove useful (see Toolbox III-1).

1.7.2 Technical Managers and Supervisors

As we go down the management hierarchy, training will need to be more technical and specific to the needs of the staff in order to bridge their *competence gap*. To be able to provide customized training such as this, a *training needs assessment* has to be carried out (see Toolbox III-1). A training needs assessment starts by specifying job requirements for a specific position, determining the level of responsibility and identifying the skills or knowledge requirements for that job and comparing them as against the personal experience and profile of a person. If there is a perfect match between the two, there is no need for training, but when there is a gap between the job requirements and existing qualifications, training needs will be identified. Here again, a simple questionnaire may support the training needs assessment of personnel at various positions in a company (see Toolbox III-1).

The purpose of *training* is to ensure that people have the appropriate knowledge and skills in order to be aware of the energy and environmental performance improvement opportunities that exist around them (Box 1.4) and to learn how to use EEMS effectively.

Box 1.4: Example of One-Day Training Program for Technical Staff and Supervisors

---

Objectives:

- *To officially introduce and update the company's energy and environmental policy*
- *EEMS organization chart*
- *To identify managers and supervisors roles and responsibilities*
- *To propose action plan on energy and environmental management program*

Suggested Duration: 09:00-16:00

Session Lists:

1. *Activities background*
2. *Review the company's Energy and Environmental Policy*
3. *Energy and environmental performance improvement*

   - *Energy costs*
   - *Environmental compliance costs*
   - *Potential for costs savings*
   - *Why is it important?*

4. *Systematic approach for effective energy and environmental management*
5. *Review EEMS organization chart*
6. *Action Plan for Energy and Environmental Management Program*

   - *Set schedule and activities*
   - *Discuss agenda of introductory workshop for all staff*
   - *Rewards and incentives*
   - *Others*

7. *Role and responsibilities*
8. *Support needed*

The principal message to all staff involved in training programs should be that it is intended to make them more effective in their jobs and facilitate the execution of their duties. At the same time it will reduce the operating costs of the company and by doing so help in securing the long-term employment prospects of its entire staff.

1.7.3 Awareness and Motivation for all Employees

The most significant problem in *achieving changes in the individual'sl attitude* towards energy and the environment is the need to change long standing, deep-seated customs and practices. In attempting to change these attitudes the aim must be to make energy efficiency and good environmental performance an integral part of the routine practices of the company by making everyone aware of the need to improve performance at their workplace.

Therefore, our experience advocates strongly the organization of awareness and motivation workshops for *ALL* employees in a company at the outset of implementing an energy and environmental management program. Awareness and motivation examines the 'whats' and 'whys' of an energy and environmental management program. A sample introductory course on awareness and motivation for all staff is illustrated in Box 1.5.

Box 1.5: Awareness & Motivation Introduction to All Employees

---

Objectives:

- *To officially introduce and declare the company's energy and environmental policy*
- *To create an understanding and awareness on opportunities for energy and environmental performance improvements*

Suggested duration: 0.5 day

Session Lists:

- *Explanation of the company's Energy and Environmental Policy*
- *Introduction to Energy and Environmental Management*

  - *What is energy?*
  - *Where are the sources of environmental impacts from our activities?*
  - *What different types of energy are used in our company?*
  - *What are the types of environmental impacts in our company?*
  - *What are the costs of environmental compliance?*
  - *What are the costs of energy?*
  - *What is the potential for saving energy and reducing environmental impacts?*
  - *What can be saved?*
  - *Why save energy and protect the environment?*
  - *Why are our efforts important?*
  - *Who is responsible?*
  - *How can energy savings be achieved?*
  - *How can environmental impacts be reduced?*
  - *What can I do?*

- *Brainstorming in groups on energy saving and environmental improvement activities and **support needed** at groups' own workplaces.*

The course structure, as shown in Box 1.5, should serve as an example only. The duration can be shorter, and the content may be covered in more than one session. However, it is very important to customize the courses to fit the real energy and environmental conditions at the place of work of a particular group of employees undergoing training, so that people can relate the course topics directly to the issues that they are facing in everyday operations. By doing so, they will recognize the opportunities for energy and environmental performance improvements at their place of work.

1.8 Bibliography

Aft, L.S. (2000) Work Measurement and Method Improvement, John Wiley & Sons.

Allaby, M. (1985) Macmillan dictionary of the environment, Macmillan Press London, 2nd edition.

Babcock, D.L. (1996) Managing Engineering and Technology, Prentince Hall International.

Daines, J., Daines, C., Graham, B. (1993) Adult learning, Adult teaching, Department of Adult Education, University of Nottingham.

Daughtrey, A.S., Ricks B.R. (1989) Contemporary Supervision – Managing people and technology, McGraw Hill International.

Davis, M.L, Cornwell, D.A. (1998) Introduction to environmental engineering, third edition, McGraw-Hill International.

De Nevers, N. (2000) Air pollution control engineering, second edition, McGraw-Hill International.

Eckenfelder, W.W. Jr. (2000) Industrial water pollution control, McGraw-Hill International.

Fayol, H. (1987) General and industrial management: Henri Fayol's classic revised by Irwin Gray. Belmont, CA: David S. Lake Publishers.

Grundy, T. (1993) Implementing Strategic Change, Kogan Page.

Investors in people (1993) Management Charter Initiative, London.

Kubr, M. (ed.) (1992) Management Consulting, ILO.

O'Callaghan, P.W. (1996) Integrated environmental management handbook, John Wiley & Sons.

Raju, B.S.N. (1995) Water supply and waste water engineering, Tata McGraw-Hill publishing company limited.

Standards for Managing Energy (1995) Management Charter Initiative, London.

Whitz, B.M. (1990) Study Skills for Managers, Paul Chapman Publishing, 1990.

2

The Energy and Environmental Management Concept

2.1 Introduction

Factories are not built to consume energy and pollute the environment but to deliver products and services. Energy is always needed for every operation in the production chain of any factory. That is why we are focusing our attention on energy performance throughout a production chain, as a driver for both energy and environmental performance improvement. Where production requires some other resources, the performance of processing these resources and the environmental impacts of their use should be related to production volume and performance analyzed in the same methodological way as that developed for energy.

It is important to emphasize that environmental impacts and pollution are the consequence of energy use and materials processing. Where no energy or materials are used, there are no environmental impacts. *If the process's use of energy and materials is optimized, the resulting environmental impacts will be minimized!* What remains to be done is end-of-pipe treatment, since the quantities of emissions, effluents and waste will be minimized as a result of the optimized use of energy and materials. Whenever some raw materials are processed, energy is required, therefore we stress again that *energy management is a driver for environmental performance management.*

A production process is a combination of people, equipment, raw materials, processing methods and environmental constraints that work together to produce an output. A process is a designed sequence of operations taking up some time, space, expertise and input resources, which are transformed into an outcome of greater value to the company than the original inputs. Hence, every process will have some inputs – raw materials, energy, labor, technology – and some outputs, i.e. products or services. A process is the base for all performance considerations, and people are the key factor for achieving good performance. The dynamics of interactions among all these factors (Fig. 2.1) will determine the energy and environmental performance of the process concerned.

It is within the production processes that additional value is created. The volume of production output determines the amount of energy and other resources needed. Consequently, the volume of production determines the type and amount of environmental impacts. So, it is within a production process that good performance is either achieved or not. It is also the source of data which can be used to quantify interactions between production outputs, consumption of resources, environmental impacts and resulting energy and environmental performance.

Applied Industrial Energy and Environmental Management   Zoran K. Morvay and Dušan D. Gvozdenac
© 2008 John Wiley & Sons, Ltd

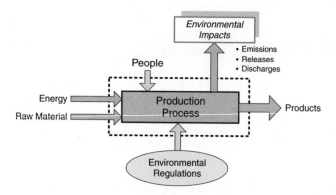

Figure 2.1   Production Process – Foundation for Development of an Energy and Environmental Management System

We will elaborate upon the basic aspects of these interactions in order to achieve the better understanding necessary for the development and introduction of the EEM concept.

2.2 Interactions between Energy and Production

Production processes present a demand side for a factory's energy system – therefore, production processes set the requirements for energy quantity and quality. Energy and environmental performance must be evaluated and improved on both sides:

Production	$\Rightarrow$	*how efficiently raw material is processed into a final product,*
	$\Rightarrow$	*how effectively energy is used to produce given amount of production;*
Utilities	$\Rightarrow$	*how efficiently input energy is converted into utilities which are required by production.*

The basic principles for optimizing energy performance are continuous monitoring of energy flows and connecting the measured amount of energy used by a process or activity and the measured output of this process or activity (Fig. 2.2). As we have already said, wherever and whenever energy performance improvements are achieved, the environmental performance will be improved simultaneously.

The essence of energy and environmental management is to *measure regularly* the use of energy and other resources, relate them to production output or activity that consumed this amount of energy and resources, express this relationship in a form of a *performance indicator (PI)*, and compare the PI with some performance standard or target.

The next step in developing EEM concept is to decentralize responsibility for energy and environmental performance along the production and energy flows and designate responsibility centers for energy and environmental costs and performance. Since energy costs can be established unambiguously by the measurement of any amount or any type of energy, the resulting responsibility centers are called the *Energy Costs Centers* (ECC) in the context of EEM.

This will enable the assignment of responsibility for performance of ECC to a nominated individual or a team of people. Holding people accountable for specified results is another critical point on the path to success.

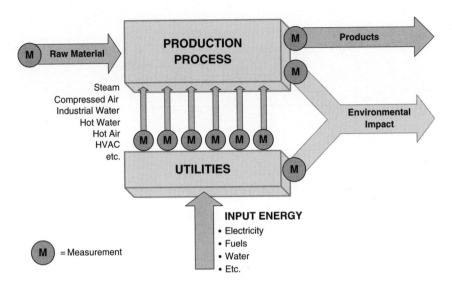

Figure 2.2   Basic Relationship of Energy and Production

2.3 Energy Cost Centers

We need to establish the important physical components or processes of both energy and production systems that play a key role in the supply and demand side of manufacturing operations. Usually, *process flow charts* and *single line diagrams* are used to establish this.

Single line diagrams present a picture of the entire plant energy system and show nominal capacities, generation, distribution and consumption of energy in forms specific to the present state of the plant. A simplified single line diagram may also be called energy flow chart.

Process flow charts consolidate data from present operations to facilitate production control, record current operating conditions and direct an analysis along a logical path, starting from raw material input, following through the production stages up to the final product packaging. They also show the sequence of tasks with related inputs, outputs, activities and opportunities for the assignment of responsibilities. Such diagrams are simplified and structured visual representations of processes that form the production chain and show the workflow of a company.

To give an overview of the various energy and mass streams, the process flow chart of a sample plant (tuna cannery) and its associated operations is presented in Fig. 2.3. When energy and process flow charts are put together, valuable information is provided on where, why and what type of energy is used. This chart can be the basis for decisions on setting up energy cost centers. *Energy cost centers* (ECC) are business segments (i.e. departments, areas, units of equipment or single equipment) where activities or production volume are quantifiable and where a significant amount of energy is used.

The same chart (Fig. 2.3) can be used to identify the sources of environmental impacts from the manufacturing operations in the areas or activities where various resources are used.

For instance, the example in Figure 2.3 indicates the ECCs in which waste water and solid waste are generated and air emissions occur. The dotted line represents the factory boundary, so where the arrows on the diagram cross the boundary, these are the instances of direct impact on the external environment.

There are no fixed rules on how to set up ECCs. An ECC can be any department, section or machine that uses a significant amount of energy or creates significant environmental impacts. As the word 'significant'

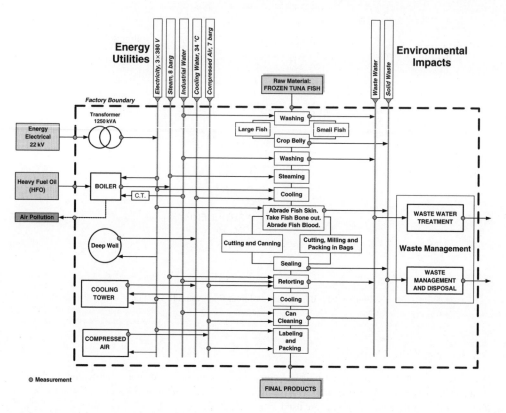

Figure 2.3 Process and Energy Flow Chart of a Tuna-Canning Factory Showing Inputs, Outputs and Interactions with the Environment

indicates, the process is quite arbitrary. However, there are several criteria that should be observed when designating ECCs:

- A process or activity that requires energy should have a measurable (preferably single) output or outcome.
- The energy consumption and/or environmental impacts of the process can be measured directly.
- The cost of measurement should be no more than 10–20 % of annual energy or environmental compliance costs related to that ECC.
- Responsibility for energy and environmental performance in the area can be assigned to the person working in or responsible for that area.
- The standardized performance indicator can be expressed.
- Realistic targets for performance improvement can be established.

The guiding principle for ECC set up is to follow the production process stages as given by the process flow chart, and try to set up the ECCs so that they coincide with existing production quantity control boundaries. On the utility side, ECC set up is quite straightforward – each utility system can be regarded as one ECC. For instance, one ECC can be a boiler house, another one a compressed air room and yet another refrigeration plant. Figure 2.4 shows the actual ECC as defined for our sample plant. Obviously, not each production phase constitutes an energy cost center, as can be seen by comparing Figure 2.3 and Figure 2.4.

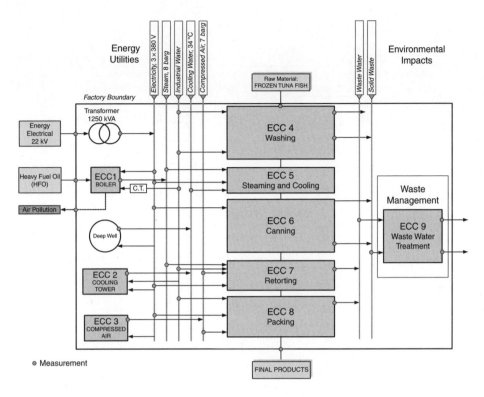

Figure 2.4 Energy Cost Centers in a Tuna-Canning Factory

Whether to include offices, warehouses and similar areas is a matter for discussion. Although they may not be significant energy users or a source of significant environmental impacts, for the sake of consistency, they should be designated as ECCs, and people working there should be responsible for the energy and environmental issues in those areas, because, as previously established, energy and environmental performance concerns everyone in the company.

As the Figure 2.5 shows, the concept of an ECC is a unifying framework that brings together all the components of EEMS:

- people;
- performance measurement;
- performance indicators;
- performance targets.

Figure 2.5 highlights the importance of accurate local measurement of energy consumption and environmental releases where applicable. Energy and environmental performance can be evaluated only against real measurement data because that is the only accurate reflection of actual performance. Assigned or allocated energy consumption and environmental impact figures cannot be the basis of performance assessment.

Figure 2.5 also highlights that there must be *a nominated individual* associated with each ECC charged with the responsibility for achieving the ECC's *energy* and *environmental performance* according to given *targets*.

The concept of an ECC extends management's responsibility for energy and environmental performance for any functional unit or business segment. It focuses attention on energy and environmental

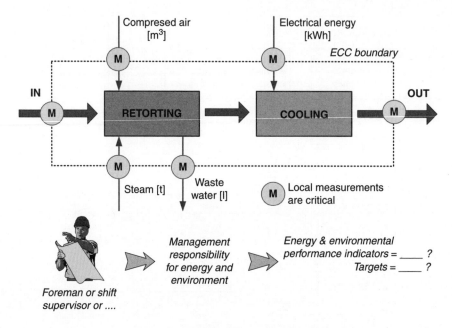

Figure 2.5   Unifying Features of ECC Concept

performance improvement opportunities through better operational and maintenance practices in the units concerned.

2.4 Assigning Responsibilities for Energy and Environmental Performance

For the success of an EEMS project, it is essential to establish the chain of responsibility for energy and environmental performance from factory floor to senior management. Responsibility for energy and environmental performance should be decentralized down to ECC level. It is not enough to assign responsibility for the energy and environmental performance of the whole department to a production department manager. Line managers or supervisors must be given responsibility for performance in their own segment or shift if there is more than one ECC in a department, or more than one shift.

To underline the importance of leadership, we usually refer to those who carry that responsibility as *ECC leaders*. They will normally have a small team consisting of local team members assigned to them as the *ECC operation team* responsible for implementing corrective actions. This team will have the main operational responsibilities for energy and environmental performance improvement at the point of use in their area of responsibilities. Therefore, *ECC leaders* may need initial support in terms of training or consultation in order to develop or upgrade their skills so that they can perform their assigned duties within the framework of EEMS, as indicated in the previous chapter.

The existing *organization chart* must be followed when assigning responsibilities (Fig. 2.6). Responsibility for energy and environmental performance should be assigned down to the people who work at the point of resources use. It means that energy and environmental performance at the production level is the responsibility of production staff, and not of utility personnel or QA (quality assurance) people.

Figure 2.6 shows that everybody in the company is involved to a degree in energy and environmental management. Top management will review energy and environmental performance reports and action plans on a monthly basis. ECC leaders will have weekly review meetings with their performance improvement teams (supervisors, foremen, operators), and the performance improvement teams will have

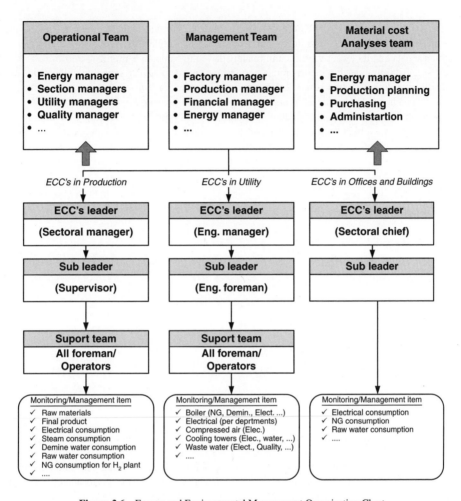

Figure 2.6 Energy and Environmental Management Organization Chart

daily responsibility for monitoring, evaluating and improving energy and environmental performance. For this particular example, the monitoring and management items are shown at the bottom of Figure 2.6. These will, of course, differ from factory to factory. The organizational structure may also differ from the example in Figure 2.6 but the fewer the changes to the existing organization required the more smoothly and easily EEMS will start to operate.

2.5 Performance Measurement System

Performance improvement starts with measurements which are aimed at quantifying past actions in order to determine current performance. Only when performance is quantified, can meaningful discussion about possible improvements begin. The basic principle of EEMS is that the use of energy and other resources must be measured regularly at the point of use and be related to activities that consume the energy or a resource. This is the foundation for energy and environmental performance monitoring and improvement. Energy and environmental managers require the means for determining whether specific energy and environmental performance improvement objectives are accomplished. A performance measurement system (*PMS*), tailored to the specific process or a part of it, provides such a mechanism and alerts managers to the problems that are developing, by measuring, collecting, processing and structuring performance data on all key activities that ultimately drive the company's profit.

Therefore, EEM will need to incorporate a *performance measurement system* that will provide *direct* and *timely* information on the *actual* consumption of resources *at the point of use*, so that needed adjustments can be made to achieve *optimized energy and environmental performance*.

Many companies produce loads of data that serve no useful purpose because they are not or cannot be converted into valuable information. Data or information is useful only if people can relate them to the performance of a process they are responsible for and if they know how their actions can influence performance. An effective PMS begins with the definition of *relevant* data, which in our case would be:

- raw materials input;
- production outputs;
- intermediary stocks;
- hours of operation;
- process parameters (pressure, temperature . . .);
- all types of energy consumption (electricity [kWh], fuels [liters, tons, m^3], water [m^3], steam [tons], compressed air [liters or m^3]).

This is then complemented by defining how data is:

- measured;
- captured;
- verified;
- organized;
- catalogued;
- secured;
- made accessible, but also
- distributed, because until data are distributed to a user, they have only potential values.

The performance measurement system is necessary in order to quantify the efficiency and effectiveness of past actions so as to determine current performance. When people know that performance is measured, it will influence their behavior and encourage the implementation of an energy and environmental policy. Therefore, PMS should include a supportive IT infrastructure and provide a framework for data:

- analysis;
- interpretation;
- knowledge creation;
- learning;
- communicating results and lessons learned.

Performance measurement will provide the means for tracking position and maintaining discipline and control over energy use and environmental processes and activities. *Performance improvement begins with measurement.* Only when performance is quantified can meaningful discussion about improvements begin.

Performance measurement system consists of:

- metrics;
- procedures for data processing;
- supporting IT infrastructure; and
- people who use and process data.

2.5.1 Metrics

The metrics must be simple and understandable to employees. The employees must be able to influence the metrics through their performance and understand how their performance will be reflected by the metrics. Metrics are defined according to performance category monitored. The main objective of EEMS in production is improvement of effectiveness of energy use and environmental performance related to production output. In the utilities, the objective is to improve efficiency of energy conversions from one form to another while minimizing release to the environment. However, the various performance categories may be covered by the same performance measurement system:

Energy	→	*Improving efficiency of generation and effectiveness of use*
Environment	→	*Reducing environmental impacts*
Material productivity	→	*Reducing the amount of input material for required product output and waste*
Quality	→	*Reducing rejects and rework*

For every performance category performance indicators must be defined and measured data sources determined.

2.5.2 Measurement Data Sources

Data reliability and relevance are of the highest importance for performance monitoring. Some data will come from daily log sheets, some from manual meter readings, and most of the rest from remote metering and automatic recording. However, the ultimate data sources are within the production process for which performance is monitored.

A single data source can support the information needs for monitoring multiple performance objectives. For the purposes of effectiveness and economics, i.e. in order to avoid excessive costs for instrumentation and measurements and to avoid overlapping and confusion, it is important at the outset of PMS design to anticipate the various potential users of data items from a single source.

2.5.3 Data Handling Procedures

Data handling routines have to be specified in order to assure *data adequacy*. In that respect, *data handling protocol*, given as an example in Box 2.1, may prove very useful. In addition to reliability and relevance, data *accuracy* is also very important. A *data verification procedure* will need to be established in order to ensure that recorded values are correct.

2.5.4 Data Verification

A good principle for data checking is to allow oneself to always have the benefit of the doubt. Ask yourself 'can this value be really correct'? If you doubt it, then a simple data verification procedure can be carried as follows:

1. *Checking accuracy of reading and recording:*

   a. Electricity – make sure that meter readings are multiplied by an adequate factor due to a current transformer (CT) ratio.

Box 2.1: Sample of Data Handling Protocol

DATA HANDLING PROTOCOL

1. Daily data recording

	Production	*Energy consumption*	*Environmental impact*
Responsible person: (manual and/or automatic)			
Time of recording:			
Source(s) of data: (daily log or instrument No.)			
Data sent for verification to:			
At what time:			

2. Data verification

	Production	*Energy consumption*	*Environmental impact*
Responsible person:			
Send back to ECC leader at what time:			

3. Data interpretation – comments and recommendations

Responsible person
(usually ECC leader):

At what time

4. Reporting on energy and environmental performance

Responsible person
(usually ECC leader):

At what time interval:

Send report to
(usually Management Team):

5. Feedback from the Management Team

Report sent back with comments
questions or approvals:

Who:

Within what time period:

    b. For electricity and water meters, the actual consumption is the difference between a current reading and previous one.

    c. Make sure that all readings are taken at the same intervals.

    d. If readings are taken at the same intervals, normally they must have the same number of digits.

    e. When summarizing and expressing monthly values, make sure that all daily readings are included in the sum.

2. *Consistency check:*

    a. Make sure that the same values are applied across all performance indicators (PIs) that use them. For example, the monthly amount of raw material processed may be used in more than one PI, but the same value must be input for all these PIs.

3. *Units check:*

    a. Make sure that the figure entered into the data log sheets is of the same unit as the one indicated at the top of the corresponding column. For instance, if the unit in the data sheet is *t/month*, do not enter the amount in *kg*, or other way around.

4. *Cross checks:*

    a. Compare related data. For instance, the ratio of fuel to steam is fairly constant, so if there is significant change in fuel data, check if the steam data values have changed similarly.

    b. Check if the figures for raw material used correspond to the final product produced.

    c. Compare electricity consumption figures from utility electricity bills with the company's recording of kWh directly from the electricity meter.

5. *Expectancy (range) check:*

    a. Each value that is monitored over a period of time appears within a certain data range. If the recorded value is unexpectedly high or small (outside expected range), this indicates an error.

6. *From time to time (i.e. six-month intervals), carry out an independent check of the whole data handling procedure.*

2.5.5 Measurement Frequency

Excessively frequent data recording can overload operators with analyses requirements. The optimal meter readings in the case of three-shift operations are at the beginning of each shift. This routine will enable the impartial assessment of energy and environmental performance for each shift independently.

    If a company operates one or two shifts a day, it is also possible to read the meters twice a day, for instance:

- at the beginning of the first (i.e. at 7:00 am) shift; and
- at the end of the first or second shift (i.e. at 11:00 pm).

The additional value of such meter readings (MR) is that companies may get the breakdown of energy use during working and non-working hours (assuming two-shift operation). For instance, energy use during working hours (E_w) can be expressed as:

$$E_w = \{MR \text{ of } 11\text{:}00 \text{ pm of day } 1 - MR \text{ of } 7\text{:}00 \text{ am of day } 1\}$$

while energy use during non-working hours (*Enw*) is:

$$E_{nw} = \{MR \text{ of } 7\text{:}00 \text{ 7am of day } 2 - MR \text{ of } 11\text{:}00\text{pm of day } 1\}$$

E_{nw} can be monitored separately to control only what is going on at night when only basic services should be operating.

All of the meters must be read at the same time in order to ensure that the energy used corresponds to the amount of products produced.

2.5.6 Supporting Infrastructure

The backbone of a performance measurement system are meters and data collection, verification and recording procedures. Most factories already have established procedures for production data collection and recording which can then also be used for EEMS purposes.

However, all data on energy and consumption of other resources at ECC level may not be readily available. Once performance indicators are defined, it will become clear where and what needs to be measured. Usually, the factory will need to install some additional meters on top of the existing ones in order to make the performance measurement system complete.

The extent of investment in new meters will vary from factory to factory depending on the scope and sophistication of existing sub-metering. If data collection is to be done remotely from a monitoring center, avoiding manual handling, investment requirements will also increase. On the other hand, remote metering and data collection will reduce manpower requirements for manual data recording and handling and the potential of human error will be reduced.

Nowadays, when local area networks (LAN) are a reality in most factories and *web based services* further reduce the costs of remote data handling, automatic data collection will prove its worth and the increased investment for most companies, excluding only the smallest, will be justified. An example of a web-based system is given in Figure 2.7.

Finally, whatever the data collection system will be, the critical factors that must be ensured are data adequacy, reliability and accuracy, as will be explained.

2.5.7 Raw Material Performance Indicators

A performance indicator quantifies information on a trend or phenomenon in a manner that promotes the understanding of performance issues in a business unit by both process operators and decision makers.

The simplest way of tracking the performance of raw material use is thorough a materials productivity indicator that is already defined as an input output ratio for a specified period of time (Fig. 2.8).

$$\mathrm{RMPI} = \frac{\mathrm{AMOUNT\ OF\ PRODUCTS}}{\mathrm{AMOUNT\ OF\ RAW\ MATERIALS}} \tag{2.1}$$

RMPI can be further tuned up by distinguishing various raw materials, including additives, where applicable, that are required to produce certain amount of products:

$$RMPI_i = \frac{AMOUNT\ OF\ PRODUCTS}{AMOUNT\ OF\ RAW\ MATERIALS\ (ADDITIVE)_i} \tag{2.2}$$

where 'i' denotes various raw materials or additives used in production.

2.5.8 Energy Performance Indicators

Performance indicators for production ECCs have to be expressed as the ratio of measured energy and resource amount or corresponding production output over a defined period of time (Fig. 2.8). The performance indicators must be defined for each designated energy cost center (ECC). They can be defined as the ratio of energy to production output, or to raw material input, or even to product monetary

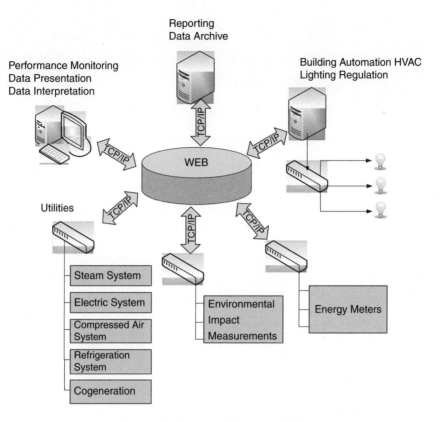

Figure 2.7 Example of Web Based EEMS

value. This will depend on the type of products and production quantity control method and can vary from one ECC to another even within the same factory.

On the utility side PIs are defined in the technically familiar way of output/input ratios, which basically express the efficiency of energy conversion from one form to another.

For each energy type used at an ECC, a separate PI has to be defined. Some energy types that may be used are listed in Figure 2.2. If the use of any energy type is significant, it has to be measured directly. The 'significance' can be determined based on the rough economic criteria that the cost of meter should be around 10–20 % of the annual cost of the relevant energy type.

When performance indicators are defined and measured as shown in Figure 2.9, they provide the basis for extraction of some additional useful information:

 (i) calculating energy and mass balance on the ECC level;
 (ii) accounting for raw material waste, rework or rejects;
(iii) monitoring quality on the ECC level;
(iv) calculating all direct costs (providing they are measured) associated with the ECC concerned;
 (v) accounting for productivity of people working at the ECC, even on a shift basis.

However, these analyses will provide useful results only if input data are adequate, reliable and relevant.

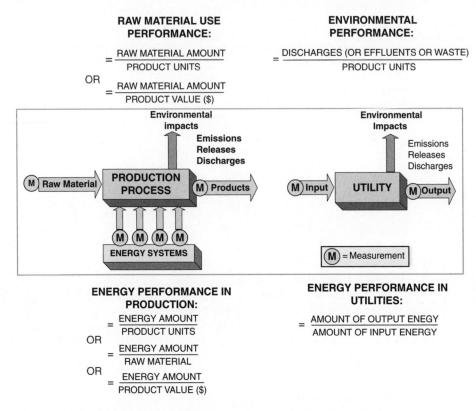

Figure 2.8   Concept of Performance Indicators in Production and Utilities

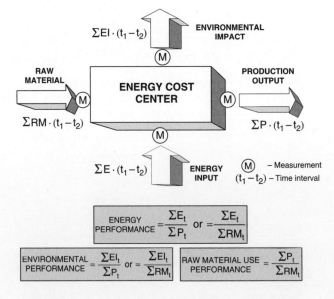

Figure 2.9   Options for PI Definition Relating to Individual ECC

2.5.9 Data Adequacy

Even if accurate, measured data can still be misleading if other aspects of performance measurement are overlooked. A very important aspect is the time interval over which energy and production are measured. This interval can be defined as a shift, a day or a batch period with an exact beginning and end. It is important to assure *time correspondence* for reading and recording energy and production amounts. It means that readings of energy and production amounts must be synchronized at the beginning and at the end of the defined time interval.

Essentially, PIs are ratios of measured input resource to a process, machine or activity and the corresponding measured output of this process, machine or activity. These amounts must correspond in time and value. Only if these conditions are fulfilled, can data be qualified as an adequate basis for expressing the performance indicator. We will refer to these conditions simultaneously as a requirement for *data adequacy*. Data adequacy has two components:

- time correspondence; and
- value correspondence.

Time correspondence means that amounts are measured over the same period of time. *Value correspondence* means that a measured amount of energy is used exactly and exclusively to produce a measured amount of output during the defined period of time. Otherwise, the performance indicator will not make much sense. It is critical that the measured amount of production and energy must correspond properly with actual consumption/production in an individual ECC cumulatively over a defined time interval. Only when this is double-checked and ensured, can performance evaluation begin.

This may sound simple and straightforward but it is of critical importance for meaningful performance measurement system that data source truly represents the intended performance indicator. If measured data do not reflect actual energy or production amounts related to the individual ECC, i.e., there is no value correspondence, every subsequent analysis will be based on a wrong input. Figure 2.10

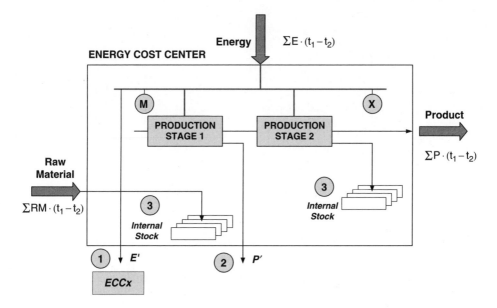

Figure 2.10   Potential Problems with Value Correspondence of Measured Data

shows the three most common potential problems with value correspondence, i.e., data source and context:

- The measured amount of energy includes some consumption by another ECC.
- The measured amount of production does not represent actual production because part of it is removed at an intermediary production stage.
- Internal stocks – if only the input of materials or output of products are measured, the potential existence of internal stocks cannot be detected.

If data adequacy conditions are fulfilled, then PI will be a meaningful basis for performance evaluation.

2.5.10 Performance Targets

Once an indicator has been defined and measured, it must be interpreted against a benchmark or a standard to determine whether the observed performance is satisfactory or not. Therefore, a target must be set for each PI and management responsibility assigned in order to achieve the given target. *Targets* must be set on the basis of *knowledge* of the representative operational circumstances, which may be affected by one or more of the following factors:

- current and baseline production volume;
- nominal production capacity (plant capacity) – production capability over the course of a day, month or year defined as number of standard shifts per day, hours per shift, days worked per year multiplied by items or volume produced (pieces, ton, l, kg, ...);
- capacity utilization – actual production divided by nominal production capacity;
- product mix.

If any of these factors change significantly over the time, targets may need to be redefined in order to reflect new circumstances. Each PI must include an operational definition that prescribes PI intent and makes its role in achieving performance targets clear.

2.5.11 Environmental Performance Indicators

The process of selecting EnPI starts from a precise understanding of the environmental problems and monitoring objectives being addressed. Why do we want to measure environmental performance? The answer is a combination of some or all of the following:

- To ensure compliance with all relevant legislation and with specific requirements to which the company has made a public commitment.
- To improve performance and reduce compliance costs.
- To better understand environmental risks and minimize future costs.
- To communicate and report on environmental performance to stakeholders.
- To fulfill mandatory statutory reporting.
- To adhere to the principles of continuous improvement.
- To achieve and retain certification in accordance with environmental management standards.

In the introduction, we described some of the most common voluntary and compulsory environmental management schemes, which are again summarized in Figure 2.11. Some of the requirements imposed by these schemes are compulsory, others are voluntary, but at the core of all of them is, once again, the production process where energy is consumed and raw materials processed. The production process is the

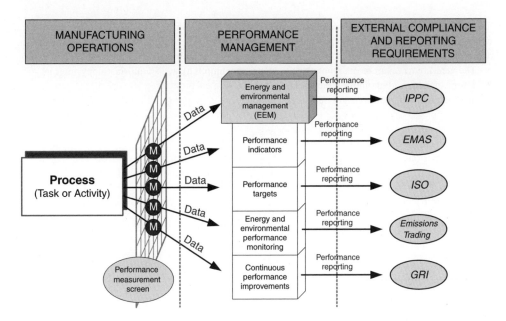

Figure 2.11 Relationship between performance measurements, performance management and external environmental compliance requirements

place where relevant data is measured, and this measured value will in turn form performance indicators, which will be used to track progress toward given performance targets.

Selected environmental performance objectives will define the indicators that have to be monitored, which will in turn identify the parameters that need to be measured or calculated in addition to those already covered by the measurement of energy performance indicators (see Table 2.1). Table 2.1 refers to our tuna cannery case study. Column 1 contains the resource types that are to be measured, column 2

Table 2.1 Measurements for Energy Performance Indicators

Resource	Energy and material performance requirements, relevant for ECC No#.., and source of data	Measurement units
Raw materials	ECC4, ECC5, ECC6, ECC7, ECC8	mass
Heavy fuel oil	ECC1	Liter
Electricity	ECC1, ECC2, ECC3, ECC5, ECC6, ECC7, ECC8	kWh
Steam	ECC5, ECC7	Ton, MJ
Industrial water	ECC1, ECC2, ECC4, ECC8	m³
Cooling water	ECC5, ECC8	t, m³
Compressed air	ECC7, ECC8	m³

indicates at which ECC the measurements have to be done, while column 3 provides the physical units in which the measurements will be expressed. While defining performance indicators for a given monitoring objective, we should take care not to duplicate measurement requirements. If required data sources have been covered by measurements (for instance water consumption), we should just copy the measured values for the required monitoring purposes.

Most commonly, environmental performance indictors that require additional measurements include the following:

- emissions to the air (quantity and quality);
- GHG emissions (quantity and quality);
- effluents and discharges into sewage and waterways (quantity and quality);
- solid waste sent to landfill (quantity and quality);
- amount of packing material used or recycled.

The intended audience for environmental reporting can result in further requirements for the measurement of environmental indicators. These audiences may be:

- management and employees;
- customers;
- the public, NGO;
- regulatory agencies;
- investors, financial institutions, insurance companies,

and depending on them, some other parameters may need to be measured or quantified. It is important to note that environmental measurements may be especially expensive in some cases. Therefore, alternative ways of quantifying particular environmental impacts can be employed, such as:

- mass balance calculation, or
- substance composition information supplied by manufacturers.

In any case, it is important to assure that data is complete, i.e. that everything is accounted for. Environmental performance indicators (EnPI) have to be expressed as the ratio of the mass of emissions, effluents and waste related to the mass of resources consumed:

$$\text{EnPI} = \frac{m[\text{kg, emitted}]}{m[\text{kg, product}]} \tag{2.3}$$

Another aspect that can be expressed, depending on the objective of monitoring, is release rate (RR), defined as follows:

$$\text{RR} = \frac{[\text{mg}]}{[\text{m}^3]} \quad \text{or} \quad \frac{[\text{mg}]}{[\text{ml}]} \quad \text{or} \quad \ldots \ldots \tag{2.4}$$

We will continue by exploring the requirements for environmental performance and compliance as imposed by individual schemes, as are shown in Figure 2.11.

2.5.11.1 Integrated Pollution Prevention Control

The IPPC Directive and similar regulations set out the conditions for an environmental permit for an industrial installation, including emission limit values (ELVs) and compliance reporting requirements. It aims at ensuring compliance with certain environmental protection standards.

Table 2.2 shows the performance indicators and designated measurements needed to comply with the requirements of energy and materials performance management, as summarized in Table 2.1, and specifies

Table 2.2 Additional Measurements of Environmental Performance Indicators for IPPC Compliance

Resource	Relevant IPPC Requirements	Relevant to ECC...	Additional measurements for IPPC compliance on top of energy performance requirements that are covered already
Air Emissions	TOTAL EMISSIONS = end-of-pipe emissions (normal operation) + diffuse and fugitive emissions (normal operation) + exceptional emissions	ECC1	Odor, Dust, Organics[1], CO_2, NO_2, SO_2, NH_3
Water Use	**Fish processing** / **Water (m^3/t raw fish)** / **COD[2] (kg/t raw fish)** Fresh fish — 4,8 — 5–36 Thawing — 9,8 Shrimp processing — 23–32 — 100–130	ECC1, ECC2, ECC4, ECC5, ECC7, ECC8	COD
Waste Water	**Parameter / Average (kg/m^3) / Range (kg/m3)** BOD — 10 000 — 5000–20 000 Fat — 12 000 — 2500–16 000 Dry matter — 20 000 — 5000–28 000 Protein — 6000	ECC4 ECC7 ECC8	– Quantity (m3/h), at point of discharge to environment – Composition – soluble organic material (BOD/COD), total suspended solids, acid/alkali, nitrate, nitrite, ammonia, phosphate, dissolved solids. – Measurement frequencies: 24h, or less, depending on case.
Solid Waste	**Solid Waste / % Solid Waste in Weight** Rejected fish — 3 Heads, offal, TAILS — 20–40 Skins, bone — 15–20 Scraps — 5	ECC4 ECC6	Weight of solid waste (kg) Type of solid waste
Energy Resources	Energy use per unit of production or per unit of raw material	All ECCs	NO
Utilities and Services	Use of water, compressed air or steam per unit of production or per unit of raw material	All ECCs	NO
Other	Consumption of specific materials, e.g., packaging per unit of production	–	Information on packaging procedures

Notes:

[1] *Organics* covers emissions containing organic material under actual processing conditions and which are present in the emission regardless of individual component's vapor pressure.

[2] *Chemical Oxygen Demand (COD):* the amount of potassium dichromate, expressed as oxygen, required to chemically oxidize, at approximately 150 °C, substances contained in waste water.

the additional requirements for the fish processing industry in order to comply with IPPC regulations. Table 2.2 also highlights the relevance of these requirements to our case study of the tuna canning factory. Column 1 contains the types of resources used or emissions and discharges occurring, column 2 describes the corresponding IPPC requirements, column 3 indicates the relevance of the requirements to a particular ECC in our case study factory, while column 4 shows which measurements are necessary for compliance with IPPC, in addition to those already introduced under the energy performance measurement system.

As it can be seen, measurements of energy, utilities and water use are already covered and the additional measurements requirements relate to the quantity and quality of air emissions, water discharges and solid waste disposal.

2.5.11.2 Eco-Management and Audit Scheme

The Eco-Management and Audit Scheme (EMAS) is a voluntary instrument of policy in the EU, which acknowledges organizations that improve their environmental performance on a continuous basis. EMAS registered organizations run an environmental management system and report on their environmental performance through the publication of an independently verified environmental statement.

Table 2.3 is organized similarly to Table 2.2 except that it now also includes additional needs for environmental performance measurement in order to comply with EMAS requirements. The meaning of 'additional' denotes these measurements which are not already covered by IPPC and by energy performance measurement (column 4).

2.5.11.3 Environmental Management Standard ISO 14000

The ISO 14000 environmental management standards represent a voluntary scheme, which helps companies to minimize the negative effects on the environment of their operations, to comply with applicable laws, regulations and other environmentally oriented requirements,and to improve continuously environmental performance.

Table 2.4, similar to Tables 2.2 and 2.3, summarizes the need for additional measurements and reporting requirements in order to ensure ISO 14000 compliance, supposing that energy performance, IPPC and EMAS measurements and reporting requirements are already provided for.

It can be seen from Table 2.4 (see column 6) that no additional measurements are required for ISO 14000 compliance if a company has already in place energy performance measurement systems supplemented by IPPC and EMAS required measurements. It means that all the required data exist and that the reports have only to be structured according to the ISO 14000 standard requirements.

2.5.11.4 Emissions Trading

Emission trading is a market based approach to control pollution by providing economic incentives for achieving reductions in the emissions of greenhouse gases (GHG). The threshold for inclusion in the scheme of 'combustion installations' (which is under the category of 'energy activities') is 20 MW rated thermal input. In other words, any combustion installation above 20 MW is included in the scheme. Any businesses or organizations that have such installations on their sites may be covered by the scheme. Companies covered by the scheme are required to keep track of their emissions.

Table 2.5 compares the requirements of EU Emission Trading Scheme (ETS) as against the requirements of previously described schemes.

Even if a company is not covered by the EU ETS scheme or any other similar scheme, it may decide to determine its GHG inventory. By doing so, it safeguards evidence of emission reductions achieved so far, so that this can be taken into consideration when emission reduction commitments are imposed.

Table 2.3 Additional Measurements of Environmental Performance Indicators for EMAS Compliance

Issues	EMAS Requirements	Relevant ECC	Covered by IPPC and energy perf. req.	Additional measurements, or reporting requirements for EMAS compliance
Direct	(a) Emissions to air;	ECC1	Y	NO
	(b) Releases to water;	ECCs 4,7,8	Y	NO
	(c) Avoidance, recycling, reuse, transportation and disposal of solid and other wastes, particularly hazardous wastes; (d) Use and contamination of land;	ECC4 ECC6	N	Information on recycling procedures, etc. Information on waste disposal management,
	(e) Use of natural resources and raw materials (including energy);	All ECC	Y	NO
	(f) Local issues (noise, vibration, odor, dust, visual appearance, etc.); (g) Transport issues (both for goods and services and employees); (h) Risks of environmental accidents and impacts arising, or likely to arise, as consequences of incidents, accidents and potential emergency situations; (i) Effects on biodiversity.	All ECC	Partly	Information on emergency procedures, Risk prevention measures, Fuel consumption for transportation,
Indirect	(a) Product related issues (design, development, packaging, transportation, use and waste recovery/disposal); (b) Product range compositions; (c) Environmental performance and practices of contractors, subcontractors and suppliers.	–	Partly	Information on contractors and suppliers.

On the other hand, the GHG emissions inventory will help to satisfy possible new statutory requirements when imposed. It can also facilitate the definition of the baseline emissions upon which the future emission reduction requirements may be imposed.

2.5.11.5 Global Reporting Initiative

The Global Reporting Initiative (GRI) aims to make reporting on economic, environmental, and social performance – sustainability reporting – by all organizations as routine and comparable as financial

Table 2.4 Additional Measurements of Environmental Performance Indicators for ISO 14000 Compliance

Type of environmental effect	Requirement (*'may include, not limited to'*)	Relevant ECC	Covered by IPPC	Covered by EMAS	Additional measurements for ISO
Direct	• Energy consumption	All ECC	Y	Y	NO
	• Water usage	ECC1 ECC2 ECC4 ECC5 ECC7 ECC8	Y	Y	NO
	• Gaseous emissions	ECC1	Y	Y	NO
	• Effluent discharges to water	ECC4 ECC7 ECC8	Y	Y	NO
	• Solid or contained wastes	ECC4 ECC6	Y	Y	NO
	• Materials handling and storage • Packaging waste disposal • Transport • Noise, odors and other irritants • Disruption of habitats and ecosystems • 'Worst case' scenarios		N	Y	NO
Indirect	• Effects on suppliers' or subcontractors' process • Effects of other business interests • Effects of use and disposal of products		N	Y	NO
Other	• Machine start-up, shut-down (boilers and furnaces), etc …	Covered by energy perform. req.	N	N	NO

Table 2.5 EU Emissions Trading Scheme Requirements

Issue	Requirement	Relevant ECC	Covered by IPPC	Covered by EMAS	Covered by ISO	Additional measurements
Air Emissions	Carbon dioxide (CO_2)	ECC1	Y	Y	Y	NO

reporting. This reporting guidance – in the form of principles and indicators – is provided free for the public good and is intended for voluntary use by organizations of all sizes, across all sectors, all around the world.

In Table 2.6, only the core performance indicators for the chapter 'Environment' of the Global Reporting Intiative requirements are shown.

Table 2.6   The GRI environmental reporting requirements

Type of report	Requirements	Relevant ECC	Covered in IPPC	Covered in EMAS	Covered in ISO	Additional measurements
Environmental Materials	• Materials used by weight or volume. • Percentage of materials used that are recycled input materials.	ECC4	Y	Y	Y	NO
Energy	• Direct energy consumption by primary energy source. • Indirect energy consumption by primary source.	All ECC	Y	Y	Y	NO
Water	• Total water withdrawal by source	ECC1, ECC2, ECC4, ECC5, ECC7, ECC8,	Y	Y	Y	NO
Biodiversity	• Location and size of land owned, leased, managed in, or adjacent to, protected areas and areas of high biodiversity value outside protected areas. • Description of significant impacts of activities, products, and services on biodiversity in protected areas and areas of high biodiversity value outside protected areas.	Information on process, plant and company procedures	N	Y	Y	NO
Emissions, Effluents, and Waste	• Total direct and indirect greenhouse gas emissions by weight. • Other relevant indirect greenhouse gas emissions by weight.	ECC1	Y	Y	Y	NO

(continued overleaf)

Table 2.6   (*continued*)

Type of report	Requirements	Relevant ECC	Covered in IPPC	Covered in EMAS	Covered in ISO	Additional measurements
	• Emissions of ozone-depleting substances by weight. • NOx, SOx, and other significant air emissions by type and weight.					
	• Total water discharge by quality and destination. • Total number and volume of significant spills.	ECC4, ECC7, ECC8,	Y	Y	Y	NO
	• Total weight of waste by type and disposal method.	ECC4, ECC7, ECC8,	Y	Y	Y	NO
Products and Services	• Initiatives to mitigate environmental impacts of products and services, and extent of impact mitigation. • Percentage of products sold and their packaging materials that are reclaimed by category.	–	N	N	Y	NO

2.6 Effective Use of Energy and Environmental Performance Indicators

Performance management begins with measurement and data collection. Without data to evaluate process efficiency and effectiveness, we are only overseeing a business unit – not managing it. If we do not know what it costs, how long it takes or how much we have produced, what can we say about the performance?

Figure 2.12 again illustrates that the foundation of energy and environmental performance is the production process which, for the sake of accountability, is broken down into energy costs centers (ECCs). Every ECC absorbs resources in order to produce services or products, thus affecting the company's business results. Performance of day-to-day operations can be discussed and evaluated only in hard data terms! Therefore, performance measurement systems should be overlaid on any business function or, unit or ECC, bearing in mind the need to avoid overlapping, i.e. multiple measurements of the same performance indicator.

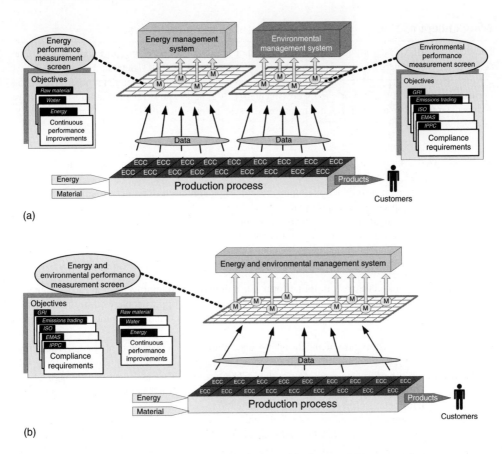

Figure 2.12 Integrated Energy and Environmental Performance Measurement System

Performance can be managed only when based on quantitative data on day-to-day operations. Energy and environmental performance measurement systems must be data rich, but not data-driven. We should be aware of the danger of 'data for data's sake' approach. There are basically three legitimate reasons for taking a measurement:

- to quantify what needs to be quantified;
- to monitor progress toward a target;
- to interpret the causes of good or bad performance.

The applied metrics must indicate the identifiable instances of non-standard performance and provide enough information to interpret what and why has happened. From that perspective, environmental performance indicators (EnPI) focus on monitoring compliance and compliance costs, while energy performance indicators (EPI) focus on operational effectiveness and direct cost savings through the more efficient use of resources or as sometimes referred to – through *eco–efficiency!* Both indicators, EPIs and EnPIs, enable the line personnel *to assess and improve* operating efficiency.

2.7 Concept of Energy and Environmental Management System

The design of EEMS starts with decentralizing responsibility for energy and environment along the actual process and energy flows in a company (see Fig. 2.3 and Fig. 2.4). The resulting responsibility centers are called energy cost centers and are illustrated earlier by the simplified diagram in Figure 2.5. For each responsibility center or energy costs center (ECC) the PIs are to be defined, responsibilities assigned, targets given and performance measurement systems established. Hence, the concept of EEMS (Fig. 2.13) is based on:

- decentralized responsibility for environmental performance, energy use and costs along the actual process and energy flows;
- regular measurement of production outputs, resources, energy used to produce them, and resulting environmental impacts, where applicable;
- calculation and evaluation of performance indicators;
- interpretation of performance level against given targets;
- implementation of corrective actions at the point of use.

Such a concept provides a framework for performance monitoring and improvement at each designated responsibility center – Energy Cost Center (ECC) – which is the task of the people working at the center. The main concern of EEMS is the continuous improvement of energy and environmental *performance* throughout a factory.

The ECCs are the basis for introducing the energy and environmental management system. A performance measurement system will be also introduced based on the ECC concept and individual targets will be assigned for each performance indicator at the ECC. Once the ECCs are identified throughout the factory (Fig. 2.4) and responsibilities are assigned, the shape of the EEMS starts to emerge and we can start to deal with the details of its design and implementation.

But firstly we should be aware of and should understand the external and internal *context* of the EEMS. There are number of factors within a factory and in the factory's environment which will cause

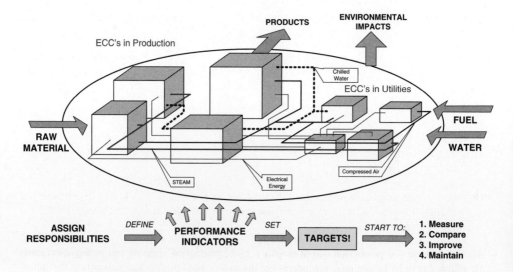

Figure 2.13   The Concept of EEMS based on Energy Cost Centers (ECCs)

performance to fluctuate. The sources of these factors should be clearly identified and their potential influence on the performance should be properly understood.

2.8 Context of Energy and Environmental Management

A common mistake by energy and environmental management projects is that the adopted approach is too narrow or that important factors that affect performance are not identified and accounted for. Often, the focus is only on the utility side, specific technical solutions or computerized control. The importance of the production side of a factory, and the importance of human factors are not sufficiently appreciated. In such cases, a number of opportunities for performance improvement will be missed, and overall results will be moderate.

For that reason, before proceeding with the considerations of how to design and implement an EEMS, we should properly appreciate the external and internal time-varying factors that influence energy and environmental performance. The EEMS must provide the tools to detect variations, to establish their causes and to minimize the effects on performance. The sources of impacts and variations have to be recognized first, regardless of whether a company can or cannot exercise control over them.

An EEMS must be designed in such a way that it can respond dynamically to changes in the environment in order to ensure that energy and other costs are kept to a minimum level whatever the variations of external and internal factors are. For energy performance specifically, the relationship between production and energy consumption must be properly understood because production is where most of the energy is used. This relationship will be elaborated upon in detail in Part I Chapter 3. We will now proceed with a consideration of the external factors influencing energy and environmental performance.

2.8.1 External Context

There are numerous external factors which are beyond control but which do influence daily operations and the company's ability to maximize profit and achieve other specified business objectives. Figure 2.14 provides a simplified overview of the main external aspects that define the context for EEMS implementation. It highlights those that are important for a business's energy and environmental performance. The relevant aspects are:

- business environment, constituting markets, customers, competitors, labor;
- conditions and limitations for purchasing natural resources such as raw materials and energy;
- climate conditions;
- environmental issues arising from the significant environmental impacts of industrial operations, and relevant regulations, etc.;
- social and political changes;
- technological advancements.

These issues will be covered in more detail in Part I Chapter 6 within the context of strategic management considerations. We will focus here on identifying the *external factors* that may have a critical impact on energy and environmental performance and the overall production costs of a company's operations. The most important external factors that affect company operational performance are

- *Market demand* – if production output drops by 25 %, production planning and management must be adjusted in order to minimize the increase of the energy and other operational costs in the unit of production. For example, after the 1997 crisis in Thailand, many companies experienced a drop in output of 30 % or more. Companies continued their operations just as before, when output was at the full capacity level. A number of them, as we have experienced, have not adjusted their *production planning and equipment utilization scheduling* to the lower overall output level, so as a consequence their unit production costs have increased by much more than necessary.

- *Weather* – energy used for heating or cooling will be added on top of energy use for production. It will also vary as a result of seasonal changes of temperature and humidity. Therefore, the weather dependent portion of total energy use should be identified prior to analysis of energy performance.
- *Energy prices and tariff system* – the impact of prices or tariff system charges must always be analyzed in order to ensure that energy is bought at minimum cost.
- *Environmental regulations* – the increasing costs of compliance can justify changing process inputs such as fuel type, raw material, waste treatment technology, etc.

Tariff systems and environmental regulations will have a so called 'one-off' impact, because they are not changing constantly and suddenly. It is easier to adapt to the impact of their changes. Market demand for products or for types of products varies most of the time; hence there should be a *cross-functional understanding* – between utility, production and sales on how these variations will impact upon overall performance and therefore the costs of production. This will be explained later in the text.

Weather conditions may have a significant impact on overall energy consumption (depending also on the type of industry) and, if this is the case, appropriate corrections should be made. The most common temperature corrective method is based on heating or cooling degree days. This method requires a collection of both weather and energy data and uses a statistical model in order to adjust to the impact of weather on total energy use.

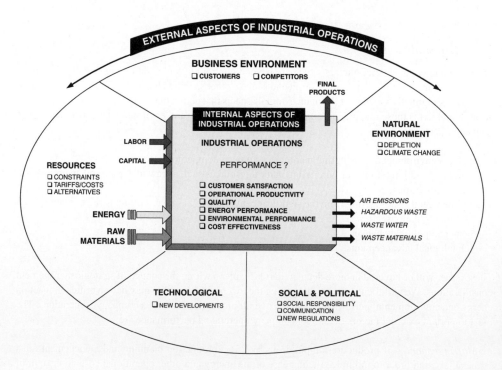

Figure 2.14   External Aspects of Industrial Operations

2.8.2 Internal Context

Figure 2.15 highlights the main internal aspects of industrial operations within a given external environment. It highlights *people*, whose skills and competencies are critical to the performance of any business function, as discussed earlier. From the energy performance point of view, any company can be seen as having two distinctive parts:

- *Production* as *energy demand* side that sets requirements for energy quantity, quality and variations over time, i.e. it determines the *energy demand profile*;
- *Utilities* as *energy supply* side that must deliver energy efficiently and when required by production.

The supply and demand sides are linked by a distribution network, with imposed metering, control and monitoring systems. The main internal factors that influence performance may be grouped as follows:

$\Rightarrow$ Production-utility coordination;
$\Rightarrow$ Production planning practices;
$\Rightarrow$ Maintenance practices;
$\Rightarrow$ Extensions of buildings or addition of equipment;
$\Rightarrow$ Changes of product mix;
$\Rightarrow$ Changes of input materials specifications;

The coordination of production and energy or utility departments in most factories presents a major challenge for cross-functional cooperation and communication. Most companies are production focused and, as such, they consider energy and environmental performance as marginal issues in relation to

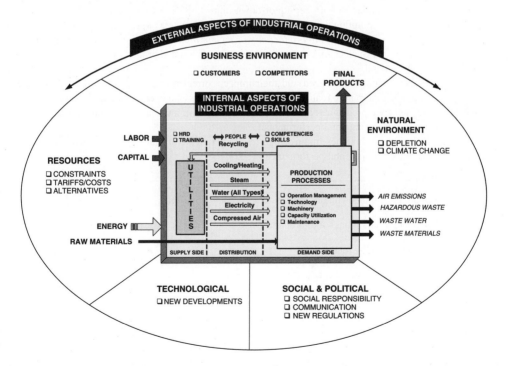

Figure 2.15   External and Internal Aspects of Industrial Operations

production output and quality considerations. But experience tells us that there are important reasons for them not to be separated operationally as they usually are, organizationally, just the opposite:

> *Utility and production departments should coordinate their operations closely in order to achieve the lowest cost for energy and for environmental compliance!*

Why? When energy performance in production is not closely controlled, actual consumption and capacity demand tend to be higher than necessary, sometimes much higher. One reason for that is the fact that energy is often seen in production departments as a peripheral necessity. Consequently, production departments' focus is on product quality and quantity, while process energy performance does not feature high on the agenda. Even more, disregarding energy and environmental performance allows for a more convenient means of production management. Since we are all human, we like convenience. But convenience always comes at a cost!

For instance, our tuna-canning factory (Fig. 2.3) had a boiler house with five boilers in operation, while only two would have been enough for average steam consumption. The analysis revealed that sterilizing retorts were operated in a way that resulted in high steam demand for a very short period, a few times a day. This resulted in excessive investment (too many boilers), inefficient operation (mostly on partial load) and excessive fuel use and emissions to the air. With more careful *production planning*, peak demand was reduced, resulting in only two boilers being sufficient for supplying all the necessary steam while the factory was operating at full load.

Maintenance practices may have a similar impact on energy and environmental performance inasmuch as inappropriate maintenance results in higher energy costs and higher rates of environmental pollution. Further, adding new machines, replacing old ones with more efficient ones, or adding more workshops or buildings all have a similar one-off step-up or-down impact on energy consumption. If any of these happen, the base line consumption data and corresponding performance targets will need to be adjusted correspondingly. If there are changes in *product mix or type of input materials* used, it will also impact upon energy consumption. Such changes must be observed and recorded properly in order to account for deviation in energy demand.

The one-off factors would be noticed and their impact corrected easily by adjusting the base line consumption. But it is important to realize that it is always people who operate machines and implement planning and maintenance procedures and their attitude, behavior, awareness and skills have a critical impact on energy and environmental performance.

EEMS must take into account these main categories of internal factors, including human factors, which influence energy and environmental performance:

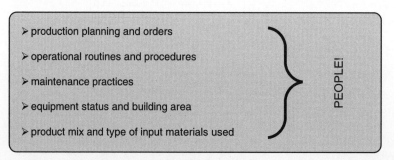

There will be a number of individual internal and external factors from these broadly outlined categories that will cause variations in energy use, sometimes on a daily basis. They should be identified and listed for every functional area in a company.

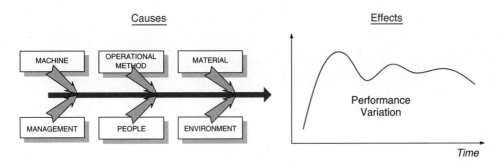

Figure 2.16   Symbolic Cause-Effect Relationship of Influencing Factors and Performance Variations

2.8.3 Factors that Influence Energy and Environmental Performance

Performance of energy use will vary depending on the presence of various internal or external influencing factors. Performance improvement involves taking action on the causes of variations and developing operational guidelines for future prevention of undesired performance levels. Developing these guidelines requires knowledge and understanding of cause-effect relationships between variations in performance and the factors that influence these variations. The general categories of *influencing factors* are summarized in Figure 2.16.

Understanding cause-effect relationships (Fig. 2.16) is important for interpreting measured data and evaluating performance. This is a knowledge-intensive process and there is no substitute for analytical thinking by a human operator. In order to *develop understanding* of the *cause-effect relationship*, it is helpful to compile a list of individual influencing factors relevant to the monitored areas. The number of influencing factors can be quite large so the relevant ones need to be recognized, based on good insight and understanding of the monitored process or activity. Usually, such a list will contain one or more of the generic factors listed in Box 2.2 and Box 2.3. Often, a combination of several factors will occur and cause particular deviations in performance level.

Box 2.2: Factors Influencing Energy and Environmental Performance

Box 2.3: Operator Induced Factors Influencing Energy and Environmental Performance

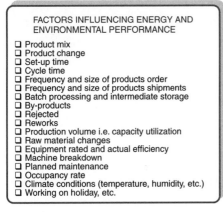

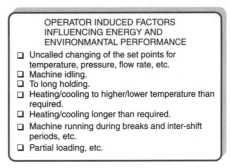

Only the people who operate machines will have hands-on experience of what the actual operational events are, and what influencing factors have occurred during the respective periods of time. They may need some initial support in order to upgrade this experience into usable knowledge that can be integrated in EEMS, with the aim of preventing undesirable performance. This knowledge is based on an understanding of the cause-effect relationship between the factors influencing the performance variations. However, the people who operate machines and work on the manufacturing floor must be the core of the performance improvement teams whose task will be to achieve enduring performance improvements.

It is vital to be aware of all the factors that cause energy use to vary before attempting to analyze actual performance data. Even though it is impossible to account accurately for every potential influencing factor, just being aware of the consequences they can avoid the wrong interpretation of data on energy and environmental performance.

Intuitively, it is clear that energy use and environmental impacts must depend strongly on production volume – the higher the output, the higher the rate of environmental releases and the more energy will be used. But how does this relationship hold in a dynamic environment where number of influencing factors mix and change with time while production output varies as well? In the next chapter we will have a closer look at the relationship between energy and production and how energy performance depends on variations in production.

2.9 Bibliography

http://ec.europa.eu/environment/emas/index_en.htm.
http://www.ec.europa.eu/environment/ippc/.
http://www.iso.org/iso/iso_14000_essentials.
http://www.weather2000.com/dd_glossary.html.
Morvay, Z., Gvozdenac, D., Sathapornprasath, M. (2003) Integrated Energy and Environmental Performance Monit-
    oring of Manufacturing Processes, 2nd Regional Conference on Energy Technology towards a Clean Environment,
    Phuket, Thailand 12–14 February.

3

Relationship between Energy Use and Production Volume

3.1 Introduction

Changes in energy or environmental performance are due to variations in one or more influencing internal or external factors in a production process (see Boxes 2.2 and 2.3 in Part I Chapter 2). When a change in performance is detected, it must be tracked back to the variation of the particular factor that has caused the performance to change.

A good understanding of and an insight into the relationship between energy use and production output is a prerequisite for performance evaluation. We know that production is an independent variable and that production output level will determine energy use. When there are variations in production output level, energy use will vary as well. Unfortunately, variability in energy use can only be partly explained by the underlying energy/production relationship. There are other influencing factors that will be the cause of of variability.

Here, we will focus only on the energy/production relationship because energy is directly related to and an easily measurable input into a production process, while environmental impacts are the consequence of the amounts of energy and materials used. However, the same methodological approach can be employed to analyze the relationship of other resources or environmental impacts to production volume variability.

We need to develop an understanding of the *causes and effects of variability* that determine the *dynamics of the energy/production relationship*. Variability is partly determined by process design and standardized operational procedures. That apart, variability will stem from the dynamic environment of everyday operations.

3.2 Energy/Production Relationship by Design

Energy and process flow charts, as described in previous chapter, provide valuable information on where, why and what type of energy is used or required. These charts are a logical first step in understanding the energy/production relationship as determined by the original process design. These flow charts should provide quantitative information on energy, material and water requirements at the designed production output capacity, and sometimes at reduced output as well.

An additional insight into the relationship between energy and production is provided by the temperature profile of the production cycle. The example of the temperature profile of a tuna-canning process is presented in Figure 3.1. It shows the duration of the whole production cycle and of individual production stages, as well as the temperature profile at each stage. Further information is obtained by providing

Applied Industrial Energy and Environmental Management   Zoran K. Morvay and Dušan D. Gvozdenac
© 2008 John Wiley & Sons, Ltd

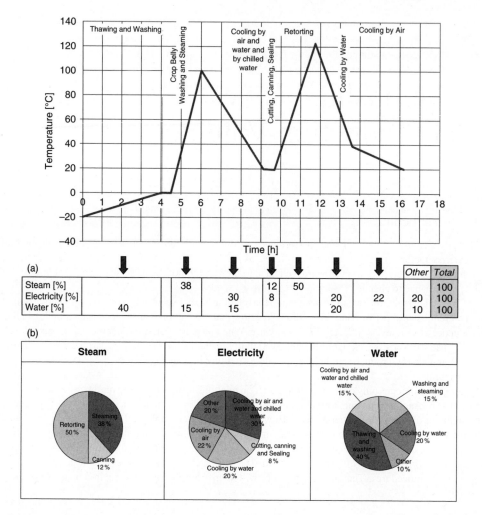

Figure 3.1   Temperature Profile during Production Cycle and Energy Used at Individual Production Stages (Tuna-Canning)

data on the type and percentage of total energy used at each production stage and correlating it with the temperature profile (Fig. 3.1a) or by showing them as a pie chart (Fig 3.1b).

The presentation style can differ, but the point is that at this stage of analyzing the energy/production relationship, we should know the types, amounts, and places of energy consumption in the process. This information will also serve as guidance for the technical analysis of process energy performance. We should focus on the parts of the process where most of a particular type of energy is used.

Such temperature profile diagrams are created by technology requirements or by process design. They present the best way to run a process in order to achieve the desired level of product quality. Consequent energy consumption is either determined, or can be calculated or measured, and then serves as a target for this production process. But, in practice, variations occur in actual temperature, in duration of process phases (which are usually too long) and in the load or capacity utilization of the production line. All deviations from the designed and desired process diagram should be followed by deviations in energy consumption.

As an initial benchmark, for any given purpose, we should establish the energy demand in an ideal process – the one determined by the original process design at a full production output level. It is also useful to have information on the energy use at reduced production output levels.

This information defines the underlying *energy/production relationship* because the process design and technology requirements determine optimal energy needs for the various production output levels.

3.3 Energy/Production Relationship by Standard Operational Procedure

The manner in which machines are operated has a strong impact on the energy/production relationship. Production lines and machines, once installed, are accompanied by blueprints, diagrams, and standardized operational and maintenance procedures. Only strict adherence to the prescribed operational procedures will result in optimal energy consumption at the designed output level and with reduced energy variability.

Standard operational procedures for a machine would provide information to fill in the blanks in Figure 3.2 and determine the nominal relationship of energy to production through the stages in production operations.

Figure 3.3 provides an example of the *operational procedure* for an injection molding machine. It emphasizes the important fact that the production machine is not necessarily a single energy-using unit. The molding press incorporates a pump, cylinder heater, materials dryer and dye heater. All of these

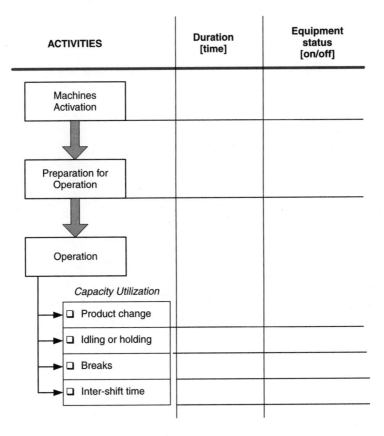

Figure 3.2   Stages in Production Operations

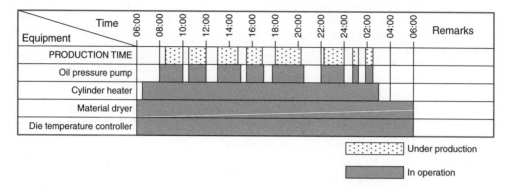

Figure 3.3 Operational Procedure for Molding Machine

components do not need to run all of the time. The purpose of a standard operational procedure is to prescribe the running status of single components during production stages.

As time goes by, the originally prescribed procedures and operational requirements often tend to be forgotten and more relaxed and convenient operational practices are applied instead.

Another important issue, for the energy/production relationship, is *capacity utilization*. Capacity utilization can be defined as the ratio between the actual production output and maximum production capacity. Even when operational procedures are strictly adhered to, if machines run at *partial capacity*, energy effectiveness will decrease, and energy cost per unit of product will increase.

This can be illustrated by an example, a hammer mill driven by an electric motor of more than 200 kW power, which displayed a particularly inefficient operation. Figure 3.4 shows the daily load profile as measured during the operation analysis. It reveals several points of concern:

1. Frequent idling (i.e. no load operation).
2. Low average capacity use – mostly at half of the nominal output level.
3. Frequent start/stop.

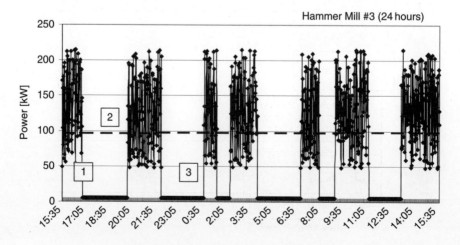

Figure 3.4 Actual Daily Load Profile of Hammer Mill

The main reason for these problems is that the operational procedures for mill operation are not followed and strictly applied. As a result, the mill uses, on average, 40 % more electricity than is required by the actual design! Of course, things would be different if the mill operated at the nominal capacity and with strictly observed operational procedures, as prescribed by the original design. Neglecting standardized operational procedures owing to convenience or other reasons will always result in higher operational (including energy) costs, and will be reflected in a changed energy/production relationship. The usual deviations from prescribed operation procedures are

- too early a start;
- operating time that is too long;
- machines left running when not required;
- operating at partial load, etc.

The issues emphasized by Figure 3.3 and Figure 3.4 may look simple and straightforward but cause most of the variations of energy use in production as well as increased energy and other operational costs. Therefore, it is of the utmost importance to have and then to understand and follow operational procedures, which must not be underestimated.

Understanding *operational procedures* and the consequences of less than strict adherence to them is at the heart of understanding variability in the energy/production relationship. This understanding provides the critical knowledge necessary for monitoring energy and environmental performance in production departments.

If standardized operational procedures do not exist, they need to be established and then followed. That is one of the aims and results of EEMS – to establish a *best practice operation* that is embodied in standardized operational procedures.

3.4 Presenting the Dynamics of the Energy/Production Relationship by Scatter Diagram

The information presented in Figures 3.1 to 3.3 is based on nominal or designed data and does not take into account actual operational conditions and circumstances. We need to assess how actual energy consumption relates to the given production output dynamically as operational conditions change and production output varies in order to evaluate process energy performance. Figure 3.4 presents an example of such a dynamic operational relationship.

Ideally, all variability in energy use should be explained by its fundamental relationship with production output. If that is the case, then it will provide an example of excellent energy performance! But this is hardly ever the case. The dynamic nature of industrial operations and various influencing factors will result in energy variability greater than that predicted by production variability, as shown in the example of the hammer mill case.

When there is an excessive *variability in energy use* (on top of variability caused by the underlying relationship), the causes need to be explored and identified. First and foremost, reliable data is required on both production output and energy consumption. Monthly data is usually available at most factories and can be used for the initial analysis. The energy data sources are monthly energy bills or utility log sheets. Production data may come either from sales or production control. Energy (E) and production (P) data of the same time interval (t) are inputs for forming a performance indicator that will provide direct indication on *energy/production relationship*:

$$PI_{Pt} = \frac{E_t}{P_t} \tag{3.1}$$

It is also possible to relate energy consumption to raw material (RM) use:

$$PI_{RMt} = \frac{E_t}{RM_t}$$ (3.2)

In this way, reworks or rejects will also be indirectly taken into account. In principle, it will be best to monitor performance of material use (MU) by relating raw material to good products (GP) output and then to monitor *energy performance* (EP) by relating the amount of good products completed over a period of time to the energy consumed over the same period.

$$PI_{MUt} = \frac{GP_t}{RM_t}$$ (3.3)

$$PI_{EPt} = \frac{E_t}{GP_t}$$ (3.4)

Table 3.1 shows a typical data set (a real case) giving monthly energy consumption (steam) and production data over a period of 12 months and the related performance indictors (PI). The PIs expressed as given by (3.1) are also frequently called *specific energy consumption* (SEC) because they present the ratio of amount of energy per unit of product.

Little can be said about underlying energy/production relationships when data is shown in a format such as in Table 3.1. In factories, energy and production data are often presented graphically, as time series (Fig. 3.5). When data points are plotted for each month and then connected, a *trend line* appears. The obvious purpose of such a graph is to show trends in energy and production data and identify swings and fluctuations over time.

Table 3.1   Data on Production and Energy – Real Case

Month	Production, $[t_p]$	Steam, $[t_s]$	PI_p, $[t_s/t_p]$
1	*2*	*3*	*4*
1	1745	1708	**0.98**
2	1347	1771	**1.31**
3	1670	1839	**1.10**
4	1825	1758	**0.96**
5	1328	1824	**1.37**
6	1297	1655	**1.28**
7	1275	1503	**1.18**
8	1249	1337	**1.07**
9	1152	1389	**1.21**
10	1012	1109	**1.10**
11	876	1093	**1.25**
12	1058	1040	**0.98**

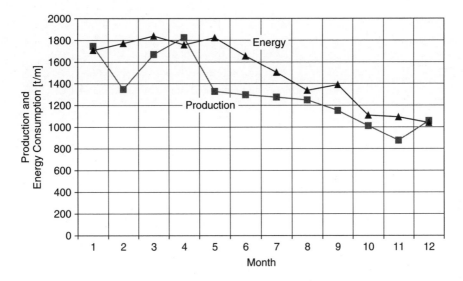

Figure 3.5 Energy and Production Data Trend Lines

In Figure 3.5, for instance, production output is generally declining over the year, and energy use tends to follow the trend of production, with the exceptions of months 2, 4, 5, 9 and 11. Also, if there are any cyclical movements, seasonal fluctuations or irregular variations in energy or production values, the time series graph will reveal them. Indirectly, the graphs may also indicate correlations between variations of energy and production values.

For this example, values of energy and production are expressed incidentally in the same units (tons) and are of similar magnitude. Therefore, it is easy to observe the correlation between the trends of both variables. When units and scales are different, that immediate visual clarity can be lost. Therefore, we cannot make effective conclusions from time series graphs about relationships between energy and production values variations.

Another graph, as shown in Figure 3.6, provides more information on the underlying relationship between energy and production.

Normally, *production* as an *independent variable* is presented on the x-axis, and *energy* as a *dependent* or *response variable* is presented on the y-axis. If production varies, it is expected that steam consumption has to vary as well. The higher the output, the more energy is needed. As we can see, there is no time dimension in this graph. The points in the graph simply indicate the amounts of energy consumption for various production outputs, irrespective of when that happened. As a result, a particular *data pattern* emerges, describing the relationship between production output and energy consumption. The points obtained are spread in an irregular pattern – therefore such a plot is referred to as *scatter diagram*.

Provided data accuracy and adequacy are verified, the emerging data pattern presents the true and actual relationship between energy use and production output over the observed period of time. It shows the degree of variability in energy consumption and the correlation to production output. The position of each data point in the scatter diagram is the effect of explainable causes and production circumstances that have actually happened during the observed period. Such variations, as displayed in Figure 3.6, are caused by factors that influence energy performance, as listed in Boxes 2.2 and 2.3 in Part I Chapter 2. When the energy production relationship is visualized in the scatter diagram, variations in performance become visible immediately and we can start acting on those variations.

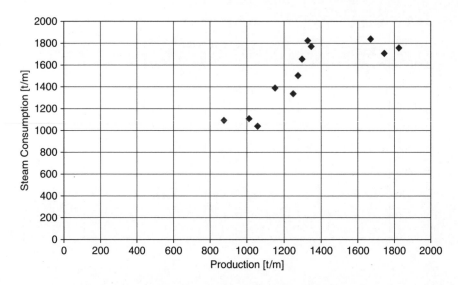

Figure 3.6   Production Output and Energy Consumption Data Pattern

Statistically speaking, we are interested in the *correlation analyses* between production output and energy use values. We are concerned about the *variability* of energy data around the anticipated trend line. However, what interests us more than the merely statistical measures of this variability are the *causes* of and *explanations* for a particular data pattern. For every significant variation in energy performance, there must be a specific cause either in production events or problems or in operators' behavior and performance.

Descriptive or enumerative statistics are of little help here. They may help by visualizing data or quantifying variability and correlation coefficients. But statistics cannot explain the causes and effects of any pattern that emerges. Quantified variability factors can be used as an indicator on performance, but not as the main measure of it. More important is the resulting *pattern of data* in the scatter diagram, and this pattern must be *interpreted*. *Inductive reasoning* and *analytical thinking* are required to *interpret data* and *extract knowledge* on performance upon which we should act.

Therefore, we will focus firstly on *qualitative analysis* and interpretation of the energy/production relationship as provided by a scatter diagram. Helpful statistical methods will be examined afterwards.

3.5 Interpretation of Energy/Production Data Pattern on the Scatter Diagram

For practical energy and environmental management, one of the main tools is *cause and effect* analysis – determining what events, practices or behavior resulted in good or bad performance. The focus is on the process and how to improve its performance. Every variation in energy performance has an explainable cause. Whether or not the variation can be reduced and avoided, the cause must be discovered and explained by the process of data interpretation.

The process of *data interpretation* must result in operational guidelines for the future. Guidelines should prevent undesired practices and help to avoid behavior that has a negative impact on process performance. The most powerful tool managers have for data interpretation is – a *QUESTION!* The managers can and should always ask '*why*' a particular data pattern occurred. *Performance improvement teams* at each ECC should seek and provide adequate answers.

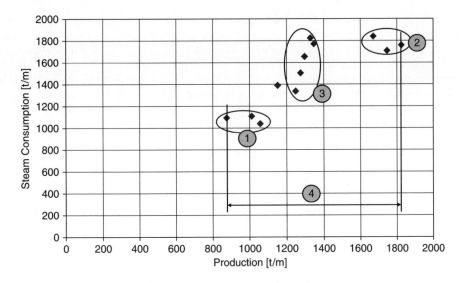

Figure 3.7   Interpreting Data Patterns

If we take a closer look at the data pattern in Figure 3.7, four areas of concern emerge, as follows:

Data sub-sets Nos 1 and 2 show that there is little variation in energy consumption for different production outputs. Moreover, for these data sets, energy consumption tends to be lower when production outputs are higher, which is a clear sign of weak energy management and control.

Insight 1: *If there is no variability in energy values that does not necessarily mean that performance is good. On the contrary, if production drops or increases, energy must drop or increase too, by the amount determined by their underlying relationship.*

Data sub-set No. 3 indicates the opposite: for almost the same production output, there are many points of recorded energy consumption, ranging from 1300 to 1800 tons of steam. If 1300 tons of steam is enough for one production level, why is it not always enough?

Insight 2: *The variability in energy values here is too extensive. If production is steady, energy use must be steady too.*

Data sub-set No. 4 indicates a range of recorded production outputs which varies from 876 to 1825 tons/month. This is very large production variability. Table 3.1 clearly indicates that specific energy consumption is higher by at least 20 % on the lower end of production output. It also means that unit production costs are higher at the lower output level. Therefore, efforts should be made to reduce variability in the production output.

Insight 3: *If production output varies often and to a great extent, it would be difficult to maintain energy use at optimum level, and greater variations in energy use will inevitably occur.*

One can argue that *production variability* is beyond factory control because the output level is set by market demand. That may be true for monthly total production or sale. But we will see later that when the daily production/energy data is analyzed a similar pattern will emerge again. By better production planning, daily production output can be controlled and large variations avoided. If production requirements are set at below maximum capacity, there is an option to operate machinery at full capacity for a shorter period of time, rather then at a partial capacity but for the full time.

Sometimes the reasons for performance variations will be legitimate, i.e. uncontrollable or unavoidable. But in many instances the causes of variations are controllable. The fact that there are always some variations in production cannot be used as an excuse for sub-optimal energy performance. When we identify the causes for variation, we are on the road to energy performance improvement.

It can be argued that, looking only into the aggregated data on past performance, it is difficult, if not impossible to think of the reasons for bad performance, say, 10 months ago. This is true, but in that case, the real question is why did we wait so long, or why don't we have more data for analysis? Monthly data may show that there is a problem with performance but can do little to explain the root of the problem. In any case, the point is that EEMS must provide *decentralized data* on energy performance – down to a business segment level, or ECC in our case, and must introduce discipline in order to interpret the data – not annually or monthly, but daily and weekly.

We suggest that data should be interpreted daily on the operational level and weekly on the managerial level, at least during the initial 6-12 months of the EEMS start-up operation. This period is necessary in order to create insight into the energy/production relationship and develop an understanding of the production process and the factors that influence energy performance.

The following example clearly illustrates the point of having daily instead of monthly data on energy and production. Figure 3.8 shows the energy/production monthly value for an ECC (a real case). A month is a standard interval for management to pay some attention to energy consumption data. What can be concluded on energy performance from a single data point, even if compared with a target or the past months? Not very much indeed.

After implementing EEMS, daily data has become available. Figure 3.9 presents a scatter diagram with daily data for the same ECC and the same month. Now there is an abundance of information on energy performance of ECC being observed and many conclusions can be made from the interpretation of this diagram. For instance, the variability in both energy and production values (point 1) in Figure 3.10 is such

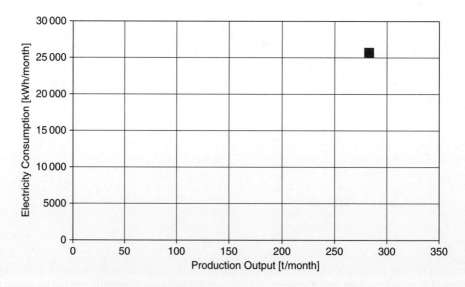

Figure 3.8   Monthly Value of Energy and Production for an ECC

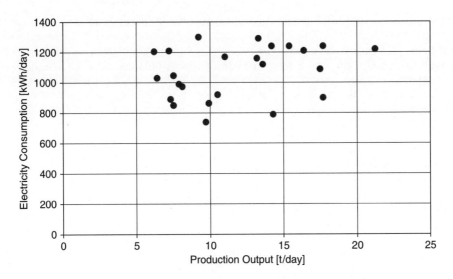

Figure 3.9 Daily Values of Energy and Production for the Same ECC from Figure 3.8

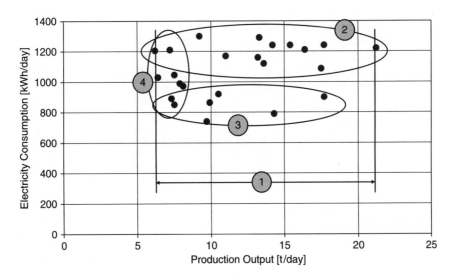

Figure 3.10 Points of Concern for Interpretation of the Data Pattern

that neither energy nor production management can be satisfied with their performance. However, interpretation has revealed some consistency in these seemingly un-correlated values (see Fig. 3.10). While production variability is too large, energy use is almost insensitive to the change in production output.

The values in subset 2 and subset 3 (Fig. 3.10) are of different product types. However, both cases demonstrate large production variability and little sensitivity in energy use to the change in production output. The values in subset 4 all happened on weekends and holidays. Apparently, production output is consistently lower outside of ordinary weekdays.

After such a qualitative analysis of the data pattern, the following questions are to be asked:

(a) Why are production variations so large for both products?
(b) Why does energy consumption for both products show so little sensitivity to production output change?
(c) Why is production level on weekends always low?
(d) What should be done to improve energy performance?

It is up to the local ECC team to seek and provide the answers by analyzing the operational events that lay behind demonstrated performance. When daily energy and production measurement and performance monitoring is a routine practice, it is simple to provide answers to the above questions and determine the real causes for performance variations. Measures can then be put in place to avoid recurrences of bad performance.

This is an example of a *qualitative and analytical study* where data interpretation raises questions about actual energy performance and seeks answers and corrective actions. It aims to uncover useful patterns (Fig. 3.10) from large quantities of data in order to *develop best practice operation*, i.e., better way of doing things. This is a process of *knowledge discovery and creating* improvement of energy and operational performance.

Please note that all considerations in this chapter require only knowledge and understanding of the process and *cause-effect relationship* for energy performance. No sophisticated statistical know-how is required. Therefore, people at the workshop level should not be put off or terrified by another complex performance monitoring system that they must master or be subjected to.

This is very important for achieving success in the improvement of energy and environmental perform-ance. We should never forget that EEMS implementation involves all of the people in an organization and not only those with a university degree, who majored in math. Therefore, at the operational level, where people have all sorts of educational backgrounds, procedures for performance monitoring and evaluation must be simple enough so that all employees can understand and implement them. In some cases, initial training and support may be required, but if the procedures are simple and practical, training requirements should not be extensive. On the other hand, managers can certainly benefit if some useful and straightforward statistical methods are applied for the evaluation of energy/production variability and their underlying relationships.

3.6 Statistical Methods for Energy/Production Variability Analysis

So far we have used a scatter diagram to visualize variability in energy and production data. The resulting data pattern is used for the interpretation of energy performance based on understanding the production process and the cause-effect relationship between energy variability and performance influencing factors. This interpretation is based only on analytical thinking and inductive reasoning, and no statistical methods are used.

A scatter diagram may indicate visually where the centralized data tendency may be, but it does not provide any quantitative information in that respect. Therefore, to introduce a certain measure of variability, to quantify the significance of deviations and to define the term 'significant', etc., we need to apply statistical methods.

A key statistical concept in characterizing variability is based on two simple parameters: one which locates the 'center' of the data and another which measures the spread (dispersion, scatter) around that 'center'. Experience has shown that the ordinary average (mean) value is generally the best indicator of central tendency, while the standard deviation gives the best measure of dispersion. These two parameters are calculated as follows:

$$\text{Average Value} = \frac{\sum\limits_{i=1}^{n} x_i}{n} = \overline{x} \qquad (3.5)$$

$$\text{Standard Deviation} = \sqrt{\frac{\sum\limits_{i=1}^{n} (x_i - \overline{x})^2}{n-1}} = s \qquad (3.6)$$

where:
x_i = individual value;
n = number of data values in sample.

Note that $(x_i - \overline{x})$ is the 'distance' of an individual point from the average and thus a measure of the scatter. A measure of average scatter is standard deviation. For the example given in Table 3.1, it is possible to find that these two parameters are as follows (Table 3.2):

Table 3.2   Average Values and Standard Deviations for the Case Presented in Table 3.1

Parameter	Production Output, [t]	Steam Consumption, [t]
Average Value	1319.50	1502.17
Standard Deviation	294.71	300.93

These figures give us separate information on the average production output and steam consumption for a period of 12 months. However, we should not forget that production output and steam consumption are not sets of random data, but reliably and accurately measured quantities, where production is an independent variable while steam consumption is a dependent one. Therefore, average values and standard deviations will not help us a great deal in characterizing the energy/production relationship. We need a tool that will describe the functional relation between energy and production.

3.6.1 Regression Analyses

Regression tells us how a dependent variable – energy – is related to the independent one – production – by providing an equation that allows for *estimating energy consumption* for the given production output. This relationship between production and energy consumption for most industries is a linear form (Fig. 3.11), which means that the relationship between the points in the graph can be approximated by a straight line and expressed by a linear equation (3.7) in a general form as follows:

$$y = a \cdot x + b \qquad (3.7)$$

where **a** and **b** are constants that need to be calculated for each data set. **x** is an independent variable, which is production output in our case and will be denoted as 'P'. **y** is dependent variable, which is energy in our case, so we will denote it by 'E'. Now the same equation can be rewritten as follows:

$$E = a \cdot P + b \qquad (3.8)$$

Such a linear equation will produce a *regression line* that 'fits' through the uneven scatter of data points. In principle, we can draw many lines through a single data scatter. It can even be done manually, which most of us have probably forgotten since a computer is now on everyone's desk.

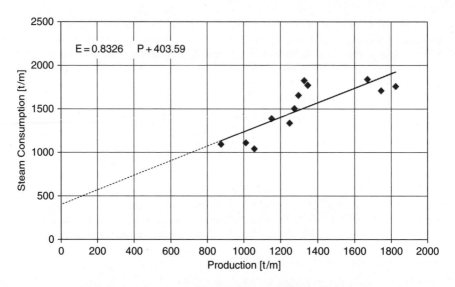

Figure 3.11   Relationship between Production and Energy Consumption

For a particular data set, if the constant values of 'a' and 'b' are calculated by the '*least square method*', the resulting line will go through the center of the data scatter and therefore is called a '*best-fit line*'. This line has a particular property – the sum of the vertical distances of data points from the best fit line is equal to zero. Nowadays, spreadsheet software will offer a 'best-fit' approximation based on the least square method, draw a line and give the equation as shown in Figure 3.11. For the presented set of data, the best-fit approximation is as follows:

$$E = 0.8326 \cdot P + 403.59 \tag{3.9}$$

The regression equation given by the least square method is the best approximation we can obtain of the underlying energy/production relationship, assuming linearity over the relevant data range.

However, we should be aware that this is just a mechanical process of fitting a regression line to the data set and which does not take into account any information about the actual environment from which the energy/production data comes. Spreadsheet software will produce an equation for any data set entered into a computer. If the data is wrong, or linearity does not hold, the software will still generate an equation. It is our – user's – responsibility to assure that input data is a correct reflection of the energy/production relationship that is analyzed. Of course, it is assumed that measurement errors are within pre-set acceptable ranges. If that is so, spreadsheet software will produce values for constants **a** and **b** that will in turn determine the regression equation and best-fit line.

If the best-fit line is extended to the y-axis (the dotted line in Fig. 3.11), it will *intercept* the y - axis at the value y = b. The *slope of the line (a)* that goes through the scatter must always be positive, which means that if production increases, energy consumption must also increase, or vice versa.

Here, we should point out that regression analysis is limited only to the range of actual observations that quantify the relationship between energy and production. No relationship exhibits the same pattern at all activity levels. It also means that linearity may apply only over the observed range of production outputs. Consequently, one should be careful when making conclusions based on an *extrapolation* of properties outside the relevant data range.

At a very low production output (say less than 30 % of nominal value), the linearity may be lost and replaced by some form of a polynomial, power, logarithmic or exponential relationship. On the other hand, we can assume that relationship will be linear in segments over a range of production output levels. Still, a very low production output is not a normal operating state, because plants are not designed and built to operate continuously at low capacity. Therefore, instead of trying to fit the data with some exponential regression line, a question should be asked firstly as to *why at all* is the plant running at such a low output?

If there are any material changes in the underlying energy/production relationship, i.e. expansion of production capacity, installation of more efficient machines, etc., a sufficient amount of new energy/production data (after relevant change) have to be collected and then the *regression equation* must be updated.

When handled with care, the regression equation proves to be a quite useful tool in the performance evaluation process. But, if the regression analysis is conducted without giving any thought to the context of the analysis, it can do more harm than good.

3.6.2 Correlation Analyses

The scatter diagram shows the relationship between energy and production values. If we are interested only in the *strength of association* between energy and production, we can use correlation analysis to provide a measure of this strength. Correlation analysis can measure how closely variations in both values move together and quantify the strength of their relationship, assuming that the underlying relationship is linear. This assumption is correct for most industries provided that production output variability is not too large.

Spreadsheet software usually has a built-in function to calculate correlation coefficients for a given data set. The symbol 'r' is normally used to denote the correlation coefficient, E_i and P_i denote energy and production, and $\bar{E}$ and $\bar{P}$ denote mean (average) energy and production for the given data set, with n equal to number of energy/production pairs ($i = 1, 2, 3, \ldots n$).

$$r = \frac{\sum\limits_{i=1}^{n} P_i \cdot E_i - n \cdot \bar{P} \cdot \bar{E}}{\sqrt{\left(\sum\limits_{i=1}^{n} P_i^2 - n \cdot \bar{P}^2\right) \cdot \left(\sum\limits_{i=1}^{n} E_i^2 - n \cdot \bar{E}^2\right)}} \tag{3.10}$$

where:

$$\bar{P} = \frac{\sum\limits_{i=1}^{n} P_i}{n} = \text{Mean value (arithmetic mean) of } P_i (i = 1, 2, 3, \ldots n)$$

$$\bar{E} = \frac{\sum\limits_{i=1}^{n} E_i}{n} = \text{Mean value (arithmetic mean) of } E_i (i = 1, 2, 3, \ldots n)$$

The calculated correlation coefficient for our case study with monthly data (Fig. 3.11) gives values of $r^2 = 0.6648$ and $r = 0.815$, which denotes a reasonably strong correlation for a real industrial environment. For the daily data set shown in Figure 3.9, the correlation coefficient has a value of $r = 0.293$ and $r^2 = 0.0860$, which denotes a very week correlation indeed.

The *correlation coefficient r* will indicate how well a best-fit line explains variations in the value of a dependent variable, i.e. energy. If $r = 1$, all data points will be exactly on the regression line. This will be a case of perfect correlation. The larger the scatter around the best-fit line, the smaller the r-value is, and the weaker the correlation between energy and production.

3.7 Meaning and Use of the Regression Line in Energy Performance Evaluation

Once the regression equation has been obtained, *estimates of past or future* energy values for a given production output can be made, and *variability* in energy use determined. Variability can be expressed as the difference between standard or reference and actual energy use. Variability analysis is a systematic process of identifying and evaluating variances in order to provide useful information for measuring efficiency and improving the performance of energy use.

A part of variance in energy use will be caused by variances in production output. This is normal or expected variability. It will be determined by the underlying relationship between energy and production, as expressed by the regression equation. For instance, let us have a closer look at month 5 in Table 3.1 which displays an apparent excessive variability in energy use. This is shown in the scatter diagram in Figure 3.6. If we want to know what should have been the energy consumption for that month according to the obtained regression equation, we should substitute the actual production output of this month (1328 t from Table 3.1) into the equation:

$$E' = 0.8326 \cdot 1328 + 403.59 = 1507.62 \qquad (3.11)$$

The underlying relationship between energy and production warrants the energy use of 1509.28 tons of steam. When we compare the estimated energy use value E' with the measured value of E, we will get the variability portion of energy that is not explained by the variability in production output:

$$\varepsilon = E - E' = 1824 - 1507.62 = 316.38 \qquad (3.12)$$

One interpretation of this difference is that energy consumption in the respective month was 314.72 tons more than required for the given production output, based on the underlying average energy/production relationship. Therefore, it comes as no surprise that this month accounted for the highest energy consumption per ton of product (see column 4 of Table 3.1).

On the other hand, we can repeat the same procedure for month 1 when the lowest specific energy consumption was recorded:

$$E' = 0.8326 \cdot 1745 + 403.59 = 1856.48 \qquad (3.13)$$

If we calculate the difference between the actual and estimated values we will get what in statistical terminology is called a **residual ε**:

$$\varepsilon = E - E' = 1708 - 1856.48 = -148.48 \qquad (3.14)$$

There is a minus sign in front of our residual. What does it mean? A quick interpretation will tell us that we have used too little energy for the given production output. But the values of energy and production have been checked and confirmed to be correct, so this production output has really been achieved with the stated energy amount.

How and why this has happened, are the questions that regression analyses cannot answer. In this case, it can only tell us that for reasons unknown we have saved about 148.48 ton of steam in this particular month compared to the expected consumption based on the average underlying relationship.

If we calculate and plot all the residues, a graph will appear that can be used for *residual analysis* (Fig. 3.12). This is yet another scatter diagram, but this one shows a *variability of energy values* at the given production level, as calculated from the regression equation. Statistics say that individual

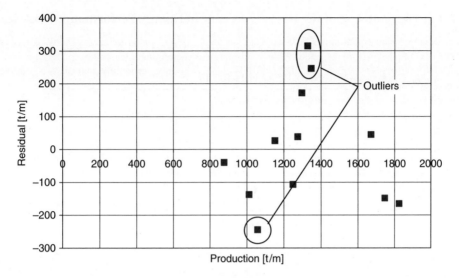

Figure 3.12   Residual Scatter Diagram

residues should fall within a narrow horizontal band, otherwise our analysis may be based on false assumptions.

In our case, we can be satisfied that the 'narrow band' is defined over ± 200 range, with three *outliers* dropping outside the range as indicated in the graph. If an identifiable cause exists for the outliers, as it should indeed exist for all the other variations, we can be satisfied with the validity of our analysis. This brings us to the task of quantifying variability in energy values. There are four issues at stake:

- How can variations be quantified?
- What is the reference for energy variation assessment?
- Whether a variation is significant or not?
  and
- If a variation is significant, what was the cause of it?

Statistics can provide the answers for the first three questions but not fthe fourth.

3.7.1 Quantifying and Understanding Energy/Production Variability

We can see that the points in scatter diagrams (Fig. 3.11) do not fall precisely at, or around, the best-fit line. They vary in a seemingly random way with smaller or greater distances from the best-fit line, similar to energy residues. This is shown by the residual diagram. The fundamental indication of variability, provided by any given energy/production data set, is the amount of scatter around the estimated regression line. The measure for this variability is called the *standard error of estimate*, which is the deviation between the actual or measured energy E_i and the estimated one E_i' as computed by using the correlation equation for the same value of production P. The standard error of estimation is given as follows:

$$S_{EE} = \pm \sqrt{\frac{\sum_{i=1}^{n} \left(E_i - E_i'\right)^2}{n - 2}}$$

(3.15)

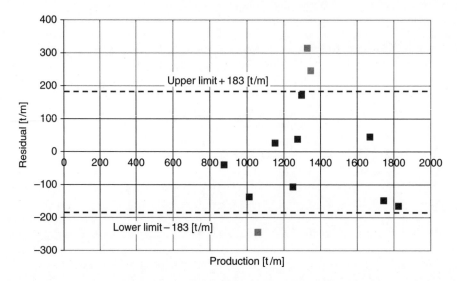

Figure 3.13   Residual Scatter Diagram of Energy Residues Variations

The particular value for the data set presented in Figure 3.13 is ± 183. One characteristic of S_{EE} is that 68 % of all data points will fall within a band of $\pm S_{EE}$. The width of that band gives a measure of variability. If we draw these lines on our residual scatter diagram as upper and lower limit, only three values will fall outside the band. It means that if we analyze energy variability around the best-fit line and use S_{EE} as criteria to establish significance of variation, there are only three instances that require exploration (Fig. 3.13).

In purely statistical terms, the described approach may be valid, but it does not satisfy the needs of energy/production variability evaluation related to energy performance monitoring.

If S_{EE} is large, the range is wider. When S_{EE} is divided by the mean energy value of the analyzed data set (1502.17 t/m from the Table 3.1), the scatter can be expressed as a percentage, which is about 12 %. This percentage represents high variability and indicates the loose control of energy in production. Therefore, the threshold for significance of variability should not be established mechanically. It must be based on knowledge of the process and the underlying energy/production relationship.

When evaluating energy performance, we are not actually interested in variability around the best-fit line, because the best-fit line represents an average of the energy/production relationship, i.e. *average* energy performance. We are interested in energy variability against a *target* that represents the best energy performance for the given production output. Another look at the scatter diagram can help us to identify three cases of best performance during the observed period (Fig. 3.14). It makes good sense for energy performance assessment to take the best-achieved performance as a reference for evaluation and as a target for future performance.

We can even calculate a new regression equation based on only three indicated points that present best performance (Fig. 3.15). The new regression equation displays changed **a** and **b** parameters:

$$E' = 0.9503 \cdot P + 36.037 \tag{3.16}$$

The correlation analysis will now yield a correlation coefficient and a coefficient of determination with values close to 1, which means almost perfect correlation. As we have said before: for any data set spreadsheet software will provide statistical indicators and equations, but which one is right or the

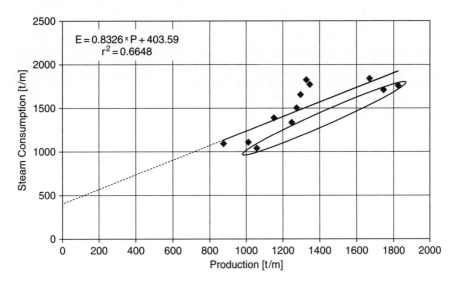

Figure 3.14 Identifying Base Line for Energy Performance Evaluation

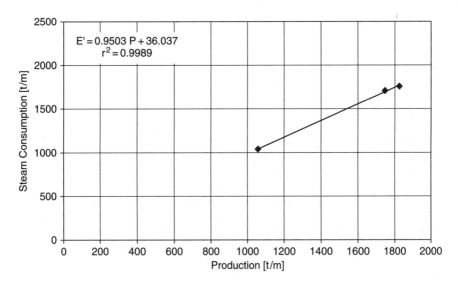

Figure 3.15 Energy Variability Evaluation – Best Performance as Base Line

best for our purposes, statistics cannot decide. This can be answered based only on the knowledge and understanding of the process and the actual energy/production relationship.

In our case it is proven that the new regression equation presents a more desirable and appropriate energy/production relationship. This is why we can use it as a *base or target line* (equation) in order to assess past or even future energy performance, as long as the context of the underlying relationship of energy/production remains unchanged. When the material changes occur within the context, the regression equation will need to be updated.

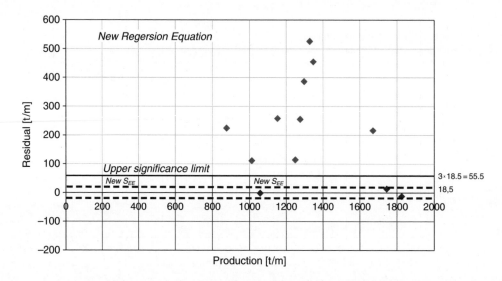

Figure 3.16   Residual Scatter Diagram against Best Performance Regression Equation

If we now redraw the residual scatter diagram with residues calculated using the new equation, the results are shown in Figure 3.16. The new value of standard error of estimate is only $S_{EEN} = 18.5$ t/m for the chosen 'best' points which represent the target for performance assessment. The variability around the 'best' (or target) performance is larger than previously around the average performance. There are now only three values within the new significance range of $+/- 1 \times S_{EEN} = 18.5$ t/m.

However, we may decide to set the upper *significance limit* at $3 \times S_{EEN} = 55.5$. It is a customary procedure when the correlation coefficient of the target line regression equation is strong (near to 1). If the correlation is week, then the significance interval determined by $3 \times S_{EEN}$ will be too wide. In other words, too many instances of substandard performance will be accepted without question. It does not make sense to establish a lower significance limit, because it may look like we are restricting the requirements for performance improvement.

In our case study, applying the $3 \times S_{EEN}$ significance limit will reveal that variations in energy use are too large in all but three months of the observed period, and individual residues are, of course, larger than in the previous case. This is as much as statistical methods can help us in performance evaluation. Finding the reasons for excessive variability and preventing recurrences is clearly beyond the power of statistics.

When performance begins to improve, the points on the *residual scatter diagram* will be closer to the significance limit line, and eventually all the points should be below the line. When all points are below the significance limit, it means that consistent performance improvements over the targeted best practice are achieved, and this requires an update of the target line, i.e., the corresponding regression equation.

3.7.2 Fixed and Variable Energy Consumption

Additional useful information that regression equation can provide is the quantification of a concept of *fixed and variable energy consumption* (Fig. 3.17). The original regression equation is:

$$E' = 0.8326 \cdot P + 403.59 \tag{3.17}$$

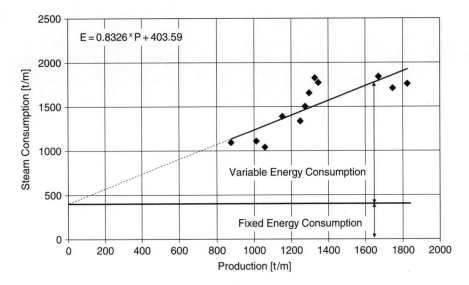

Figure 3.17   Fixed and Variable Energy Consumption

The intercept value of 403.59 will be roughly equivalent to the fixed energy consumption, and '$0.8326 \times P$' will show the rise in energy consumption for additional units of production P. If $P = 1$, then the additional justified variation in energy will be 0.8326 for an additional unit of product.

For instance, when the production output is 1347 tons, the estimated energy use by regression equation will be 1525.1 tons of steam, and the actual energy use was 1771 tons (from Table 3.1). The energy/production relationship, as described by the regression equation, can provide the following breakdown of this energy value:

- Fixed energy consumption contributes to 403.59 tons of steam use.
- The variable energy portion contributes to $0.8326 \times 1347 = 1121.51$ tons of steam use.
- The energy residue $\varepsilon = 1771 - 1525.2 = 245.9$ tons cannot be explained by the underlying energy/production relationship, as it is caused by the factors that influence energy performance.

Again, we should be careful when coming to any conclusions based on extrapolation and on a mechanical interpretation of statistical parameters. We have shown already that for the data subset based on best performance, quite different statistics emerge and provide a different breakdown of energy variability. Which data sets are representative and over what range can be decided only on a case-by-case basis and by relying on an understanding of the process and of the energy/production relationship.

3.7.3 Specific Energy Consumption

Another way to present data from Table 3.1 is shown in Figure 3.18. Here, instead of energy, the specific energy consumption (SEC) is plotted against production. It is again an approximate linear relationship, but in this case the slope is negative, i.e. the line is declining. It means that with increased production, energy consumption per unit of product is decreasing and vice versa.

This brings us back to the impact of market demand variations. If production output is falling, it means that we are moving back along the x-axis in Figure 3.18, which in turn means that energy consumption

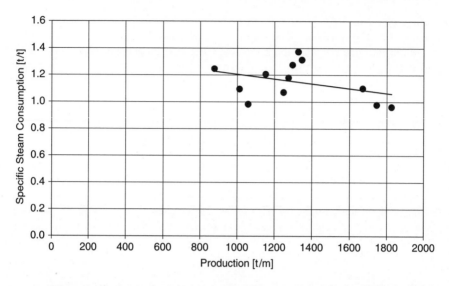

Figure 3.18   Relationship between Production and Specific Energy Consumption

per unit of product is increasing. As a result, the production at *low capacity utilization* is more expensive and should be avoided.

3.8 Summary of Presenting and Analyzing the Energy/Production Relationship

The key to energy and environmental performance improvement is to understand the energy/production relationship and the impact of influencing factors on energy/production variability. In this chapter, we have explored the ways of expressing and analyzing this relationship. Let us review them briefly. Energy/production relationship is determined by:

- process design and technology requirements;
- operational procedures;
- capacity utilization;
- a combination of influencing factors affecting day-to-day operation.

A prerequisite for the analysis of the actual energy/production relationship is the availability of reliable data. Monthly data at the factory level are suitable only as an indication of potential problems. For performance improvement we need daily energy/production data for each designated ECC. A *scatter diagram* with the resulting data pattern is the best visual representation of the energy/production relationship.

Correlation analysis provides a measure of strength for the association between energy and production along an assumed linear relationship, given by the correlation coefficient **r** which takes values between 0 and 1. If **r** is close to 1, it implies a very strong correlation, and indicates good energy management practice and control of energy use in production.

Regression analysis provides the equation which is the best approximation of the underlying energy/production relationship given by available data. The regression equation can be used to estimate the energy value for any given production output. When coefficients of regression equations are calculated by the least square method, the result will be the *best-fit line* that will go through the center of the data scatter. The measure of variability in energy values is given by the *standard error of estimate*.

To evaluate the *significance of variations*, we need to establish a meaningful reference or base line against which individual variations can be judged. A common sense approach suggests selecting several instances of achieved best performances in the observed period, checking beforehand that these have happened under normal circumstances. A new regression line can then be calculated using these data points only and the resulting equation can be used as a baseline for assessing the significance of variability in energy values.

We should be careful not to rely too much on the statistical indicators, and they should *not be used mechanically* to draw conclusions. They must be used with care to support the analytical study of a particular data pattern that represents an energy/production relationship. In any case, statistics cannot *explain causes* for particular variability patterns. Only *inductive reasoning*, based on an understanding of the underlying energy/production relationship and on factors that affect energy performance, can provide explanations for particular deviations and suggest corrective actions.

In practice, data interpretation of energy performance is based on *analytical thinking* supported by correlation and regression analyses. Such data interpretation is the process of creating knowledge of the causes and effects of variability in the energy/production relationship that will result in operational guidelines, which will ultimately bring the factory to achieve the best practice operation.

3.9 Bibliography

Morvay, Z., Gvozdenac, D., Košir, M. (1999) Systematic Approach to Energy Conservation in the Food Processing Industry, Manual for Training Programme 'Energy Conservation in Industry' (training course performed in Thailand), supported by Thai ENCON Fund, EU Thermie, ENEP program.

Olson, C.L. (1987) Statistics: Making Sense of Data, W C B/McGraw-Hill, December.

4

Evaluating the Performance of Energy and Environmental Management Practice

4.1 Evaluation of Past Performance

In previous chapters, we have developed a framework for the implementation of energy and environmental management system (EEMS). We have emphasized the importance of dealing with people, decentralizing responsibility for energy and environment to individual energy cost centers (ECC), introducing a performance measurement system, and understanding the energy/production relationship. The diagram in Figure 4.1 summarizes these organizational and procedural considerations in a simple step-by-step guideline that integrates all of the actions, structures and procedures as necessary prerequisites for a successful EEMS operation. The diagram also reminds us that EEMS continuously seeks to identify significant interactions within an organization, which exercise combined influence on energy and environmental performance, and which provide the basis for decisions on performance improvement measures.

The prerequisite for the introduction of EEMS should be the evaluation of current energy and environmental performance and existing energy and environmental management practice, in order to establish the potential for improvement and assess to what extent proper management structures and procedures are already in place. The evaluation needs to focus on two aspects of energy and environmental management:

- *Organizational*, i.e. checking to what extent adequate management structures are in place, and how effective they are;
- *Operational*, i.e. establishing current energy and environmental performance in the factory, based on actual energy consumption and production data and related environmental issues.

Such an evaluation will have four main objectives:

- to establish the current position;
- to set directions and targets for performance improvement;
- to establish the base line for progress evaluation;
- to prepare an energy and environmental management action plan.

Applied Industrial Energy and Environmental Management   Zoran K. Morvay and Dušan D. Gvozdenac
© 2008 John Wiley & Sons, Ltd

Figure 4.1   Energy and Environmental Management Process

Performance evaluation should start with a top-down approach. Usually, more data are available on the overall or aggregated energy and environmental performance in a factory than on individual energy cost centers (ECC). Therefore, this data should be evaluated first. However, the evaluation methodology will be the same for the specific data on ECCs as it is for the aggregated data on a factory as a whole. Besides data evaluation, the suitability of energy and environmental management procedures throughout the factory has also to be assessed.

This methodology is usually referred to as energy and environmental auditing and it covers both organizational and operational aspects.

4.2 Energy and Environmental Auditing

An energy and environmental audit (EEA), if conducted in a systematic and a comprehensive manner, is a powerful tool for evaluating current or past energy and environmental performance and management practice. The ultimate aim of energy and environmental audit is clear – identifying opportunities for reducing the cost of energy and of environmental compliance. Prior to the introduction of EEA process, it is useful to remind ourselves of the following definitions:

- *Energy Cost Center (ECC):*
  A department, a process, or unit of equipment, where the use of raw materials, energy consumption and environmental impacts can be related to ongoing activities and be continuously monitored.
- *Significant Environmental Impacts:*
  Environmental impacts are categorized according to the effects on the environment that they produce in terms of air emissions, waste water discharges, solid waste (hazardous and non-hazardous), land contamination, underground water pollution, noise and odors. Their 'significance' is established based on the quantity and quality of environmental impacts, and in the regulatory, socio-political and economic context of a company's sphere of activity.
- *Baseline Energy Consumption*:
  The calculated or measured energy usage at energy cost centers (ECCs) over a period of time.
- *Baseline Environmental Impacts:*
  The calculated or measured amounts of waste, emissions and discharges at energy cost centers (ECCs) over a period of time.
- *Standard Energy or Environmental Performance:*
  This relates the amount of waste, emissions and discharges or energy use to production output.
- *Energy or Environmental Conservation Opportunity (ECO):*
  The opportunity for reducing energy consumption or environmental impacts by implementing organizational or technical improvement measures.
- *Energy or Environmental Conservation Measures (ECM):*
  Technical measures for the realization of ECO, designed and costed for implementation in a specific plant, and at a particular ECC, usually with a short payback period.
- *Performance Improvement Projects (PIP):*
  PIPs are key practical solutions to the problems of energy and environmental performance, identified during an energy and environmental audit, requiring some investment and presented with full technical and financial evaluation.
- *Post-Installation Energy Usage or Environmental Impacts:*
  Measured and verified energy usage, amount of waste, releases or discharges of ECC after the implementation of ECMs and PIPs.
- *Performance Improvement:*

$$= \text{(Baseline Performance)} - \text{(Post-Installation Performance)}$$

- *Techniques for Continuous Performance Monitoring:*
  - direct performance measurement;
  - utility billing analysis;
  - mathematical models and engineering calculations;
  - suppliers' specifications.

The EEA is carried out through a sequence of activities aimed at determining the energy and material flows in a company, calculating energy and mass balances, including emissions and discharges, and establishing opportunities for reducing the costs of energy and of environmental compliance (Fig. 4.2). The term *energy and environmental auditing* denotes a study of a facility which has to provide the following:

(a) To determine *how* and *where* energy and raw materials are used or converted from one form to another, and the performance thereof.
(b) To establish a *base line* for energy and raw materials consumption, and for the types and amounts of waste generated.
(c) To *provide* a comprehensive analysis of all of the environmental issues and impacts of a company's activities.
(d) To *identify* the relevant legal and regulatory requirements.
(e) To *quantify and qualify* significant environmental impacts.
(f) To *identify* the opportunities for improving energy and environmental performance and to define measures for improvement.
(g) To *evaluate* the economic and technical viability of implementing these measures and to recommend performance improvement projects.
(h) To *create* prioritized recommendations for implementing performance improvement projects with the aim of reducing the cost of energy, raw material and environmental compliance.

It is important to clearly distinguish EEA from energy and environmental performance monitoring. Auditing is a *periodic activity*, sometimes carried out by an independent service provider, particularly in a case when it needs to verify conformance. It is based on a data sample, aimed at establishing the company's overall position regarding energy and environmental issues.

Performance monitoring is a *continuous activity*, part of management process, based on regularly measured data, carried out as an operations responsibility, aimed at assessing ongoing energy and environmental performance.

In practice, unfortunately, EEA is often reduced to checking the efficiency of main utilities, such as boilers, chillers or air compressors and the environmental impact thereof. It is obvious that a simplified form of an audit can yield only partial results and cannot provide full benefits and inputs to a comprehensive energy and environmental management program. Often, the audit neglects the role and impact of the human factor in energy and environmental performance.

Therefore, to avoid a piecemeal approach, the extent of the audit is shown in Figure 4.3. It is obvious and a matter of common sense that EEA covers all aspects of industrial operations. However, the scope of an audit may vary and will be defined by the actual need that triggers the requirement for carrying out the audit.

4.2.1 Scope of Energy Audit

An energy audit is aimed at identifying the opportunities for reducing energy, water and raw material costs. It starts from understanding the processes at a company's individual plants and departments and recognizing issues that are relevant for energy consumption.

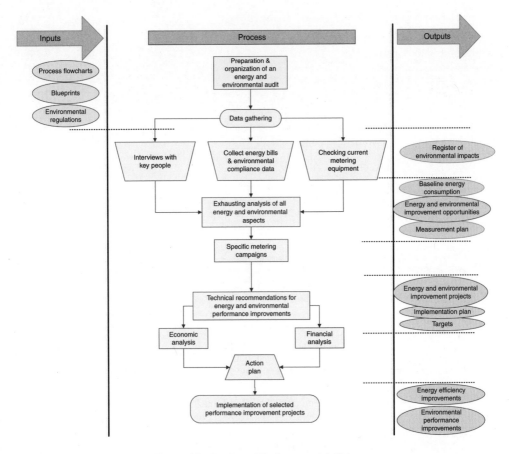

Figure 4.2   Energy and Environmental Auditing

· A *systematic approach to energy auditing* in a facility requires the study and understanding of interactions among the main factors influencing energy performance, i.e.:

- production;
- utilities;
- people; and
- technology.

The energy audit is important because it identifies how energy performance improvements can be achieved. The energy audit is focused equally on the physical aspects of energy consumption and on the human aspects related to the operation and maintenance of facilities. Generally speaking, the scope of an energy audit includes an examination of the following:

- energy conversions (in boilers, transformers, compressors, furnaces, etc.);
- energy distribution (of electricity, steam, condensation, compressed air, water, hot oil, etc.);
- energy end-users' efficiency (equipment and buildings);
- production planning, operation, maintenance and housekeeping procedures;
- management aspects (information flow, data collecting, data analysis, feedback of results and achievements, education of employees, their motivation, etc.).

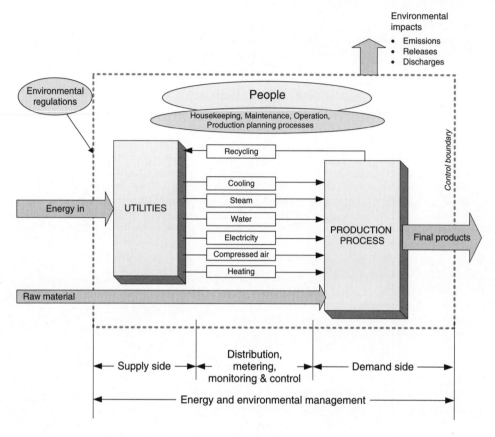

Figure 4.3   Extent of Energy and Environmental Audit

When conducted thoroughly and systematically an energy audit provides recommendations for improvement that tie together various aspects of energy performance:

- people and their motivational and training needs;
- housekeeping, operation, and maintenance procedures improvements;
- equipment and technology efficiency improvements;
- efficiency of energy supply, distribution, and consumption.

Therefore, the emphasis is not only on the single aspect of the efficiency of a particular energy utility generation and distribution but also on the integration of the energy needs of a particular production process, the supply side options available to satisfy these needs and management techniques, which tie together all the resources *(people, energy, materials, technology)* in a cost effective way.

Another important area considered by an energy audit is waste (energy, water, raw material) minimization, and reuse and recovery or recycling of these resources. The elaboration of effective recommendations for performance improvement in such a context requires a basic understanding of industrial operations and main production processes.

4.2.2 Scope of Environmental Audit

Unlike the energy audit, the scope of an environmental audit may vary depending on the requirements of specific schemes for which the audit is requested. These may include:

- pure regulatory requirements (like IPPC);
- emission trading requirements (like EU ETS, or CDM);
- voluntary environmental schemes (like ISO 14000, EMAS, GRI, etc.);
- investment, insurance or money borrowing related requirements;
- customers or community requirements.

Regulatory requirements are usually embodied in the form of an environmental permit, and an environmental audit in that context aims at demonstrating compliance with the conditions specified in the permit.

Emission allowances that are traded have an economic value. For this reason, it is important that the parties in the market can be assured that emissions are determined with a high degree of certainty. Level of assurance means the degree to which an auditor, in this context also called a verifier, is confident in that the information reported for a plant is free from material misstatement.

In the context of ISO 14000 and similar voluntary schemes, the environmental audit is also referred to as an 'Initial Environmental Review'. Such a review focuses on three key areas:

1. environmental practices and procedures;
2. identification of significant environmental impacts;
3. identification of legal and regulatory requirements.

It is worth mentioning that in the context of the ISO standards the term 'Audit' is used to describe activities related to the periodical checking of the functions of an established environmental management system according to the ISO Standard's procedure. It is based more on a documentary review, rather then on a plant and facility reviews and measured data analyses.

When an environmental audit is required in the context of an investment, banks will require a detailed analysis of all environmental issues related to the company's operations (see the example in Box 4.1). Such environmental audits are concerned with past and ongoing environmental pollution at the site and the environmental status of the company.

In any event, any environmental or energy audit will have two components, i.e. organizational and operational, as explained at the beginning of this chapter.

4.3 Evaluating Organizational Aspects

4.3.1 Qualitative Evaluation

The usual method for evaluation is an *energy and environmental management audit*. It reviews the adequacy and effectiveness of the management structures, working relationships, utilization of personnel, adequacy of the lines of communication and clarity of company objectives related to the implementation of energy and environmental policy, if any.

Various tools are used in carrying out an energy and environmental management audit but the most common are interviews, questionnaires and reviews by checklists or an energy management matrix.

Some of these tools are provided in Toolbox III-1. The focus here is on the results that an energy and environmental management audit is expected to provide. It is possible to follow the steps in Figure 4.1 and use the questionnaires in Toolbox III-1 as a check list for a quick review in order to see which of the required structures, actions and procedures are fully or partially implemented and which ones are missing completely.

Box 4.1: Sample Scope of Work for Environmental Audit Required by Development Bank

1.   Nature of the Project to be Supported
1.1   Description and Context of the Proposed Project
2.   The Existing Company/Facility Conditions
2.1   Description of Processes, Facilities and Assets
2.2   Key Environmental, Health and Safety Aspects of the Company/Facility (if any)
2.3   Facility Location and Description of Natural Environment
2.4   Facility and Site History
3.   Corporate Environmental, Health and Safety Management
3.1   EHS Policies and Practice
3.2   Organization of EHS Management
3.3   Contingency Planning and Emergency Procedures
3.4   Staff Training and Supervision
3.5   Internal and External Stakeholder Dialogue
4.   Company's Environmental Performance
4.1   Local/National Regulatory Requirements
4.2   Applicable EU (World Bank/Other) Requirements and Standards
4.3   Inputs, Products and Releases
           Raw Materials Source and Consumption (where appropriate)
           Water Source and Consumption
           Energy Source and Consumption
           Greenhouse Gas Contribution
           Intermediate Products Arising
           Wastewaters/Effluent Amounts and Quality
           Air Emission Amounts and Quality
           Wastes Amounts and Characteristics
4.4   Process Efficiency
4.5   General Housekeeping Issues
4.6   Product-Related Issues
4.7   Material Handling and Storage
4.8   Hazardous Materials Management
4.9   Waste Management (inc. PCBs and Asbestos)
4.10 Soil, Surface and Groundwater Contamination
4.11 Current Environmental Expenditure
5.   Health and Safety Performance
5.1   Local/National Regulatory Requirements
5.2   Applicable EU/International Requirements and Standards
5.3   Key Health & Safety Issues
5.4   Control of Major Accident Hazards
5.5   Current Health and Safety Monitoring Practice
5.6   Summary of Regulatory Compliance Status
5.7   Summary of Health and Safety Expenditure
6.   Other Worker Related Issues
6.1   Forced Labor
6.2   Harmful Child Labor
6.3   Discriminatory Practices
7.   Conclusions and Recommendations
7.1   Summary of Regulatory Compliance Status
7.2   Key Risks and Liabilities
7.3   Process Efficiency and Environmental Opportunities
7.4   Environmental Action Plan

Source: EBRD, http://www.ebrd.com/about/policies/enviro/procedur/procedur.pdf, Annex 3.

Table 4.1 Performance Measurement and Evaluation Review

Which performance aspects need to be monitored?	
What related values are measured?	
Why are they measured?	
Is there any performance target?	
What is the frequency of measurements?	
Who measures?	
Where is the source of data?	
Who acts on data?	
What do they do?	

The key areas to be assessed are whether appropriate performance measurements are performed at all and secondly whether appropriate structures for performance evaluation are in place. These can be checked effectively if a simple table (Table 4.1) is completed for each performance aspect that has to be managed.

Such a table should be complemented by the assessment of the completeness of performance measurement systems. This requires determining whether all relevant performance indicators are measured, and if not, which ones are missing. This assessment should be done during the course of EEA.

A thorough analysis of organizational aspects related to the appropriateness of structures for energy and environmental management will reveal which aspects of the existing energy and environmental management organization require improvement. In short, improvements can be required in one or more of the following areas:

- top management commitment to energy and environmental performance improvement;
- a declared energy and environmental policy with clear objectives and targets;
- assigned responsibilities for energy and environmental performance;
- adequate levels of staff awareness, motivation and skills;
- an appropriate metering system for performance monitoring;
- a reporting and communication system for checking progress, providing feedback and communicating results.

How to implement the required improvements or set up management structures from scratch will be explained in the next chapter by elaborating upon the steps in Figure 4.1 and illustrating them by real case studies.

4.3.2 Quantitative Evaluation

In order to quantify the performance of current energy and environmental management practice, we need to use real data on production, energy and raw material consumption, emissions and waste. Further,

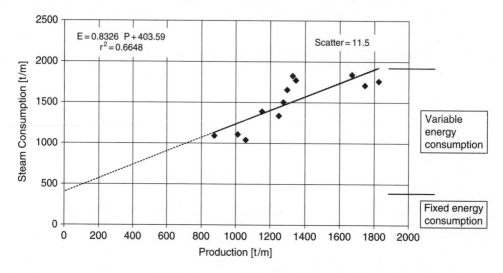

Figure 4.4   Sample Scatter Indicative on Energy Management Performance

we need to prepare a scatter diagram in order to take a look at emerging data patterns, as elaborated in detail in Part I Chapter 3. Owing to variations of internal and external factors, the different values in the scatter diagram will not fall neatly on the 'best fit' line but will be spread above or below the line.

The extent of the scatter will indicate how strong or weak the current energy and environmental management practice is. A small scatter (less than 2 %–3 %) generally indicates good management practice, while larger values like 11.5 % that characterize the case study from Table 3.1 in Part I Chapter 3 (see Fig. 4.4) will indicate that there are significant opportunities for improvement.

However, scatter is only an approximate indicator of performance because it is possible to find small scatter values where, after analysis, performance may turn out to be quite bad. Typically, this may be the case when production output varies considerably while energy consumption remains almost constant. Statistically, scatter will be small, but actually this is a very wasteful operation, and energy management is weak or nonexistent.

The interpretation of the data pattern from a scatter diagram can indicate the performance of an existing management practice. Looking again at our tuna-canning case study, even though only monthly energy/production data are available, we are still able to identify a number of areas of concern that provide opportunities for performance improvement. Even if limited information is available, it can still be used to set the general direction for performance improvement across the business based on the qualitative assessment of energy and environmental management performance as shown before.

This can be summarized along the following lines:

(1) Production planning should be focused on avoiding large variations and low daily production output, thereby reducing the cost of energy and environmental compliance in the unit cost of products.
(2) Energy consumption in production should be tightly controlled so that it varies in accordance with production variations. This way, variable energy consumption can be reduced.
(3) Utilities should improve the efficiency of energy supply thereby contributing to the reduction of fixed energy consumption and environmental releases. It is worth mentioning that fixed energy

consumption is not only caused by utility sections. It may be also the consequence of operational practice in production. For instance, if water or steam flows are controlled by overflow or bypass instead of by a stop valve, the energy consumption, as well as resulting environmental impacts, will be constant whatever the production activities are.

Such qualitative assessment is a useful first step in establishing the current position and giving the direction for performance improvement but it cannot provide proposals for specific technical measures for performance improvement. To obtain such proposals, we need to evaluate the ongoing operational practices in a factory.

4.4 Evaluating Operational Aspects

The actual level of energy and environmental performance is determined by how people operate machines, equipment and processes, under given external and internal circumstances characterized by influencing factors, as identified earlier in Part I Chapter 3. An energy and environmental audit (EEA) is the process that establishes the current performance level and identifies projects and measures for energy and environmental performance improvement (Fig. 4.2).

The inputs for EEA are process and energy flowcharts, related blueprints and applicable environmental regulations. At the heart of EEA is a measuring campaign that complements (and verifies) existing data on energy and environmental performance. A prelude for a measuring campaign is preparation of a work plan for thorough technical data collection and for carrying out on-site activities and for planning the necessary measurements.

4.4.1 Planning and Preparation for Data Collection

For any project planning and preparation is a crucial phase. Planning includes at least the following:

- lay out a time table and the scope of all activities;
- set-up the project team and assign specific tasks;
- establish a relationship with the plant's management and personnel;
- establish effective lines of communication and co-ordination between the project team and the plant's personnel;
- initiate data gathering;
- kick off the project successfully.

The planning activities should be documented in the form of a work plan, which should also be shared with the plant's personnel prior to visiting the plant in order to carry out the intended work. The work plan will outline which part of the plant is to be visited, the purpose of the visit, documentation and data required and the plant personnel to be involved (Box 4.2).

4.4.2 Data Collection on Energy Consumption

Toolbox III-2 contains a number of forms that may help in structuring the process of data collection, which can be used as checklists so that important information is not omitted and which can provide guidelines for personnel involved in an audit on the scope and the steps to be taken in the work ahead. This toolbox can be used to support the preparation of the work plan. The work plan will facilitate the collection of relevant and available data from invoices and other available sources based on:

Box 4.2: Sample Work Plan for EEA

WORKPLAN

A) Sterilization plant

Purpose:

(**1.**) To establish available data on process flow parameters; raw material and utilities within the plant:

(**2.**) To establish existing measuring points;

(**3.**) To establish actual control system arrangements and practice;

B) Utilities

Purpose:

(**1.**) To establish available and relevant data on the compressed air system (process requirements and operational practices)

(**2.**) To establish available and relevant data on the boiler plant (process requirements and operational practices)

(**3.**) To establish available and relevant data on the steam distribution and condensate return system (process requirements and operational practices)

(**4.**) To establish existing measuring points

(**5.**) To establish actual control system arrangements

C) Background documents

(**i**) Detailed single line diagrams

(**ii**) Process flow charts

(**iii**) Daily log book forms

D) Company representatives included

(**a**) Sterilization plant representative (production manager or shift supervisor)

(**b**) Utility representatives

- monthly amounts of energy consumption for at least last 12 months (electricity, fuels):
  - quantity used
  - amount of money paid
  - tariff system

- raw water and treated water:
  - quality, quantities, costs

- weather data;
- information on electricity, steam or water meters around the site:
  - type, age, last calibration date, intervals of readings and reporting.

The collected data will provide inputs for the calculation of energy balances, and determination of specific energy costs in a unit of production.

4.4.3 Data Collection on Production Processes

To develop an understanding of production processes, the following data needs to be collected:

- plant layout and description;
- production processes and products;
- original design specifications;
- process flow diagrams;
- operating and production records:
  - capacity utilization,
  - hours of operation,
  - monthly production output data

- operating manuals;
- maintenance records;
- major plant disruption incidents, frequent breakdowns;
- plant availability; on-line times, off-line times (scheduled and unscheduled);
- Expansion: in hand, planned or anticipated;
- raw materials: types, quantities, costs;
- raw material waste;
- rejects, reworks;
- finished products: types, quantities, prices.

Based on this data, we should be able to assess the performance of material use, to comment on quality issues, to assess adequacy of manufacturing planning regarding optimizing costs of production and capacity use and to understand ongoing maintenance and operational practices.

4.4.4 Specification of Major Utilities and End-Use Equipment

Toolbox III-2 also contains the forms for data collection on main utilities. When planning data collection on utilities and major equipment, it is important to approach the task systematically. This means not only performing checks on the utility side, but also establishing what the main users of this energy/utility type are, the distribution and control arrangements and the actual demand and losses. This is the essence of a systematic approach to EEA.

An example of a work plan for a systematic approach to detailed analysis of a compressed air system is shown in Box 4.3.

Box 4.3: Sample Work Plan for Analysis of Compressed Air System

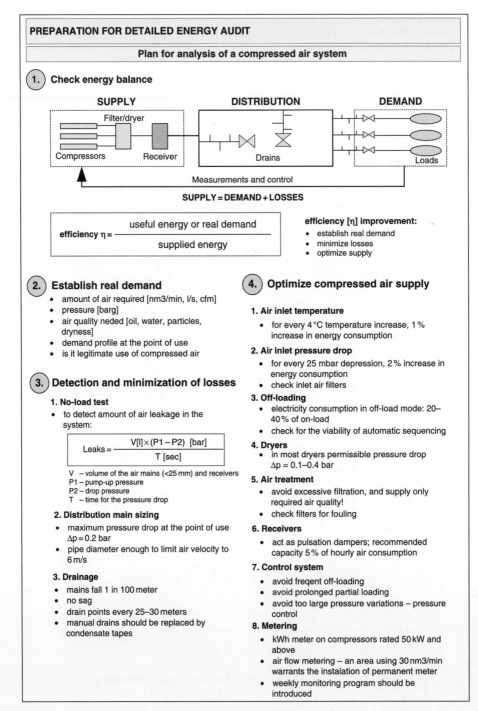

PREPARATION FOR DETAILED ENERGY AUDIT

Plan for analysis of a compressed air system

1. Check energy balance

SUPPLY   DISTRIBUTION   DEMAND

Filter/dryer

Compressors   Receiver   Drains   Loads

Measurements and control

SUPPLY = DEMAND + LOSSES

$$\text{efficiency } \eta = \frac{\text{useful energy or real demand}}{\text{supplied energy}}$$

efficiency [η] improvement:
- establish real demand
- minimize losses
- optimize supply

2. Establish real demand
- amount of air required [nm3/min, l/s, cfm]
- pressure [barg]
- air quality neded [oil, water, particles, dryness]
- demand profile at the point of use
- is it legitimate use of compressed air

3. Detection and minimization of losses

1. No-load test
- to detect amount of air leakage in the system:

$$\text{Leaks} = \frac{V[l] \times (P1 - P2) \ [bar]}{T \ [sec]}$$

V – volume of the air mains (<25 mm) and receivers
P1 – pump-up pressure
P2 – drop pressure
T – time for the pressure drop

2. Distribution main sizing
- maximum pressure drop at the point of use $\Delta p = 0.2$ bar
- pipe diameter enough to limit air velocity to 6 m/s

3. Drainage
- mains fall 1 in 100 meter
- no sag
- drain points every 25–30 meters
- manual drains should be replaced by condensate tapes

4. Optimize compressed air supply

1. Air inlet temperature
- for every 4 °C temperature increase, 1 % increase in energy consumption

2. Air inlet pressure drop
- for every 25 mbar depression, 2 % increase in energy consumption
- check inlet air filters

3. Off-loading
- electricity consumption in off-load mode: 20–40 % of on-load
- check for the viability of automatic sequencing

4. Dryers
- in most dryers permissible pressure drop $\Delta p = 0.1$–0.4 bar

5. Air treatment
- avoid excessive filtration, and supply only required air quality!
- check filters for fouling

6. Receivers
- act as pulsation dampers; recommended capacity 5 % of hourly air consumption

7. Control system
- avoid freqent off-loading
- avoid prolonged partial loading
- avoid too large pressure variations – pressure control

8. Metering
- kWh meter on compressors rated 50 kW and above
- air flow metering – an area using 30 nm3/min warrants the instalation of permanent meter
- weekly monitoring program should be introduced

4.4.5 Data Collection on Environmental Impacts

The scope of environmental data collection is determined by the purpose or scope of the environmental audit itself, as elaborated earlier in this chapter. However, the key environmental data are common to most requirements and these are summarized in Toolbox III-1. The logical first step is to identify the relevant environmental regulations that a company is required to comply with. This should be complemented by a review of any complaints from the neighboring communities and of warnings or fines issued by relevant authorities.

The second step is to determine all the physical source points of environmental impacts (Fig. 4.5). For air emissions, these will be stacks and ventilation exhaust ducts. For waste water discharge, these are the points where waste water is released into the sewage, canals or waterways outside the factory perimeter. For solid waste, these are waste collection areas around the site and for soil and underground

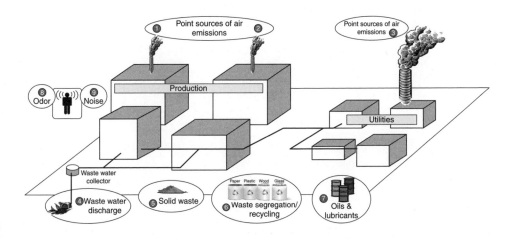

REGISTER OF ENVIRONMENTAL IMPACTS

1: Ventilation shaft		2: Ventilation shaft		3: Boiler house		4: WW discharge		5: Solid waste	
Quality	Quantity	Quality	Quantity	Quality	Quantity	Quality	Quantity	Quality	Quantity
VOC	mg/yr	Dioxin	m^3/yr	CO_2	m^3/yr	BOD	kg/yr	Scraps	%
		SOx	m^3/yr	NOx	m^3/yr	Fat	kg/yr	Hazardous	kg/yr
		NH_3	m^3/yr	SOx	m^3/yr	Dry matter	kg/yr	material	kg/yr
				NH_3	m^3/yr	Detergents	kg/yr	Sludge	

6: Waste segregation		7: Oils & lubricants		8: Odor		9: Noise	
Quality	Quantity	Quality	Quantity	Quality	Quantity	Quality	Quantity
Paper	kg/yr	Waste oill/yr	n/yr	Odor	yes/no	Noise level	dB
Plastic	kg/yr	Oil barrels					
Wood	kg/yr						
Glass	kg/yr						

Figure 4.5 Determining Sources and a Register of Environmental Impacts

water contamination these are the areas for storage of chemicals, oils, lubricants, fuels, and temporary disposal sites for other waste materials.

The next step is to establish the annual quantities and qualities (compositions) of emissions, releases and discharges at identified points of impact, packaging waste disposal and the presence of any noise, odors or other irritants.

This is to be followed by analyses of materials and waste handling and storage procedures, environmental risks prevention (like spills, accidents, fire, etc.) and emergency procedures.

When an environmental audit is conducted in concurrence with an energy audit, then data on energy consumption including transport, and water and raw material use will already have been collected as a part of the energy audit, as shown above.

The considerations described so far referred mostly to what is known as *direct impacts*, i.e. those which arise directly from the business operations over which the company is able to exercise a high degree of control.

Indirect impacts are, by their nature, more difficult to assess because they generally arise from activities remote from the company site, and therefore are largely beyond the direct control of the company. Although the company cannot control them directly, it should recognize their existence and significance. One way to do that is to engage suppliers and subcontractors for waste disposal services and require their statements on environmental impacts.

Another category for which data are required is *environmental compliance costs*. This is not such a straightforward task because these costs are usually distributed over several accounts and some are even hidden because nobody is made explicitly aware of them or they are not used in business decisions. The main components of environmental costs are:

- conventional (or direct) compliance costs;
- potentially hidden or indirect costs such as administration, legal, transportation, etc.;
- health and safety costs;
- contingent or future costs;
- intangible costs.

The ultimate aim of an environmental audit is the identification of all environmental issues at a factory and the preparation of an environmental pollution inventory or register. This register should include all the significant environmental impacts arising from the company's operations (see Fig. 4.5); it should describe the procedures employed in identifying and quantifying the impacts and screening out those which are insignificant.

4.4.6 Assessing On-Site Metering and Control Equipment

On-site metering is important because it provides first hand information on energy and environmental performance but only if it is accurate enough! If metered data are not used regularly, it is very likely that the meters are not calibrated regularly either or that the log book entries are not very reliable. The truth is that if log books are not checked, the entries are at best approximate, and at worst quite random.

Therefore, the following needs to be checked:

- how energy consumption and environmental impacts are metered and monitored;
- how the operation of the equipment/processes is controlled;
- the accuracy and reliability of existing metering equipment.

Depending on the extent of reliable on-site metering, an EEA metering plan has to be prepared in order to provide a complete picture of energy and mass balances based on complementary on-site measurements performed during the course of EEA.

Often, the existing metering system will not be sufficient to provide all necessary data for EEMS implementation and additional metering will be required (see Toolbox III-2). One rule of thumb suggests that in plant metering should be installed where annual energy and environmental costs exceed five times the cost of the meter. The extent of additionally required instrumentation for EEMS implementation has to be established during energy and environmental auditing along with appropriate data collection and handling procedures.

4.4.7 Measurement Plan

When conducting a detailed energy and environmental audit, it is usually required to carry out on-site measurements with portable instruments in addition to stationary ones. There is a direct relationship between the extent of data collection and the subsequent evaluation of performance improvement opportunities. While insufficient data may prevent the identification of some performance improvement opportunities, a very extensive audit may prove unnecessary and wasteful by diverting funds and time from more rewarding improvement opportunities.

To ensure that neither too many nor too few measurements are taken, a measurement plan has to be prepared taking into account the following factors:

* the purpose of measurement;
* foreseeable performance improvement measures (see Toolbox III-2).

These factors should be identified during the initial stage of an audit and plant inspection. A well prepared measurement plan will ensure that measuring results will serve as a good basis for identifying performance improvement opportunities. The example of the sterilization plant of our tuna cannery case study is shown in Figure 4.6. This presents a simplified chart of the sterilization process (retorting) together with related equipment (cooling tower and water tanks). The figure illustrates the measuring points and provides a table with quantities to be measured. The purpose of the measurement is to establish energy and mass balance at the sterilization process, and to identify opportunities for performance improvements through better controls, waste heat recovery and improved operational procedures.

4.4.8 Timing

It is important to choose a representative time for the execution of measurements, especially for manual measurements. Generally, the measurement results of interest are those which represent the normal operation of the plant. Therefore, while preparing the measurement plan, we should satisfy ourselves that production circumstances will be normal and that all units that are usually in operation will be in operation during the measuring campaign. Also, the plant should be running undisturbed for a period long enough to be in equilibrium before measurements begin (i.e. to avoid start-up transients during measurements).

Depending on the type of operation, measurements should extend over a week, a day, or at least over a couple of full batches in the case of batch operations.

It is also important to note that the results of measurements from the selected time interval and the subsequent analysis results will be used to extrapolate the effects of any performance improvement

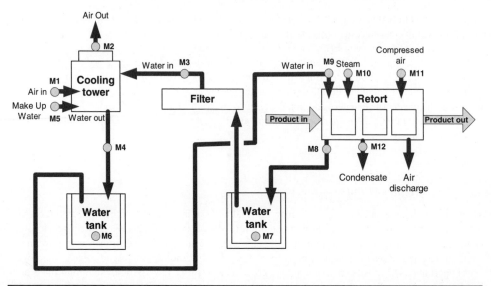

Measuring point	Description	Value	Unit	Duration of measurements	Frequency of readings
M1	Air in temperature Relative humidity Mass flow rate	$t_{A,in}$ $\phi_{A,in}$ $m_{A,in}$	°C % kg/s	7 days	15 min
M2	Air out temperature Relative humidity	$t_{A,out}$ $\phi_{A,out}$	°C %	7 days	15 min
M3	Water in temperature Water mass flow rate	$t_{W,in}$ $M_{W,in}$	°C kg/s	7 days	1 h
M4	Water out temperature	$t_{W,out}$	°C	7 days	1 h
M5	Make up water temperature Mass flow rate of make up water	t_{MUW} M_{MUW}	°C kg/s	7 days	15 min
M6	Water temperature	t_W	°C	7 days	1 h
M7	Water temperature	t_W	°C	7 days	1 h
M8	Water mass flow rate	M_W	l/s	7 days	1 h
M9	Water mass flow rate	M_W	l/s	7 days	1 h
M10	Steam mass flow rate	M_S	m3/s	7 days	15 min
M11	Air mass flow rate	M_A	m3/s	7 days	15 min
M12	Condensate mass flow rate	M_C	l/s	7 days	15 min

Figure 4.6   Sample Measurement Plan

measure cumulatively over a period of one year. This is an additional reason to be careful when deciding on the time period for carrying out measurements.

4.4.9 Records and Note Keeping

It is very important to keep good records on *where, how* and *what* has been measured in order to be able to use measurement results correctly during the back-to-the-office analysis. A good practice is to use prepared measurement plans (as in Fig. 4.6) and related forms for the different tasks to be performed and insert the measurement results later along with comments and observations made during the measurements.

We should be aware that EEA at a large company is a complex task, consisting of a number of activities condensed over a short period of time, and without proper note keeping, one might not be able even to decode one's own scribbling when back in the office some weeks after the field work has been completed.

4.4.10 Instrumentation

There is no reason to use instruments which have an unnecessary high degree of accuracy when it comes to manual or portable measurements. Precision instruments are expensive and are often easily damaged when used in industrial conditions. The accuracy of results is normally determined more by the manner in which measurement equipment is used, the manner in which measurement is executed and measurement results are processed, than by the accuracy of measurement instruments. Measurements should always be as direct as possible.

4.4.11 Calculations

After the measurement campaign is completed according to the plan, measurement results become inputs for the extensive analysis of the operational performance of the observed processes. Part II of the book will elaborate on how calculations and subsequent analyses are carried out.

It is important to note that performance improvement opportunities are evaluated based not only on calculations, but also on other qualitative data and observations from the audit, such as capacity utilization, control procedures, production planning routines, maintenance practices employed, etc.

Specifically for *environmental impacts*, where measurements may sometimes be too expensive or too complex, calculations can serve as an indirect or *surrogate method* for quantifying the amounts of pollutants. Generally, there are quantitative, qualitative and indicative surrogates. Quantitative surrogates give a reliable quantitative picture of, for instance, flue gas amounts, based on combustion mass balance calculations (see Toolbox III-5 and III-13). Qualitative surrogates give information on, for instance, flue gas composition, based on combustion process analysis and the suppliers' provided composition of the fuel (see Toolbox III-5). Indicative surrogates give information about the operation of the plant or process through parameters such as relative temperature difference, pH factor, flow rate, etc.

4.4.12 Economic and Financial Analysis

Economic analysis in the context of an energy and environmental audit is actually a cost-benefit evaluation of the proposed measure or project. On the one hand, we have to estimate the costs of identified performance improvement measures or projects (PIP) while, on the other, we have to forecast cost reductions consequent to their implementation. As a quick cost-benefit analysis, simple payback period (SPP) can be calculated as:

$$SPP = \frac{\text{Annual cost savings} \left[\$/y \right]}{\text{Total investment} \left[\$ \right]} \, [y]$$

If the payback period is very short, from a couple of months up to three or even five years, this information may suffice for management to decide in some cases to approve the project. But if the payback period is longer and particularly when the required investments are larger, then additional analysis may be required. The usual analysis is a *discount cash flow* method (Toolbox III- 3) which provides additional *financial performance indicators*: *internal rate of return* (IRR), and *net present value* (NPV) that enable decision makers to select which projects are to be financed and subsequently implemented.

Cost estimate is required for all payments necessary to implement suggested PIPs. It is based upon engineering analysis and the solution to the problem of cost reduction. In order to be able to prepare sufficiently accurate cost estimates, it is required that energy and environmental auditing delivers technical results at the level of conceptual engineering design. The conceptual engineering design will provide sufficient details to enable *cost estimate* for economic analysis purposes. An item-by-item approach to the cost estimate is suggested, which includes the following:

- detailed design;
- equipment and installation;
- spare parts;
- insurance and handling;
- duties, tariffs, taxes and other equipment related charges;
- project management cost for project implementation;
- O&M cost for new equipment;
- operator training costs;
- license, patent-rights costs.

Financial analysis deals with the costs of specific ways to secure financing for the implementation of performance improvement projects. It needs to determine which capital sources can be used for project implementation and what it takes to get them. The company's financial resources may come from two sources:

- available cash at hand;
- access to external capital.

The important considerations are effective lending rates, lending terms, methods of calculating interest rates, methods of repaying the principal, grace period, tax breaks, etc. The amount of cash outflows based on the lending terms is to be developed for every year, to provide a basis for the calculation of the taxes, taxable profit and finally net cash flow resulting from PIP implementation.

Toolbox III-3 provides a guideline on how to carry out economic and financial analysis and includes simple software that can be used to calculate basic financial indicators (Fig. 4.7).

4.4.13 Presentation of Audit Results

A comprehensive *energy and environmental audit* involves extensive work, which comprises measurements, technical calculations and economic and financial analyses, where required. However, no

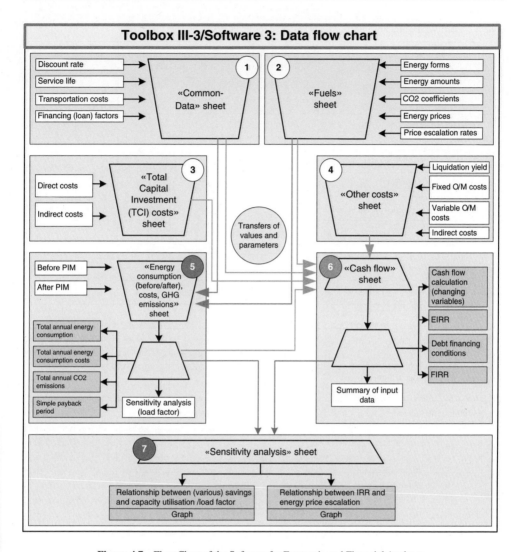

Figure 4.7 Flow Chart of the Software for Economic and Financial Analyses

matter how many calculations are performed and how good they are, an audit is only as good as the quality of its final report. The main purpose of the *final report* is to convince the top management of the factory to invest money in performance improvement measures and projects, and to prove that these measures can compete with other investment opportunities in terms of their own return on investment.

While the report itself may be quite comprehensive, it is important that it has an executive summary at the beginning where the main recommendations are described briefly and condensed into a table, accompanied by the key financial indicators of the proposed projects. Such a table may look like the example in Table 4.2.

Table 4.2 Performance Improvement Projects Proposed for Implementation

	Proposed Project (Title and short description)	Energy Savings	Energy savings	Expected Investment	Pay Back Period	EIRR**	FIRR**
		% of TEC*	US$/year	Euro	Years	%	%
1	**Energy and Environmental Management System**	3 % (over 3 years)	275 000	72 000	0.3	N/A	N/AZ
	Performance measurement system						
	Waste minimization						
	Training for best practice operation						
2	**Chilled Water System (12/20 °C)**					N/A	N/A
	– PIM*** #1: Increasing the chilled water temperature	1.97	175 000	–			
	– PIM #2: Additional heat exchanger	6.01	535 000	265 000	0.5		
3	**Chilled Water System (5/10 °C)**						
	– PIM #1: Increase the chilled water temperature	0.17	165 000	–			
4	**Steam System – distribution, users, control**						
	– PIM #1: Steam system – Steam Blow Losses Elimination	1.37	125 000	50 000			
	– PIM #2: Steam/Condensate System Recovery	0.52	70 000	55 000		N/A	N/A
	– PIM #3: Plate Heat Exchangers Efficiency Improvement (FFE)	0.12	10 000	5000	0.5		
	– PIM #4: Plate Heat Exchangers Efficiency Improvement (AF)	0.20	17 500	5000	0.3		
	– PIM #5: Plate Heat Exchangers Efficiency Improvement (CC)	0.64	55 000	15 000	0.3		
	– PIM #6: Waste Heat Recovery	0.24	21 500	20 000	0.9	N/A	N/A
5	**Boiler House**		21 500			N/A	N/A
	– PIM #1: Tuning the burners	0.96	87 500				
	– PIM #2: Economizer installation	6.83	610 000	125 000	0.2		
	TOTAL:		**216 8000**	**612 000**			

* *TEC – total energy costs*
** *EIRR, FIRR – Economic and Financial Internal Rate of Return in case not calculated because payback period for all measures is less than a year*
*** *PIM – performance improvement measure*

4.5 Setting a Baseline for Monitoring Performance Improvements

Every improvement process starts from a known performance level against which a clear target and time frame for improvement is given. To establish the performance level prior to any improvements is relatively easy. Available monthly data may be considered to represent a *baseline data set* that determines past energy performance (usually over the last year or two) against which future improvements will be assessed.

The nature of energy and environmental performance improvement projects and its results (energy and other cost savings) is somewhat hypothetical because the *actual savings* and performance improvements are defined as follows:

> ⇒ post-improvement consumption or costs,
>   subtracted from agreed *base line consumption*,
>   that would have occurred had the project not been implemented,
>   while all other factors are held constant.

Figure 4.8 illustrates this definition. The left part of the graph represents energy or environmental compliance costs in the previous year or two. '0' stands for present time when we are starting the implementation of the EEMS project. We assume that the project implementation period would be 20 months and over that period the costs would be gradually reduced to reach a target level. After that, if no other performance improvement projects are implemented, the costs will stabilize at the new, lower level. Now cost savings are the difference between past and current cost levels, and will accumulate over time.

This definition assumes a static approach which can explain the nature of cost savings in a dynamic industrial environment. But it cannot help to actually quantify or verify whether there are any savings at all against a background of varying production output and energy use. Therefore, we have introduced the regression equations (see Part I Chapter 3) calculated from the baseline data set given in Table 3.1, Part I Chapter 3:

$$E = 0.8326 \cdot P + 403.59 \tag{4.1}$$

which represents the best fit line for this data set. This equation is the usual first choice for the *base or reference line* against which performance will be monitored and assessed and the future cost savings calculated and verified.

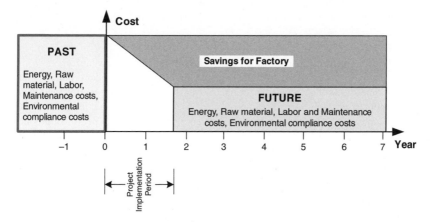

Figure 4.8   Evaluating Results of Performance Improvements

4.6 Setting Initial Targets for Performance Improvement

Performance improvement *targets* are important for an energy and environmental management program because they are clear statements of what management wishes to achieve and when. Targets must mobilize people toward the achievement of management objectives, hence they must be:

- measurable;
- achievable;
- time-based;
- capable of being monitored;
- capable of maintaining standards.

The starting point for target setting is the knowledge of current position. Looking back to our tuna-canning case study, even with only monthly energy/production data available, we are still able to identify instances of good performance which may be used for establishing targets at the company level. However, we have said that targets must be measurable and capable of being monitored. For that purpose, we must use some of the tools provided by regression analysis in order to quantify performance results and compare them with given targets. At the outset of the discussion on target setting, we have to emphasize that there will be always at least two levels of targets:

(i) targets at the company level;
(ii) targets on the ECC's level.

The ECC's targets need to be broken down further and established for different products and for various energy types used at the ECC.

4.6.1 Setting Targets at the Company's Level

The targets at the company's level are usually aggregated and based on available monthly data on achieved production and raw materials and energy used over a period of at least least 12 months. Such data, when checked for adequacy, represent average performance over the selected period, and provide the basis for setting overall targets for performance improvement.

The realistic specific initial target can always be based on the best performance already achieved in the past. In this case, this can be a target line that passes through the three points representing best performance in the respected period, as shown in Figure 4.9. In Part I Chapter 3, we have already calculated regression equation for the best-fit line based on these three data points:

$$E = 0.9503 \cdot P + 36.037 \qquad (4.2)$$

Again, before deciding firmly that this equation can present realistic (achievable) *target*, we must satisfy ourselves that particular performances at these three months are achieved under ordinary circumstances.

The 'best-fit' line shows the average performance over the corresponding period, and *the target line* provides a clear objective for performance improvement at any ordinary production output. It is important to note that the target cannot be just one figure, for instance one ton of steam per ton of product, because energy consumption cannot be fixed irrespective of actual production level. Therefore, we speak about the 'target line' which is expressed by *target equation*, so that for any production output that may occur, a corresponding energy target value can be calculated and compared with actual energy used.

As said earlier, environmental impacts are consequent to the energy, water and raw material use, therefore there is no need to set specific target equations for environmental issues in operations, because when energy, water and raw material use are at their lowest for the desired level of production, the resulting environmental impacts will also be at their lowest. Only for end-of-pipe treatment of effluents and discharges, can specific targets be set, that will be achieved by applying particular environmental abatement technologies.

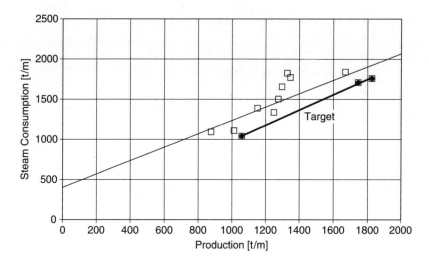

Figure 4.9   Setting Target Based on Best Past Performance

4.6.2 Setting Targets on ECC Level

The targets for ECCs are set in the same way as for the company as a whole. But when trying to establish targets for an ECC, we almost invariably stumble on the problem of a lack of data. We have set initial targets for the factory based on the aggregated data on energy and production, but reliable data on production and energy use at the ECC level are usually not available prior to implementing an EEMS project .

Later, when the energy and environmental management system provides data for an individual ECC, we will need to set individual targets for ECCs in the similar manner as the aggregated ones. The overall target and individual targets should not necessarily have the same value.

Why? It is because targets must always be established from the position of knowledge of the actual performance at respective ECCs. For a factory, this knowledge is derived from the given monthly data set and subsequent analyses. If similar data for an individual ECC are not available and if we set a target of say 10 % improvement across the factory, that may be easy to reach for some ECCs because their actual potential may be larger, but for others it may be impossible because their actual performance may already be quite good.

Usually, individual ECC targets can be established when the initial operation of the EEMS provides daily energy/production data for at least two to three months. These initial targets should be checked periodically for adequacy and reset if necessary.

4.6.3 Targets for Different Products

When a company or ECC produces more than one product, targets must be established for each product separately. At the first glance, it may look like a complex and complicated procedure, but actually it is quite simple. The principle for target setting is always the same as described above and it must always be based on measured data. An operational EEMS will provide daily data on energy and production and, if necessary, a further break-down into data by shifts. This enables the separation of energy and production data on a product by product basis. Data for individual products is consolidated into one scatter diagram, so we will have the same number of scatter diagrams as there are individual products where performance needs to be monitored separately. These *scatter diagrams* then provide the basis for *performance assessment* and *target setting* for *individual products* in the same way as has been already described.

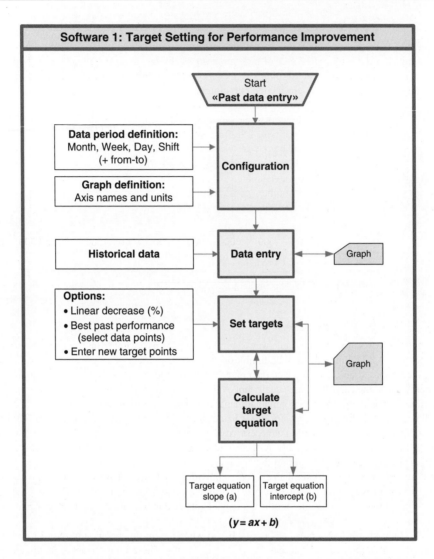

Figure 4.10 Flow Diagram of the Software for Performance Improvements Target Setting

To help illustrate target setting from past performance data, a sample software tool (see Part III Software 1) has been developed. It provides the tool for presenting the energy performance of the industrial process in a scatter diagram. Once entered, data can be analyzed and targets determined either as uniform linear reduction, or based on selected best past performance points, or based on new discretionary target points. As the final result, the software provides target line equation. The software flow diagram is shown in Figure 4.10.

4.7 Monitoring Energy and Environmental Performance

A *monitoring procedure* will need to be introduced to follow up and check regularly on the progress of performance improvements towards targets. The information on current performance level must be provided by the performance measurement system through continuous measurement of key performance indicators.

The usual monitoring practice in factories is to record measured data in a tabular form into log sheets. However, this is the least informative way of data presentation for the purposes of performance monitoring. A much more informative way of presenting data is by means of the already described scatter diagrams. This is particularly true for monitoring energy consumption and environmental impacts from manufacturing operations, where production output figures are set out as an independent variable.

There are also other means of data presentation, which may be more appropriate for performance monitoring where there is a direct input/output relationship of the variables involved. These include load profile diagrams, which present variations of load over a period of time (an hour, a day or a week) and performance control charts that emphasize variations around a given target value.

Any presentation style is acceptable as long as it flags *substandard performance* promptly and indicates the potential causes of performance deviations.

Software 2, provided within the Toolbox, entitled 'Performance Monitoring and Control', offers an example for visualizing data and structuring reports on performance (Fig. 4.11). The inputs are a target

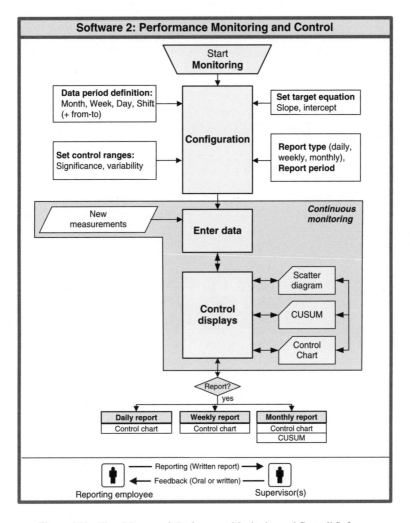

Figure 4.11 Flow Diagram of 'Performance Monitoring and Control' Software

equation determined earlier and data on performance measurement received from performance measurement system. At any moment of time a user can select the style of *visualizing performance data*, as an input to the performance evaluation of the process and an interpretation of the data pattern that should identify possible corrective actions. The software also provides a structured means for reporting performance results and providing proposals for corrective measures, as well as generating daily, weekly or monthly reports.

4.8 Verifying Performance Improvements – CUSUM Technique

The nature of energy performance improvement projects and their results (energy and other cost savings) is somewhat hypothetical because *actual savings* and performance improvements are defined as post-improvement consumption or costs, subtracted from agreed base line consumption, that would have occurred had the project not been implemented, where all other factors remained constant.

When this is applied against the changing operating environment, where other factors are not constant, it is not surprising that in many cases people actually wonder how to determine how much, if any, they have improved their performance. In fact, this is the question that has puzzled a number of managers we have worked with. Usually, they produce two annual data sets on production and energy and all kind of colorful time-series graphs, but the question: 'How much did we save on the year to year basis?' remains unanswered!

Let us have a look at a real set of the data (Table 4.3, Table 4.4) representing monthly energy use and production output over two consecutive years and a set of diagrams (Fig. 4.12) that the factory has produced in order to try and answer the question: 'How much did we save?'

Table 4.3 Year-On-Year Data for ECC 1

Month	Production	Electricity	Month	Production	Electricity
	Units	MWh		Units	MWh
Apr-00	8584	1003	Apr-01	8787	848
May-00	10 877	1199	May-01	10878	1153
Jun-00	10 880	1203	Jun-01	10918	1134
Jul-00	10 211	1175	Jul-01	9710	1059
Aug-00	10 567	1195	Aug-01	11434	1227
Sep-00	10 741	1155	Sep-01	10114	1124
Oct-00	10 161	1108	Oct-01	10344	1126
Nov-00	10 784	1205	Nov-01	10538	1135
Dec-00	8728	1021	Dec-01	8228	1033
Jan-01	10 020	1068	Jan-02	9778	1042
Feb-01	10 007	1071	Feb-02	8146	962
Mar-01	implementation		Mar-02		

Table 4.4　Year-On-Year Data for ECC 2

Month	Production	HFO	Predicted	Difference	CUSUM
	Units	l	m³	m³	m³
Apr-00	8584	172	**Apr-01**	8787	181
May-00	10 877	206	**May-01**	10 878	202
Jun-00	10 880	218	**Jun-01**	10 918	207
Jul-00	10 211	173	**Jul-01**	9710	190
Aug-00	10 567	140	**Aug-01**	11 434	210
Sep-00	10 741	138	**Sep-01**	10 114	206
Oct-00	10 161	133	**Oct-01**	10 344	188
Nov-00	10 784	138	**Nov-01**	10 538	208
Dec-00	8728	123	**Dec-01**	8228	180
Jan-01	10 020	118	**Jan-02**	9778	185
Feb-01	10 007	127	**Feb-02**	8146	178
Mar-01	implementation		**Mar-02**		

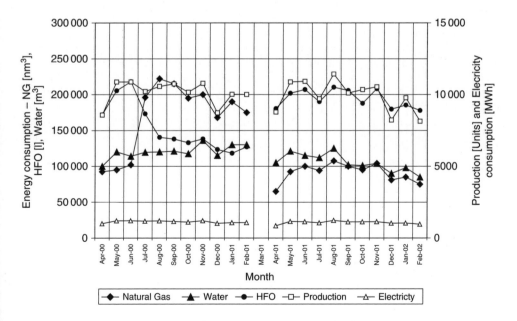

Figure 4.12　Year-On-Year Energy Use and Production Output Data

This is a very common way of presenting data that we have seen in many factories, but it can be misleading because different values are shown on the same y-scale diagram. Apart from some swings in natural gas consumption (which, by the way, require an explanation) all other lines look nice and tidy and the first impression is that everything is under control.

Following such presentations, the energy team was unable to provide any definite and quantitative answer to the important question: 'Do we save, and if yes, how much?'

Why?

We have already said that a target cannot be a single figure, but must be an equation, because energy consumption depends on the actual production output. Correspondingly, base line consumption also cannot be represented by a single figure (like average annual consumption) but also must be expressed as a *base line equation*, which is calculated from the agreed base line data set.

An appropriate base line equation is given by the regression equation representing the best-fit line. The best-fit line goes through the center of the data scatter, hence represents an average energy/production relationship over the observed period of time (see Fig. 4.13). For this example, the base line equation for the year 1, for heavy fuel oil and production relationship, is expressed as follows:

$$E = 12.865 \cdot P + 22860 \qquad (4.3)$$

where P is in production units and E in liters.

What is the meaning of this equation? For any actual production output 'P', it will give us energy consumption 'E'' that would have happened if no improvements were made. This consumption 'E'' we can then compare with actually measured consumption E_m in the subsequent year, to determine if there are any savings:

$$\text{Energy Savings} = E_m - E' \qquad (4.4)$$

If the above equation yields a negative value this means that energy performance improvements have been achieved since actual consumption is lower than one as given by the base line equation and calculated for the same production output.

Now, if we calculate energy savings in the same way month-by-month, we can sum the monthly savings and maintain cumulative sum of the savings achieved. This is shown in Table 4.5 and Table 4.6. Values of cumulative savings can be plotted in a so-called *CUSUM graph* (see Fig. 4.14a and Fig. 4.14b). Such a presentation makes any savings in the year-to-year data visible immediately! For instance, we can read from the graphs that cumulative increase of heavy fuel oil are around 450 $\mathbf{m}^3$, while electricity consumption has decreased by around 350 MWh.

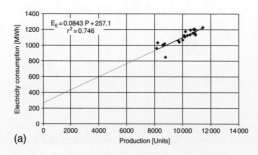
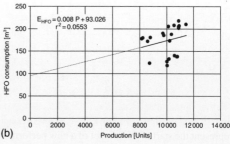

(a)   (b)

Figure 4.13   Baseline Equations

Table 4.5 CUSUM Data for ECC 1

Month	Production	Electricity	Predicted	Difference	CUSUM
	Units	MWh	MWh	MWh	MWh
Apr-01	8787	848	1010	−162	−167
May-01	10 878	1153	1192	−39	−206
Jun-01	10 918	1134	1196	−62	−268
Jul-01	9710	1059	1090	−32	−300
Aug-01	11 434	1227	1240	−14	−313
Sep-01	10 114	1124	1126	−1	−315
Oct-01	10 344	1126	1146	−20	−334
Nov-01	10 538	1135	1163	−28	−362
Dec-01	8228	1033	962	72	−291
Jan-02	9778	1042	1096	−54	−345
Feb-02	8146	962	954	8	−337
Mar-02					

The CUSUM graph has another useful feature: whenever the slope or direction of the line changes at a particular data kink, it means that something has happened in the monitored process at that instance of time. Again, the causes of change cannot be explained by the CUSUM graph. Analytical reasoning skills must be applied to analyze operational conditions and circumstances at the time of change and extract reasons for the detected change. These are the permanent tasks of *performance improvement teams* – to keep the performance under control and provide explanations for any detected variation, to apply corrective actions and strive for continuous improvements.

The CUSUM graph is a very good tool to make these objectives clear and their achievements immediately recognizable.

If there is a permanent change in the underlying energy/production relationship (such as the addition of a new machine), that will affect the CUSUM graph, then the base line equation must be adjusted. The CUSUM method offers a basic and effective mechanism for verification of performance improvements and monitoring and for actual savings verification at individual ECCs.

The value of CUSUM as a management tool is presented in Figure 4.15. If there are no changes in consumption patterns, the CUSUM curve will simply fluctuate around a horizontal line. However, any change which produces saving or waste will cause the CUSUM curve to change direction. It will then continue in this direction until there is another change in consumption when the curve changes direction again. The common CUSUM curve consists of a number of straight sections. Intersecting points represent the start of the impact of a new performance improvement measure or of some events that make performance worse. Extrapolating these segments to the present day shows immediately the cumulative savings (or wastes) to date from respective performance improvement measures. The advantage is that

Table 4.6   CUSUM Data for ECC 1

Month	Production	HFO	Predicted	Difference	CUSUM
	Units	m^3	m^3	m^3	m^3
Apr-01	8787	181	136	20.8	20.8
May-01	10 878	202	163	42.4	63.2
Jun-01	10 918	207	163	47.0	110.2
Jul-01	9710	190	148	41.6	151.8
Aug-01	11 434	210	170	51.4	203.2
Sep-01	10 114	206	153	53.3	256.5
Oct-01	10 344	188	156	43.2	299.7
Nov-01	10 538	208	158	50.6	350.3
Dec-01	8228	180	129	35.5	385.8
Jan-02	9778	185	149	36.1	421.8
Feb-02	8146	178	128	34.9	456.7
Mar-02					

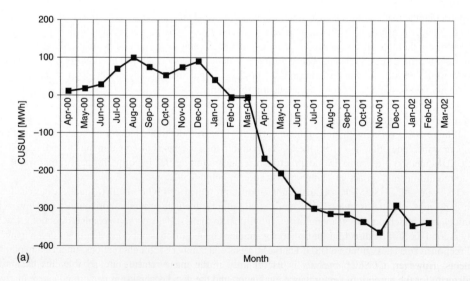

(a)

Figure 4.14   CUSUM Graphs for Savings Verification

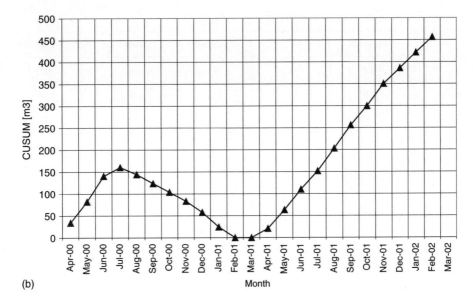

(b)

Figure 4.14   (*Continued*)

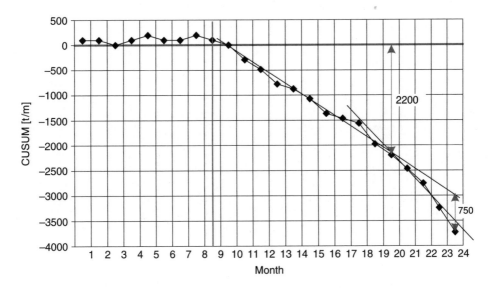

Figure 4.15   Evaluating Energy Savings Using CUSUM

values on CUSUM graphs are measured in physical units and can be converted simply in money terms by multiplying the measured amounts with corresponding unit price.

This example illustrates how the CUSUM technique shows the results of performance improvements. However, *CUSUM analysis* is useful only if the measurements are reliable, the data are adequate for the purpose of performance evaluation and the data processing is performed carefully and consistently.

4.9 Moving Toward Targets – Process of Change

The energy and environmental performance improvement process starts from a known performance level and moves toward a clear target over a given time frame. The base line or reference performance level is determined by the best-fit line and corresponding regression equation for the reference year or another selected period of time.

A reasonable target for improvements over the next 12 month period can be based on the best performance achieved during the reference year and expressed by the target line and corresponding equation based on the selected data points. For our original case study, the base line and target line are given by the following equations:

- *Base line equation:* $E = 0.8326 \cdot P + 403.59$
- *Target equation:* $E' = 0.9503 \cdot P + 36.037$

The target line can be used to evaluate progress in performance improvement and variability in energy and raw material use. It can also be used to quantify overall energy savings related to the given target. For instance, if we calculate the estimated monthly energy consumption by using the target equation and substituting P for each month from the reference period, and then take the sum of the resulting monthly Es, we will get the value of:

$$E'_\Sigma = 15,479 \text{ ton of steam} \tag{4.6}$$

The difference between actual annual energy use E_Σ and the value E'_Σ, gives the targeted annual amount of energy that needs to be saved:

$$\text{Targeted Annual Energy SAvings Amount} = E_\Sigma - E'_\Sigma = 18,026 - 15,479 = 2547 \text{ ton} \tag{4.7}$$

The equivalent *emission reductions* consequent to this *energy performance improvement* are as follows:

$$\Delta CO_2 = (710.37 kg\, CO2/h - 612.19 kg\, CO2/h) \cdot 5000 h/a = 490900\ [kg/a]\ or\ 491\ [t/a]$$

This is calculated by using *Software* No. 7 in Toolbox on the accompanying website.

To achieve these savings, we will need to improve energy management practice and performance so that at the end of a given time interval (a year) all the points on the scatter diagram are below or around the target line, which is based on only three data points from the reference period! This will be, of course, an ideal case.

The CUSUM method based on the base line equation from a reference year will give us the actual amount of cumulative energy savings over the observed period. However, if we want to press for more performance improvements after first 12 months of EEMS operation, we can draw a new base line for this new 12 month data set and then establish a new target line for the next period. Or alternatively, we can decide to change only the target line and to evaluate the results against the original base line from the reference year.

Every improvement process is at the same time a process of change. Operational practices need to be adjusted; energy and material use controlled more strictly; data evaluation done regularly; two-way communication maintained continuously; etc. All of these issues require a change in people's attitude, behavior and habits. And these are the most difficult aspects of change. Therefore, the process of introducing an energy and environmental management system in a factory will face similar challenges to the introduction of any other change.

We will go through the practical steps of this process in the chapter that follows.

4.10 Bibliography

Bannister, K.E. (1999) Energy Reduction through Improved Maintenance Practices, Industrial Press, Inc.

Behrens, W., Hawranek, P.M. (1991) Manual for the Preparation of Industrial Feasibility Studies, UNIDO, Vienna.

Environmental Engineer's Handbook (1999) Ed. D.H.F. Lui, B. Liptak, CRC Press. http://www.iso.org/iso/iso_14000_essentials.

Gvozdenac, D., Morvay, Z., Kljajić, M. (2003) Energy audit of cooling towers, PSU-UNS International Conference 2003: Energy and the Environment, Prince Songkla University, Hat Yai, Thailand, 11–12 December.

Ministry of Energy, Philippines (1985) Guides to quick estimates of energy costs for industrial use.

Morvay, Z. et al. (1991) Energy Audit in Industry, published in proceedings: Environmental Management in Ukraine, held in Geneva, Switzerland (English)

Morvay, Z. et al. (1993) Energy Auditing Handbook, UNIDO, (English)

Morvay, Z., Gvozdenac, D., Košir, M. (1999) Systematic Approach to Energy Conservation in the Food Processing Industry, Manual for Training Programme 'Energy Conservation in Industry' (training course performed in Thailand), supported by Thai ENCON Fund, EU Thermie, ENEP program.

Schnell, K.B., Brown, C.A. (2002) Air Pollution Technology Handbook, CRC Press.

The Engineering Handbook (1998) Editor-in-Chief, R. C. Dorf, CRC Press.

UNEP (1990) Environmental Auditing, UNEP, Technical report series No. 2.

UNEP (1991) Audit and reduction manual for Industrial emissions and wastes, UNEP, Technical report series No. 7.

Welch, T.E. (1998) Moving Beyond Environmental Compliance: A Handbook for Integrating Pollution Prevention with ISO 14000, CRC-Press; February 1.

Welsh Office (1989) Environmental Assessment – A guide to procedures, Department of Environment, , HMSO, London.

Welsh Office (1990) Integrated Pollution Control – A Practical Guide, Department of Environment, Welsh Office, HMSO, London.

Whitelow, K. (2004) ISO 14001 Environmental Systems Handbook, Second Edition, Butterworth-Heinemann; October 19.

5

Implementation of the Energy and Environmental Management System

5.1 Introduction

The implementation process transforms decisions into activities and requires the cooperation of all of the actors involved. Accountability and performance are assigned and measured at all levels. Practical implementation plans with tangible phases, targets and documentation of goals and realization schedules need to be prepared in order to help monitor progress.

However, when one tries to introduce energy and environmental management as everyday practice in a factory, one will face a number of difficulties caused by:

- lack of motivation and awareness of energy and environmental performance improvement potential and techniques;
- inadequate organization of the energy and environmental sector;
- lack of insight into where, how and why energy is consumed and the environmental impacts generated;
- lack of experienced personnel;
- limited resources (financial, manpower, know-how, equipment).

One will often encounter a prevailing culture which can be summarized as a 'why it can't be done' *attitude*, which must be changed. Therefore, this background, common to many factories, requires a phased approach to the introduction of energy and environmental management. There is a sequence of evolving energy and environmental management activities which may be visualized as a series of phases (Fig. 5.1):

  (i) preparation and planning;
 (ii) implementation;
(iii) operation;
(iv) learning.

Each phase can be viewed as a process that takes time and resources, requires some inputs, and delivers some results that the next phase builds on.

It is important to realize that energy and environmental management is an ongoing process, which deals with both technical and human aspects, requires continued support from top management, the high quality of energy and environmental management staff and adequate funding.

Applied Industrial Energy and Environmental Management   Zoran K. Morvay and Dušan D. Gvozdenac
© 2008 John Wiley & Sons, Ltd

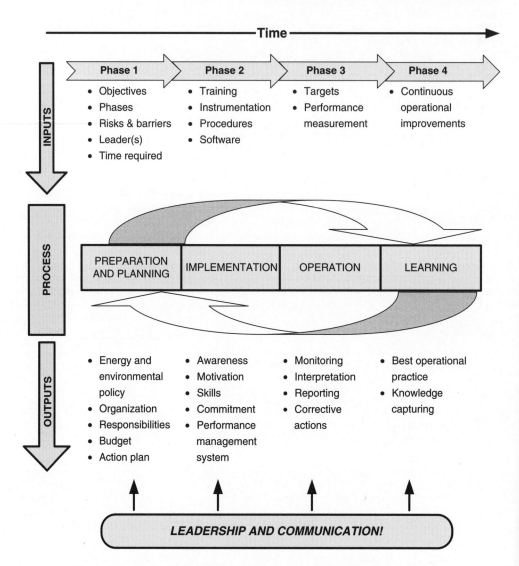

Figure 5.1   Phased Implementation of Energy and Environmental Management System

The cornerstones for the successful implementation of energy and environmental management practice are:

- conspicuous commitment and support from top management;
- accountability of line managers;
- trained and motivated staff;
- metering and monitoring instrumentation in place;
- established reporting and communication channels;
- culture of continuous improvement practice embedded in the workplace; and
- integration of EEMS into the fabric of the organization.

With these firmly in place, energy and environmental cost savings of 10 % or more are easily achievable with little or no further investment.

5.2 Phases of EEMS Implementation Process

Our experience suggests that successful implementation of EEMS is characterized by multiple influencing factors and their interrelation evolving through the phases of the *implementation process*. Therefore, particular attention should be paid to careful preparation and planning. There is not much to be gained by jumping into implementation if not all of the prerequisites (Fig. 5.1) are in place. This may only cause subsequent delay, loss of interest or diminished enthusiasm.

We have to be aware that completion of any phase takes a certain amount of time. This brings us to another key dimension of the implementation of EEMS – which is *time*. Time is a resource that should not be underestimated because a change cannot happen and become sustainable overnight. Hence, implementation of EEMS requires persistence and patience. For large companies (more than 1500 employees) our experience has shown that up to 24 months of intense effort is required for the change to be absorbed, for continuity to be assured, and for effective learning to actually begin. This also emphasizes other important aspects for EEMS implementation: *leadership and communication*. Leaders are required to instill persistence, and well planned and executed communication will ensure the patience of and staying the course by the workforce and managers alike.

Therefore, good *preparation and planning* at the outset of implementation of EEMS is of critical importance.

5.3 Preparation and Planning

Preparation and planning starts with enlisting the support of *top management*. The issues to be discussed at top management meetings and the decisions to be made are listed in Box 5.1.

Box 5.1: Top Management Meetings – Issues and Sample Decisions Required

---

POLICY

– *Prepare, approve and declare energy and environmental policy.*

RESPONSIBILITIES

– *Nominate persons responsible for controlling energy consumption throughout the factory – ECC leaders and members.*
– *Nominate person(s) responsible for environmental compliance issues.*
– *Nominate person(s) responsible for coordinating energy and environmental management activities.*

STRUCTURE

– *ECC teams will monitor energy and environmental performance daily and report to the energy/environmental manager and to the operational team weekly.*
– *The energy/environmental manager will make monthly reports to the top management team.*

LINES OF COMMUNICATION

– *Formal communication on matters relating to energy and environmental performance by end-users or budget holders will be directed through the energy/environmental manager who will, where appropriate, bring it to the attention of the operational or top management teams.*

Box 5.1: Continued

RESOURCES

– *The number of staff employed in energy and environmental management, their mix of skills and the amount of investment corresponding to the demands of these activities.*
– *Performance measurement system is to be secured.*
– *Performance improvement projects with an internal return of more than [15 %] per annum will be considered for funding approval.*

REVIEW

– *All energy and environmental management activities will be subject to monthly reviews by the top management team.*
– *Action Plan*
– *Indicate energy and environmental management actions to be undertaken by designated personnel.*

Once implementing EEMS is under way, other preparatory activities may follow, such as:

- declaring an energy and environmental policy;
- identification of related requirements for jobs at all levels and preparation of adequate training programs;
- determining targets and assigning responsibilities;
- establishing reporting and communication channels;
- designing an internal awareness campaign;
- defining a performance measurement system;
- development of an agreed-upon action plan that all concerned will understand.

It goes without saying that adequate financial, manpower and 'time' resources have to be secured and, if required, external support ensured.

5.3.1 Energy and Environmental Policy

Energy and environmental issues involve everybody in an organization, from top management to operators and caretakers. The successful implementation of energy and environmental performance improvements in an organization must begin with top management's awareness of the potential and its commitment to the program.

As a first step, management awareness and commitment can be demonstrated by publishing the organization's energy and environmental policy. Policy statements are very important because they set targets and provide guidance and direction. They also communicate basic requirements and standards regarding EEMS accountability.

The aim of energy and environmental policy should be to *establish the goals* of the program and make them known to everybody in the organization by publishing a short and clear policy statement. A generic example is given in Box 5.2.

However, every company will have a slightly different approach to formulating and declaring its energy and environmental policy, which may depend on its culture, priorities, technologies, etc. Therefore, some examples of 'real' *energy and environmental policies* are provided in Box 5.3 and Box 5.4. The policy example in Box 5.3 is focused on energy issues and use of raw materials, while that in Box 5.4 is focused on productivity improvements through waste minimization, recycling and energy efficiency improvements.

The specific performance improvement targets that these policies are referring to come from energy and environmental audits that were carried out as a prerequisite for EEMS program implementation.

Policy making is an arbitrary task that companies approach in their own way, emphasizing the operational aspects of performance that are most important to them. It is important that policies are declared and that the employees are made aware of them, and that the consequent action plans are implemented.

Box 5.2: Generic Energy and Environmental Policy

---

1. Policy

The policy of this organization is to manage energy and environmental issues in order to:

- *avoid unnecessary expenditure*
- *improve cost-effectiveness, productivity and working conditions*
- *protect the environment*

2. Objectives

The long-term objectives are to:

- *buy fuel and raw materials at the most economic costs*
- *use them as efficiently as possible*
- *minimize impacts on environment*

3. Immediate aims

In the short term, immediate aims are to:

- *gain control over energy and environmental aspects of the business by reviewing and improving metering, operation, maintenance, motivation and training practices*
- *invest in the rolling program of energy and environmental performance improvements which will maximize returns on investment in order to generate funds that can be re-invested into continuous performance improvement activities*

4. Action plan

During the coming years, the following energy and environmental management activities will be prepared and undertaken:

- *program of housekeeping and maintenance work*
- *detailed timetable with specified milestones*
- *indication of actions to be undertaken by designated personnel*
- *promotion plan to heighten awareness among employees*
- *devise a plan to monitor and evaluate achievements*
- *review on a regular basis what is working, what is not*
- *review and supplement the promotional campaign*
- *keep employees fully informed as to joint achievements*
- *devise a reward system to recognize efforts of both individuals and groups of employees*

The aim is that over the next three financial years we will reduce energy and environmental compliance expenditure by a minimum of 3 % each year.

Signature: _____

Box 5.3: The Start of New Energy Management Project (an original statement)

Throughout past operations, we have been working on various energy saving projects and have been making significant progress. In order to make further strong gains in energy savings; we are going to work with a professional energy consultant in order to re-start the ENERGY MANAGEMENT PROJECT. On the occasion of this new start, I would like to explain the background and the substance of the project.

I. '5W 2H' FOR ENERGY MANAGEMENT PROJECT

'WHY' – Energy Management now?

Energy Management is the current preoccupation of the world, and it should be our mission for society

1. *Energy Management is implemented throughout the world. If we continue to waste energy as we do now, the Earth will perish, if we continue to waste energy as we do now, OUR CITY will become part of the ocean.*
2. *The Government has laid down the law in the 'ENERGY CONSERVATION ACT' which encourages enterprises to execute energy saving strategies. The nation can no longer accept this increase. In order for our nation to grow and stay strong worldwide, we must implement an energy and environmental management system*
3. *It is the duty of the company to promote energy saving, and we must do this also in order to strengthen our competitiveness by achieving cost reduction.*

'WHAT' – all consumable energies

1. *Direct energy – electricity, oil, compressed air, steam; water.*
2. *Indirect energies: raw materials, consumable materials*

'WHERE' – any place within our premises

   Not only the production departments but also any area within the boundary of the factory.

'WHO'

   Company-wide, b ythe participation of all people working in the factory premises.

'WHEN' – a project over a period of three years

• *Do not run out of stamina in maintaining momentum throughout the project.*

'HOW MUCH' – *our challenge is to make our factorya model for minimized consumption by achieving a 3 % saving each year*

'HOW TO'

1. *Obtain advice from a professional energy consultant.*
2. *Establish Energy Management Teams.*

II. POLICY TO PROMOTE THE ENERGY SAVING ACTIVITIES

1. *We must be convinced that these targets can be achieved,*

   • *A strong belief is the origin for discovering as many energy saving opportunities as exist in the factory*
   • *Never give up. Power comes from persistence.*

2. *We must have a sense of duty and responsibility to promote energy saving.*

   • *A strong sense of duty and responsibility is the source of the power to promote the energy saving*
   • *A person who leaves matter to take their own course orwho ignores these activities committing a crime against society.*

3. *We will promote and achieve energy savings through the efforst of the entire staff.*

   • *All those working for our company will board the 'Energy Savings' boat as her crew, and all of the crew will row that boa tin order to go with the flow of the world.*

4. We will strive enhance the abilities of each staff member through the activities of Energy Management System project.

- To learn the proper techniques for data gathering, analysis, how to act, how to follow-up, etc., and apply and develop such techniques to the other issues that surround us.

The world is always changing. If we do not respond to these changes, we will gradually lose the value of our identities. Energy saving is one of the great causes of the modern world. Do not think, 'my job has nothing to do with energy', and do not thin kthat the task is 'too difficult'. We welcome your contribution even the result is very small! 'Many a little . . . . !.'

Please always keep Policy items 1 to 4 above in mind, and always with the positive attitude, 'yes we can'. With this in mind, let us begin our energy saving!

FACTORY MANAGER

Box 5.4: 3R – Reduce – Reuse – Recycle

Policy:

The policy of this organization is to introduce a new Energy and Environmental Management System project to make further energy and environmental performance improvements, in addition to the significant progress made in the past. By implementing the new Energy and Environmental Management System, we aim:

- to cut energy costs and improve profits
- to improve cost-effectiveness, productivity and working conditions
- to protect the environment

Target:

To define implementation programs in order to save 10 % of the 1999 production costs over next three years by optimizing existing facilities and implementing an Energy and Environmental Management System (EEMS) which utilizes the skills of the factory staff in order to establish the best practical operation for maintaining the long-term performance improvement gains.

Period:

1st year –March–December 2000
2nd year – January–December 2001
3rd year – January–December 2002

Key success factors:

1. We have to avoid unnecessary expenditures by the 3Rs – Reduce, Reuse, Recycle.
2. We have to improve cost- effectiveness, productivity and working conditions.
3. We have to be convinced that the target can be achieved.
4. We should have a sense of duty and responsibility in order to promote the energy and environmental performance improvements.
5. We are to achieve the improved profitability b ythe efforts of the entire company.

We should always keep our above commitments in mind and, always with a positive attitude.

Chief Executive Officer

Defining objectives, setting performance targets, assigning responsibilities, measuring achievements over time and establishing a regular reporting and reviewing system within the management structure are essential steps in converting a performance improvement policy from a raw concept into a detailed plan and ultimately into results.

5.3.2 Assigning Management Responsibilities for Energy and Environment

The *responsibilities* for energy and environmental management are assigned to the entire management structure, as described earlier (Part I Chapter 1):

  (i) top management team;
 (ii) operational team(s);
(iii) ECC teams.

ECC teams will be given individual targets for energy and environmental performance. The achievement of overall performance or profitability improvement is the task of the operational team(s), while the top management team should facilitate and support the continuous improvement process so that it gradually becomes a standard operational routine.

An important task is to assign leadership responsibility for energy and environmental matters to dedicated person(s) – *energy and environmental manager (EE manager)*. He or she should be a driving force behind the implementation of EEMS and a champion of change. Because the EE manager crosses the functional boundaries of a company, he or she should also be a repository of knowledge. Internally, the EE manager should know what the company knows about EE issues and externally he or she should know what issues the company should be aware of.

The common basic duties for an energy and environmental manager are:

- promote and support improvement of energy and environmental performance;
- verify performance information regularly and prepare it for internal and external purposes;
- supervise handling, interpretation, reporting and record keeping of performance data;
- verify the results of corrective actions and the achievement of targets and plans;
- provide technical support for ECC leaders and operators if required;
- periodically inspect the adequacy of performance measurement systems;
- reset targets when required.

The specific responsibilities of the EE manager do not release line management from their performance responsibilities and the operational requirements of the EEMS. For example, line managers may be delegated responsibility for detailing the operating procedures which their operational staff must follow and to upgrade and document them into the best operational practice.

5.3.3 EEMS Organization

For EEMS to be effective, individual roles and responsibilities must be clearly defined, including how they relate to the achievement of energy and environmental objectives and targets and to the overall operation of the EEMS. Holding people accountable for specified results is another critical point on the path to success.

The organizational structure should be defined in writing. An *organizational chart* will be helpful for illustrating the energy and environmental responsibilities that must be addressed. The organizational structure and responsibilities should involve all relevant functions and all levels throughout the organization. All the employees of the organization should clearly understand their energy and environmental performance related roles and responsibilities, as well as the importance of the energy and environmental targets and objectives that they can effect.

Two examples of organizational charts are given in Figure 5.2 and Figure 5.3. These examples illustrate that each company defines its organizational structure in its own way – the one that is most

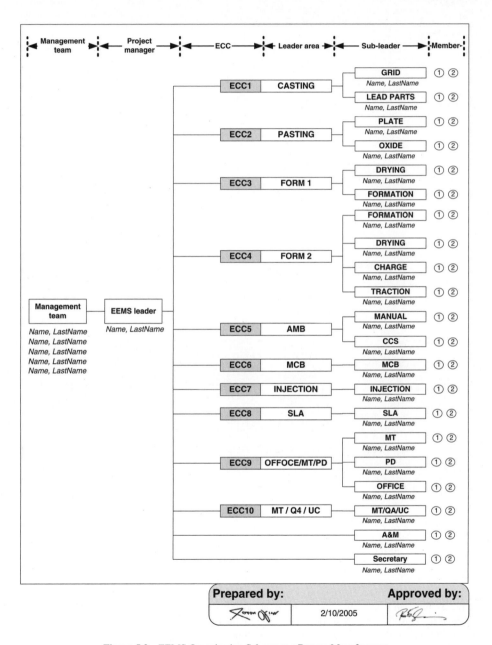

Figure 5.2   EEMS Organization Scheme at a Battery Manufacturer

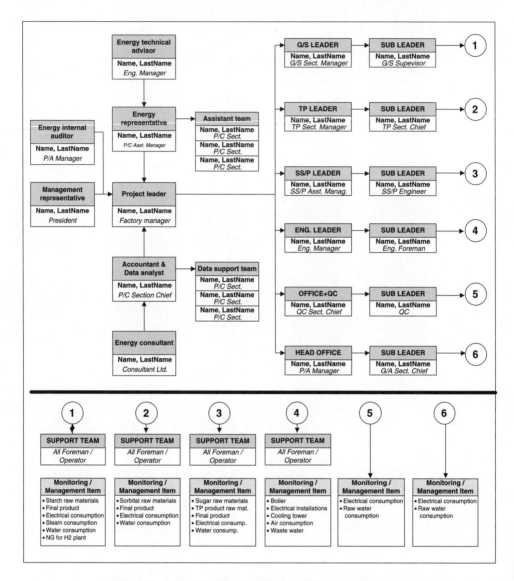

Figure 5.3   EEMS Organization Scheme at a Chemicals Manufacturer

familiar to the workforce. The format itself is not important. The important point is that the employees can identify clearly their own role and responsibilities for the implementation of the EEMS in their company.

The key roles that affect energy and environmental performance should be included in the employee's job description, where possible, and could be included in the employee's performance evaluation. These roles and responsibilities may be documented by human resources in a consistent format that includes the person's name, title, organizational responsibilities, key energy and environmental tasks, authority, and interrelation with other key tasks and other personnel.

5.4 Implementation Plan

After all of the preparatory and planning activities have been executed, implementation follows. Even the best designed EEMS is worthless unless it is integrated efficiently and effectively into the organization's routine activities, and is fully accepted and supported by those who should make it work. Therefore, an *implementation action plan* needs to be prepared that will contain the detailed steps to accomplish specified objectives and which will show what is to be done, by whom, by when, inclusive of the key events or milestones appropriate for progress reporting (Fig. 5.4). The example plan in Figure 5.4 outlines the

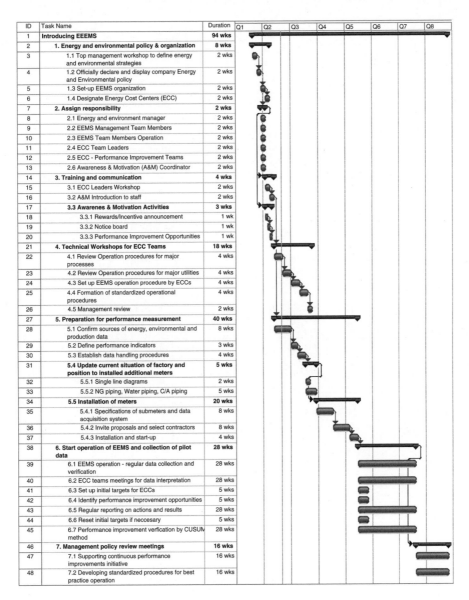

ID	Task Name	Duration
1	**Introducing EEEMS**	**94 wks**
2	**1. Energy and environmental policy & organization**	**8 wks**
3	1.1 Top management workshop to define energy and environmental strategies	2 wks
4	1.2 Officially declare and display company Energy and Environmental policy	2 wks
5	1.3 Set-up EEMS organization	2 wks
6	1.4 Designate Energy Cost Centers (ECC)	2 wks
7	**2. Assign responsibility**	**2 wks**
8	2.1 Energy and environment manager	2 wks
9	2.2 EEMS Management Team Members	2 wks
10	2.3 EEMS Team Members Operation	2 wks
11	2.4 ECC Team Leaders	2 wks
12	2.5 ECC - Performance Improvement Teams	2 wks
13	2.6 Awareness & Motivation (A&M) Coordinator	2 wks
14	**3. Training and communication**	**4 wks**
15	3.1 ECC Leaders Workshop	2 wks
16	3.2 A&M Introduction to staff	2 wks
17	**3.3 Awarenes & Motivation Activities**	**3 wks**
18	3.3.1 Rewards/Incentive announcement	1 wk
19	3.3.2 Notice board	1 wk
20	3.3.3 Performance Improvement Opportunities	1 wk
21	**4. Technical Workshops for ECC Teams**	**18 wks**
22	4.1 Review Operation procedures for major processes	4 wks
23	4.2 Review Operation procedures for major utilities	4 wks
24	4.3 Set up EEMS operation procedure by ECCs	4 wks
25	4.4 Formation of standardized operational procedures	4 wks
26	4.5 Management review	2 wks
27	**5. Preparation for performance measurement**	**40 wks**
28	5.1 Confirm sources of energy, environmental and production data	8 wks
29	5.2 Define performance indicators	3 wks
30	5.3 Establish data handling procedures	4 wks
31	**5.4 Update current situation of factory and position to installed additional meters**	**5 wks**
32	5.5.1 Single line diagrams	2 wks
33	5.5.2 NG piping, Water piping, C/A piping	5 wks
34	**5.5 Installation of meters**	**20 wks**
35	5.4.1 Specifications of submeters and data acquisition system	8 wks
36	5.4.2 Invite proposals and select contractors	8 wks
37	5.4.3 Installation and start-up	4 wks
38	**6. Start operation of EEMS and collection of pilot data**	**28 wks**
39	6.1 EEMS operation - regular data collection and verification	28 wks
40	6.2 ECC teams meetings for data interpretation	28 wks
41	6.3 Set up initial targets for ECCs	5 wks
42	6.4 Identify performance improvement opportunities	5 wks
43	6.5 Regular reporting on actions and results	28 wks
44	6.6 Reset initial targets if neccesary	5 wks
45	6.7 Performance improvement verfication by CUSUM method	28 wks
46	**7. Management policy review meetings**	**16 wks**
47	7.1 Supporting continuous performance improvements initiative	16 wks
48	7.2 Developing standardized procedures for best practice operation	16 wks

Figure 5.4   Sample EEMS Implementation Plan

generic activities for EEMS implementation. Any company embarking on the implementation of EEMS will need to produce a customized version of that plan, reflecting the tasks, milestones and responsibilities specific for their own circumstances.

The first step in implementation should be to present EEMS policies, procedures, responsibilities and individual work instructions to the staff through training sessions. Training should highlight the individual requirements under the EEMS. Since the implementation process may require a significant portion of organizational resources for a prolonged period of time, it may be best to have the management team oversee this phase in larger organizations.

5.4.1 Initiate Awareness and Training Programs

5.4.1.1 Awareness and Motivation

In order to facilitate the effective implementation of EEMS, it is essential to have the full and wholehearted cooperation of the workforce. This is the main objective of *awareness program*. The employees must understand what is happening and not feel threatened by it. Everyone involved in implementation should understand what EEMS is intended to achieve and how it should be operated. Everybody should be motivated to make the best possible use of it. This ideal state of affairs can only be approached if the awareness and motivation (A&M) process starts at the beginning of EEMS implementation. For large organizations that employ thousands of people, the A&M activities will require dedicated action plan. A sample action plan is provided in Figure 5.5.

The principal message to all staff involved in the EEMS project should be that the project is intended to make them more effective in their jobs and facilitate the execution of their duties. At the same time, the

Activities Month	1	2	3	4	5	6	7	8	9	10	11	12	Status
Officially declare company energy policy	X												☑ One-off / ☐ Continue
Display energy policy statement		X											☑ One-off / ☐ Continue
ECC leaders workshop on A&M		X											☑ One-off / ☐ Continue
Rewards/incentive announcement			X										☑ One-off / ☐ Continue
Awareneww & Motivation (A&M) training to all staff			X		X		X		X				☑ One-off / ☐ Continue
Assign A&M action group			X										☑ One-off / ☐ Continue
Notice board			X	X	X	X	X	X	X	X	X	X	☐ One-off / ☑ Continue
Energy conservarion opportunities suggested by staff			X	X	X	X	X	X	X	X	X	X	☐ One-off / ☑ Continue
Energy saving slogan contest				X									☑ One-off / ☐ Continue
Newsletter					X			X			X		☐ One-off / ☑ Continue
Energy exhibition							X						☑ One-off / ☐ Continue

Figure 5.5   Sample Action Plan for A&M Activities

project will reduce the operating costs of the organization and so help secure the long-term employment prospects of entire staff.

The most significant problem in achieving *changes in the individual attitude* towards energy and environmental issues is the need to change long standing, deep-seated customs and practices that regard energy and environment as marginal and as vague issues of no concern to individual workers. In attempting to change these attitudes, the aim is to make energy and environmental performance improvement an integral part of the routine practice of the organization by making everyone aware of the needs and ways how to improve performance at their own place of work.

With small to medium size enterprises, having several hundreds of employees, this is a moderately difficult task. However, in large companies with several thousands of employees and working three shifts a day this is a real challenge! Since everybody should feel included and affected by the ongoing EEMS program, the problem is how to reach them all over a reasonable period of time.

An effective approach could be by 'train-the-trainers' first, and then spreading well planned A&M activities throughout a factory (Fig. 5.5). The staff best positioned to become trainers are ECC leaders. They are already assigned the responsibility for performance in their own areas, where they need to work with other staff members in order to achieve results. Therefore, besides the technical training that they are undergoing anyway, they should be also trained for A&M activities in order to mobilize the entire workforce for EEMS project. Box 5.5 suggests an agenda for such training.

Box 5.5: Train-The-Trainer Workshop for A&M Introduction to All Staff

---

Participants: *ECC Leaders/A&M Coordinator*

Objectives:

- *To provide participants with an overview of the 'Awareness and Motivation Introduction for All Staff' course*
- *To provide participants with an understanding of important factors for developing awareness and motivation*
- *To encourage participants to identify specific A&M activities in their own area of responsibility*

Duration: *1 day*

Training Topics:

- *Content of 'Awareness & Motivation introduction for EMS Staff'*
- *Important factors for development Awareness and Motivation*
- *Adult Learning*
- *Techniques and Activities for training*
- *Support Activities to Create Awareness and Motivation*
- *Basic communication skills*

EEMS implementation will only be effective if those who need to act have been *motivated* to become involved. The efforts and performance of individual employees should relate directly to the energy and environmental performance in the area of their operational responsibilities. Training should promote this awareness.

The individuals should also have the time required to actually perform tasks that they have been assigned relating to the EEMS. Performance should be *acknowledged and rewarded*. During implementation, both top management and operational team members play crucial roles. Since implementation on the shop floor requires constant attention, the ECC team leaders must be relied upon to supply that effort, support and motivation.

The EEMS project must be visible all over the factory. Notice boards can be used to update information on EEMS activities and indeed on the results when achieved (Box 5.6).

Box 5.6: Notice Board

Objective: *to use as a medium to update monthly energy and environmental information to all staff*

Responsible person: *A&M coordinator*

Location: *at each ECC*

Content:

- *monthly energy, raw material and water consumption charts, and environmental compliance data*
- *update energy and environmental management information*
- *energy and environmental activities of the month*

Another communication channel can be the company's regular newsletter, or a special supplement on EEMS (Box 5.7).

Box 5.7: Newsletter

Objective:

1. *To use as a medium to communicate and update information and results on energy and environmental management program to all staff.*
2. *To use a sa medium for all staff to communicate their opinions, comments or ideas to the othesr and to the management.*

Responsible person: *A&M coordinator or newsletter editor team*
Date to be issued: *every three months*
Distribution: *Distribute to all departments and all rest areas.*

Content:

- *Message from management*
- *Summary of energy and environmental performance over the last three months and costs saving results*
- *Energy and environmental tips and useful knowledge*
- *Question and answer column (to clarify any questions that staff may have on energy and environmental management)*
- *Interviews with people who take part in EEMS activities*
- *Games/puzzles/cartoons concerning energy and environment*
- *Ideas/suggestion/comments from readers*

At the beginning of the EEMS implementation, the A&M activities may include some *promotional events* like an EEMS slogan contest, energy and environmental exhibition, screening of relevant films, etc.

However intensive A&M activities are there will always be some resistance to change. Therefore, A&M activities must be carefully planned (Fig. 5.5) to extend over the period of EEMS introduction, implementation and operation, until such time as EESM is fully integrated into the company's daily routines.

5.4.1.2 Training

While A&M activities address the entire workforce, aiming to inform them on and mobilize them for EEMS implementation, training is planned for technical personnel that will be involved with the procedural aspects of EEMS implementation. Box 5.8 suggests an agenda for the *on-the-job training* of technical staff members responsible for operating EEMS within the company.

Box 5.8: Training for EE Managers and ECC Leaders on EEMS operations

---

OBJECTIVES:

This training should upgrade the skills of the staff to operate the energy and environmental management system on a continuous basis.

METHOD:

Training will consist of presentations, practical assignments, specific tasks at the factory, review, analysis and report preparation, on the job, or outside the factory if required.

DURATION:

1-2 weeks

SCOPE

Task 1

Introduction to EEMS procedures and equipment
Clarify Energy Cost Centers (ECCs)
Performance Indicators (PI) description
Clarify single line diagrams for installation of additional meters:
Electrical – check single line diagram and positions to install kWh meters
Water – check water piping and position to install water flow meters
Natural gas (NG) – check NG piping and position to install NG flow meters
Compressed air (CA) – check compressed air piping and position to install CA flow meter s
Steam – check steam piping and position to install steam meters
. . . .
Production control process – check production control method and records keeping

Task 2

Preparation for EEMS pilot operation
Check utility and maintenance logs and production records of each ECC
Set data handling procedures and forms for each ECC,

Task 3

Pilot data collection and analysis for all ECCs

Task 4

Setting targets for individual ECC

Task 5

Proposing corrective actions and report making

The purpose of *training* employees in this area is clear enough. It is to ensure that they have the appropriate knowledge and skills to carry out daily performance monitoring activities and to identify the energy and environmental performance improvement opportunities that exist around them.

To support a proactive attitude of the workforce toward proposing performance improvement opportunities (PIO), the ECC leaders should be trained to describe such opportunities in an appropriate format that will facilitate consideration and approval of proposals. Box 5.9 suggests a format that ties together all the relevant aspects of a particular performance improvement measure.

The newly learned skills and knowledge should be periodically reinforced as the results start to show, by structured 'review workshops' attended by ECC leaders, EE manager(s), operational and top management teams.

Box 5.9: PIO 1: Automatic Turning ON/OFF of Injection Molding Machine

Purpose: Save electricity used by the cylinder for the injection-molding machine and build and dissipate awareness of energy minimization

Overview: Improvement measures are to be implemented in order to reduce energy loss based on a new concept that the cylinder heater is automatically switched off using a timer.

Target:

■ Electricity	■ Production equipment	During operating time
	Non-production equipment	■ During non-operating time (lunch break, between shifts)

Description:

Current status	**Improvement measure**
• switching off the heater is forgotten when production is complete for the day	• a timer circuit is to be designed so that it automatically turns off at the end of production as well as automatically turn on (unlike the traditional timer circuit that considers the temperature rise time only). This is a fool-proof measure to keep the heater off during breaks • For overtime work, the switch-off circuit is to be temporarily reset when the overtime button is pressed

Energy consumption before improvement	5640 kWh/year	**Energy consumption after improvement**	4794 kWh/year
Reduced energy amount	846 kWh/year	**Reduction of CO2**	419 t/year

	(effect by horizontal application: 846 · 40 lines = 33 840 kWh/ year)		(effect by horizontal application: 16 760 t/year
Environmental improvements			
Investment costs		**Cost savings**	

Approved by: _____.

5.4.2 Installation of Performance Measurement System

The main aim of PMsS is to provide control over energy and environmental costs by developing an understanding of where, why and how energy is used and the environmental impacts that occur. PMsS also provides a way to determine whether specific objectives are being accomplished. The *performance measurement system* (PMsS), tailored to the needs of energy and environmental performance monitoring of specific business processes, provides such a mechanism and alerts managers to the problems that are developing, by collecting, processing and structuring performance data on all key activities that ultimately drive the company's profits. The prerequisites for this are the availability of reliable on-site metering and monitoring instrumentation and a system for record keeping.

In each factory a kind of metering system exists with the ultimate goal of assisting plant personnel in managing energy and environmental performance while producing a specific product or service. Another goal of such a metering system is to provide historical energy and environmental impact data to aid in projecting future loads and developing standards for the next year. But often the existing metering system would not be sufficient to provide all necessary data for an EEMS operation. If that is the case, *additional instruments* will be required. The extent of additionally required instrumentation is established during energy and environmental auditing, and appropriate data collection and handling procedures will be specified.

Unless energy and environmental data is measured, it is almost impossible to know where to direct performance improvement efforts. A metering system provides the vital ingredient in a successful energy and environmental management program. It is obvious that unless we do not measure, we cannot manage and control. Essentially, PMsS provides framework for:

- analyses;
- interpretation;
- knowledge creation;
- learning, and
- communication of results and lessons learned, on energy and environmental performance.

All data comes from the measuring instruments, or sometimes indirect measurements, that are read and collected regularly at given time intervals at all ECCs according to prescribed procedures and verified and documented by trained personnel. Figure 5.6 shows how the real ECCs are set up at our tuna canning

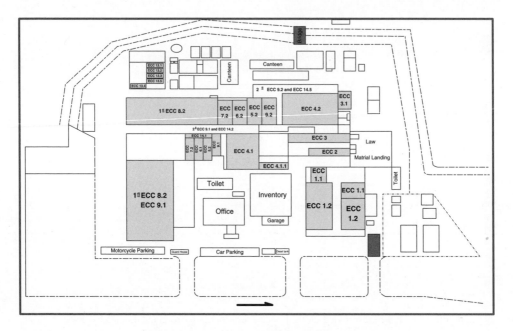

Figure 5.6 Map of ECCs at the Tuna Canning Factory

factory. Table 5.1 contains a list of required additional instrumentation which is a cornerstone of the PMsS at the tuna canning factory. Figure 5.7 shows the ECC arrangements at a textile factory.

These figures remind us that ECCs are not an imaginary concept but real, physical parts of industrial plants, identified and defined during the preparatory phase, together with the necessary instrumentation that is required to enable effective energy and environmental management and continuous performance monitoring and improvement.

In reality, the PMsS is only operational when all required instruments are installed at all ECCs at designated places, and when daily routines of data recording, verification and performance indicators calculations are established. When all of this is truly achieved, captured data will be useful, information will be processed and PMsS will be an effective tool for supporting an organization's energy and environmental performance improvement objectives. An operational PMsS will provide data on performance of individual ECCs, which will in turn enable setting up *specific targets* for each ECC.

5.4.3 Target Setting

Targets can be set only from a position of knowledge on actual performance at every ECC, based on measured data. During the initial three to six months pilot-run, some errors may be detected and corrected, people will get into the routine of continuous performance monitoring, and base line data for all ECCs will be accumulated. This data will be used to determine the individual performance targets for each ECC. New skills developed in training should be applied during this period by regularly recording, monitoring and interpreting data on energy and environmental performance.

With an operational PMsS in place, it becomes possible to establish targets for all performance indicators related to various energy types used (electricity, gas, fuel, . . .), resources consumed (raw material, water, . . .) and environmental impacts generated (waste, air emissions, effluents, . . .). The given targets must be explained to the workforce affected, and accepted by them. *Example target setting procedures are illustrated in Box 5.10 and Box 5.11.*

Table 5.1 Summary List of PMsS Instruments for the Tuna Canning

ECC	Measured	Adding	no.	Range
ECC 1.1 Freeze room	Temperature	Temp. meter	4	−20–50 C
ECC 1.2 Cold storage	Temperature	Temp. meter	5	−20–50 C
ECC Thawing	Electrical / Water	kWh meter / Water flow meter	1 / 2	13.43 kW, 24.01 A / dia. 2 and 4"
ECC 3.1 Butchering & Washing Recking Precooking Cooling	Electrical / Steam / Water	kWh meter / Steam flow meter / Water flow meter	4 / 1 / 3	6–38 kW, 10–68 A / dia. 4" / dia. 2 and 4"
ECC 3.2 Fish extract plant	Electrical / Steam / Water	kWh meter / Steam flow meter / Water flow meter	1 / 1 / 1	48.86 kW, 87.34 A / dia. 2" / dia. 2
ECC 4.1 Cleaning Smoking Loin slicing Filling in can Media filling	Electrical / Water	kWh meter / Water flow meter	1 / 1	42 kW, 75.08 A / dia. 4"
ECC 4.2 Cleaning Smoking Loin slicing Filling in can Media filling	Electrical / Water	kWh meter / Water flow meter	2 / 1	2.3–38 kW, 4–11 A / dia. 4"
ECC 5.1 Seaming Can cleaning	Compressed air / Electrical / Steam / Water	Air flow meter / kWh meter / Steam flow meter / Water flow meter	1 / 2 / 1 / 2	dia. 1", p = 7 bar / 15–48 kW, 27–85 A / dia. 3" / dia. 4"
ECC 5.2 Seaming Can cleaning	Compressed air / Electrical / Steam / Water	Air flow meter / kWh meter / Steam flow meter / Water flow meter	1 / 2 / 1 / 2	dia. 1", p = 7 bar / 48–60 kW, 108–170 A / dia. 3" / dia. 4"
ECC 6.1 Basket loading Holding retort	Compressed air / Electrical / Steam / Water	Air flow meter / kWh meter / Steam flow meter / Water flow meter	2 / 2 / 2 / 1	dia.1 ad 3", p = 7 bar / 3 kW, 5.36 A / dia. 5" / dia. 5"
ECC 6.2 Basket loading Holding retort	Compressed air / Electrical / Steam / Water	Air flow meter / kWh meter / Steam flow meter / Water flow meter	3 / 1 / 2 / 1	dia.1 ad 3", p = 7 bar / 3kW, 5.36 A / dia. 4" / dia. 6"
ECC 7.1 Air cooling (Post process area)	Electrical	kWh meter	1	25 kW, 44.5 A
ECC 7.2 Air cooling (Post process area)	Electrical	kWh meter	1	12 kW, 21.45 A
ECC 8.1 Palletizing Storage	Electrical	kWh meter	1	51 kW, 91.16 A
ECC 8.2 Palletizing Storage	Electrical	kWh meter	1	10.1 kW, 18.05 A
ECC 9.1 Lanbelling & Packing Human/Ped food	Electrical	kWh meter	1	44 kW, 78.7 A
ECC 9.2 Lanbelling & Packing Human/Ped food	Electrical	kWh meter	1	20 kW, 35.8 A
ECC 10.1 Compressor No.1	Electrical	kWh meter	1	130 kW, 232.4 A
ECC 10.2 Compressor No.2	Electrical	kWh meter	1	130 kW, 232.4 A
ECC 10.3 Compressor No.3	Electrical	kWh meter	1	130 kW, 232.4 A
ECC 10.4 Compressor No.8&9	Electrical	kWh meter	1	149.2 kW, 266.7 A
ECC 11.1 Compressor No.4	Electrical	kWh meter	1	74.6 kW, 133.3 A
ECC 11.2 Compressor No.5	Electrical	kWh meter	1	74.6 kW, 133.3 A
ECC 11.3 Compressor No.6	Electrical	kWh meter	1	90 kW, 160.9 A
ECC 12 (Ante room) Compressor No.7	Electrical / Temperature	kWh meter / Temp. meter	1 / 1	30 kW, 53.62 A / −20–50 C
ECC 13.1 Boiler No.1	Steam / Water	Steam flow meter / Water flow meter	1 / 1	dia. 6" / dia. 4"
ECC 13.2 Boiler No.2	Steam	Steam flow meter	1	dia. 6"
ECC 13.3 Boiler No.3	Steam	Steam flow meter	1	dia. 8"
ECC 13.4 Boiler No.4	Steam	Steam flow meter	1	dia. 8"
ECC 13.5 Boiler (new)	Steam	Steam flow meter	1	dia. 8"
ECC 14 Compressor room	Compressed air / Electrical	Air flow meter / kWh meter	2 / 6	dia.4", p = 7 bar / 0.75–15 kW, 1–20 A
ECC 15.1 Old chiller for CMC 1	Electrical / Water	kWh meter / Water flow meter	1 / 1	0.75–15 kW, 1–20 A / dia. 3"
ECC 15.2 New chiller for CMC 1&CMC2	Electrical / Water	kWh meter / Water flow meter	1 / 1	0.75–15 kW, 1–20 A / dia. 4"

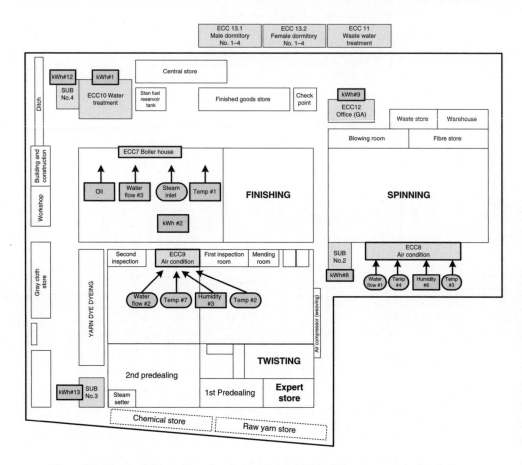

Figure 5.7   Mapping ECCs and Locations for Installation Additional Instruments at the Textile Factory

Just to remind ourselves, the performance indicators should always be related to the environmental parameters, energy and resources used to measure production output. Therefore, when more than one product is being produced at ECC, then separate *performance indicators* and related targets have to be defined for each of the product types. The PMsS data recording system will distinguish data according to product change cycles, so that only the performance data representing certain product types are grouped and analyzed together.

With targets set, accountability established and PMsS in place and operational, we can start the continuous process of performance improvement through an effective operation of EEMS.

5.5 EEMS Operation

> *Confidence – critical for success!*

To actually operate EEMS means practicing a daily routine of measurement readings and analysis at each ECC. Initially, the biggest challenge is usually to recognize a connection or a cause and consequential relationship between measured performance data on the one side and actual operational events on the other and to develop an *understanding of the reasons for performance variations*. It is a creative process that

Box 5.10: Agreed Target & Baseline Dataset – Electricity Data Sample

Subject: Electricity

Baseline data period: From: ___October 2000___ To: ___September 2001___

Table with baseline data set:

Month	Production	Electricity	P_{ELEC}
	(Lbs)	**(kWh)**	**(kWh/Lbs)**
Oct-00	686 693	2 156 760	3.141
Nov-00	662 321	2 089 080	3.154
Dec-00	551 368	1 764 000	3.199
Jan-08	780 993	2 363 400	3.026
Feb-08	813 163	2 343 960	2.883
Mar-08	869 165	2 425 320	2.79
Apr-08	464 220	1 557 360	3.355
May-08	576 638	1 881 900	3.264
Jun-08	860 045	2 485 260	2.89
Jul-08	734 915	2 051 460	2.791
Aug-08	764 023	2 128 680	2.786
Sep-08	688 542	1 992 060	2.893
Total	**8 452 086**	**25 239 240**	

Resulting average energy/production graph & equation

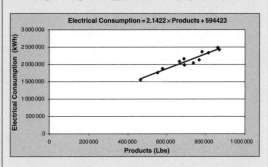

Electrical Consumption = $2.1422 \times$ Products + 594423

Oct 2000 – Sep 2001 : Average Performance Indicator

$PI_{ELEC} = 3.01$ **kWh/lbs**

Resulting average PI/production graph & equation

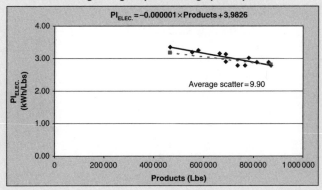

$PI_{ELEC.} = -0.000001 \times$ Products + 3.9826

Average scatter = 9.90

Agreed target

Performance improvement for electricity = 3 %

We accept the data set from this period to be a base line for performance improvement verification.

Approved by: _____.

Box 5.11: Agreed Target & Baseline Dataset – Water Data Sample

---

Subject: Electricity

Baseline data period: From: October 2000   To: September 2001

Table with baseline data set:

Month	Production (Lbs)	Water (m³)	PI_WATER (m³/Lbs)	Resulting average energy/production graph & equation
				$Water\ Consumption = 0.0038 \times Products + 9231$
00-Oct	693 686	17 757	0.026	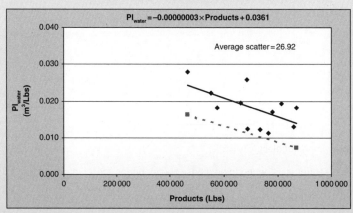
00-Nov	321 662	12 939	0.020	
00-Dec	368 551	12 266	0.022	
01-Jan	993 780	13 363	0.017	
01-Feb	163 813	15 723	0.019	
01-Mar	165 869	15 861	0.018	**Oct 2000 – Sep 2001: Average Performance Indicator**
01-Apr	220 464	12 997	0.028	$PI_{WATER} = 0.017\ m^3/lbs$
01-May	638 576	10 493	0.018	
01-Jun	045 860	11 185	0.013	
01-Jul	915 734	9065	0.012	
01-Aug	023 764	8687	0.011	
01-Sep	542 688	8584	0.012	
Total	**8 452 086**	**148 920**		

Resulting average PI/production graph & equation

$$PI_{water} = -0.00000003 \times Products + 0.0361$$

Average scatter = 26.92

Agreed target

Performance improvement for water = 20 %

We accept the data set from this period to be a base line for performance improvement verification.

Approved by: _____

requires analytical thinking, as explained in Part I Chapter 3 and a deep understanding of the energy-production relationship. Therefore, ECC teams may need intensive external support during that period until they develop *confidence* that the system works well and that their 'reading' of performance data is correct.

Regular and prompt feedback from top management will be essential for developing staff confidence and motivation for a continuous search for performance improvement. A structured way of reporting should be introduced in order to facilitate communication between management and EEC teams. An example form in Box 5.12 ties together all the EEMS operational issues:

- recording;
- monitoring;
- interpretation;
- corrective actions;
- reporting and communication.

The actual performance data is recorded and presented in a graphical format that supports *monitoring* and enables the quick identification of unusual performance variations. The point is to keep the process simple and neither time consuming nor statistically heavy. *Simplicity* is the key for positive acceptance of performance monitoring as a routine practice and it is a short cut to confidence.

5.5.1 Performance Data Interpretation

Organizations are usually great collectors of human, financial, production and technical data. But only the best organizations know how to use that data! Once an indicator has been defined and measured, it must still be interpreted in order to provide useful and actionable information.

Interpretation is a process where people are actively looking for new insights from available performance data. Interpretation involves:

- review of an operational event with a substandard performance;
- reconstruction of what happened;
- agreement on corrective actions;
- corrective actions implemented;
- the effects checked and performance improvement confirmed.

The interpretation requires an understanding of the operational events that are reflected in the measured data and a scatter diagram presentation (the diagram in Box 5.12). This understanding has to be developed through professional, dynamic and open discussion focused on the causes of good or bad performance that enable people to discover for themselves what has happened, why it has happened, how to sustain strengths and how to remove weaknesses.

5.5.1.1 Evaluation of Operational Practice

The proposed table in Box 5.12 contains daily data on aspects of operations at a particular ECC. Columns 2 and 3 contain measured data on daily raw material and energy use, while column 4 provides the performance indicator of the day. Columns 5, 6 and 7 offer a choice of targets based on past average performance, best performance, and linear 3 % improvement over the base line case. The scatter diagram visualizes data on daily performance and raises concerns on particular days when performance was substandard or very good. The rest of the fields in the form are reserved for the communication of evaluation results for these instances of good or bad performance.

Structured and purposeful communication is about operational practices and resulting performances and everyone's opinion is valued. An open discussion has to be established with sufficient detail on performance and the active participation of all operational personnel concerned without any hierarchical barriers. Therefore, the conclusions from *interpretation* discussions should be reported to the

Box 5.12: Sample of Monitoring and Reporting Form on ECC Performance

Time	Raw material Starch (t)	Steam (t)	PI =M^t/ Mt Today	Target From Best Fit Average	Best	-0,03
1	62, 00	148	2, 4	exceptional		
2	100, 50	134, 00	1, 3	1, 19	1, 05	1, 15
3	97, 00	130, 00	1, 3	1, 24	1, 09	1, 20
4	95, 00	110, 00	1, 2	1, 27	1, 12	1, 23
5	91, 50	123, 00	1, 3	1, 32	1, 16	1, 28
6	100, 00	129, 00	1, 3	1, 20	1, 06	1, 16
7	97, 50	142	1, 5	exceptional		
8	91, 50	123, 00	1, 3	1, 32	1, 16	1, 28
9	103, 00	123, 00	1, 2	1, 15	1, 02	1, 12
10	102, 50	124, 00	1, 2	1, 16	1, 02	1, 12
11	105, 00	97	0, 9	exceptional		
12	106, 00	121, 00	1, 1	1, 11	0, 98	1, 07
13	66, 50	119, 00	1, 8	1, 68	1, 48	1, 63
14	51, 00	86, 00	1, 7	1, 91	1, 67	1, 85
15	63, 00	114, 00	1, 8	1, 73	1, 52	1, 68
16	71, 00	105, 00	1, 5	1, 62	1, 42	1, 57
17	77, 50	120, 00	1, 5	1, 52	1, 34	1, 48
18	104, 00	127, 00	1, 2	1, 14	1, 01	1, 10
19	109, 00	113, 00	1, 0	1, 07	0, 94	1, 03
20	95, 50	111, 00	1, 2	1, 26	1, 11	1, 22
21	36, 50	77, 00	2, 1	2, 12	1, 86	2, 05
22	90, 50	150	1, 7	exceptional		

ECC1: Plant 1
ECC Leader _____
Date issued --.--.----

Remarks of concern

- Product was changed and so steam consumption exceptionally bad on days 1, 7, and 23. Steam consumption exceptionally good on Day 11.
- Why was steam consumption on Day 11 lower than other days ?
- How to keep steam consumption levels as on Day 22, 14 and 17?

Recommended or Answered by ECC Leader

Recommended by Production Advisor / EEMS Manager / EEMS Advisor:

Return this comment to EEMS Center within _____ .

higher management and their comments or agreement on the corrective actions communicated back to ECC teams.

The successful realization of performance improvements is the result of consecutive decisions for improvement of operational effectiveness and of actions based on active search for knowledge and understanding rather then the result of an isolated and instantaneous event. Therefore, energy and environmental performance improvements come as a cumulative result of an ongoing process of improving operational efficiency.

The main components of operational efficiency are:

Cost	– unit cost of a product
Time Productivity	– time to produce a product
Material Productivity	– raw materials input to product output ratio
Quality	– rejects or rework over acceptable items ratio.

Cost efficiency and material productivity are directly affected by the efficiency of material use. Fast changeover is an absolute requirement for time productivity. Waiting to receive parts or send products to subsequent operations is lost opportunity to *create value*. Even more, waiting means machines idling which causes unnecessary energy consumption and environmental releases, so on top of lost opportunity to create value, there is also a direct cost for non-productive consumption of energy. Reduced quality, i.e., increased amount of *rejects and rework*, will have the direct measurable consequence of increased energy consumption and waste generation.

Therefore, energy and environmental performance indicators can help line personnel to *assess and improve operating efficiency* by identifying operational decisions that will generate improvement of one or more of the important characteristics of the process, for instance:

- reduced scrap or rework:
- shorten cycle time:
- shorten set-up time:
- increased productivity:
- optimized frequency of product order;
- optimized frequency of product change over.

A direct consequence of such operational efficiency improvements will be reduced energy consumption and environmental impacts. Therefore, incremental operational practice improvements will in turn result in incremental energy and environmental performance improvements and cost reductions.

The operation of EEMS supports seeking knowledge while working, i.e., *purposeful learning* as we go, thus providing a dynamic loop where capability is enhanced and performance improved.

5.6 Learning Through EEMS Operation

Knowledge is nowadays widely recognized as the only sustainable source of competitive advantage. EEMS provides a framework for stimulating productive processes for seeking, sharing and refining practical knowledge. It enables companies to achieve extraordinary results at the operating level by building a culture of continuous performance improvements and creating, sharing and applying new operating knowledge throughout the company.

But, how do organizations learn? All people in an organization that want to excel in performance need to develop analytical, decision making and communication skills. It requires employees to be curious and highly motivated in order to become *multi-skilled* and valued staff members. By supporting that, organizations will broaden their leadership base and create an environment in which people can self-actualize in the process of performing their tasks.

5.6.1 Learning Cycle

On-the-job learning goes through a cycle of knowledge discovery, documentation, dissemination and integration into daily practice and routine. *Asking questions* is the key tool for *knowledge discovery* while working. People are a valuable source of knowledge if they pay attention to what is going on and if they are able to ask effective questions that shape thinking and actions in the area where they are responsible for energy and environmental performance.

Asking questions, analyzing causes and consequences of operational events and interpreting data patterns is a qualitative approach that synthesizes implementation oriented knowledge and expertise in the performance improvement process.

Learning starts with *discovering* the causes, patterns, and events leading to good or bad performance. When successful approaches are discovered, the lessons learned need to be discussed, corrective actions confirmed and new knowledge captured. Such discussions increase the amount of information retained and individual motivation to accept the agreed-to lessons and changes in personal behavior in the future, even if that means admitting some errors in the past.

Captured knowledge that is gathered through this experience should be documented in the form of guidelines for Best Operational Practices (BOP). To achieve permanent energy and environmental performance improvements, the company has to move gradually from a process of discovering knowledge to implementing it as routine EEMS practice, which will ultimately be transformed into the best operational practices. These are the operational measures that make better use of raw material, energy, and water, thus reducing environmental impacts by generating less solid waste, waste water and emission, at the same time developing better working conditions and communication. These practices must be documented in order to *institutionalize the lessons learned* and retain the discovered knowledge.

An example of such a documented *best operational practice* is given in Box 5.13.

Box 5.13: Best Operational Practice for Ambient Air Treatment in Textile Industry (In Tropical Climate!)

BEST OPERATIONAL PRACTICE – Air washers operation

STEP 1: Adjusting the set points for air quality in premises

- *The flow rate of supply air must be set at the maximum by adjusting inverters or dampers of all fans.*
- *Check in each set parameters (dry bulb temperature and relative humidity) in each premises. If there are deviations, then adjustment has to be done by regulating the flow rate of chilled water and flow rate of recirculation of water (opening or closing the valves).*
- *Adjust the air washers starting with the one where the room temperature and relative humidity are closest to the set point.*
- *Calculate the enthalpy of air in premises using HYGROMETER No#2 readings at 01:00; 05:00; 09:00; 13:00; 17:00 and 21:00 hours) and following the procedure:*

  Input:
  > *Dry bulb temperature of ambient (fresh) Air [oC]*
  > *Relative humidity [%]*

  Output:
  > *Enthalpy of ambient air [kJ/kg]*

  OR
  Input:
  > *Wet bulb temperature of ambient air [oC]*

  Output:
  > *Enthalpy of ambient air [kJ/kg]*

STEP 2: Managing fresh air

  IF
  > *the enthalpy of fresh (ambient) air is lower than enthalpy of the air in premises*

  THEN
  > *open the fresh air damper 100 %*
  > *close the return air damper 100 %*

  IF NOT
  THEN
  > *fresh air damper must be closed*

  AND
  > *return air damper must be opened.*

> **Regularly check the following:**
>
> 1. *Spray nozzles (not clogged or broken)*
> 2. *Spray pumps (no leak and properly operating)*
> 3. *Filters (air and water, keep clean)*
> 4. *Dampers (clean and operating easily)*
> 5. *Fans (vanes clean and bearings greased, belts tight)*
> 6. *Accuracy of instrumentation*
>
> *If any of these items is not in good operating condition, it has to be repaired immediately.*

Dissemination is aimed at sharing lessons with others and encouraging them to try new approaches. The mechanisms for recording and disseminating mutually understood lessons institutionalize the lessons and set the stage for further performance improvements. Accumulating and disseminating knowledge is too important to be left to chance. There must be a specific person designated to ensure that new knowledge is documented and treated as something of value. It cannot be assumed that knowledge discovery, documentation and dissemination may just happen. It must be the result of conscious effort. Companies routinely fail to formalize and institutionalize what individuals know, mostly because it is hard work to ensure that knowledge is recorded and becomes part of the company know-how.

An implemented *learning cycle* ensures that valuable knowledge is captured, *core competencies* retained and *continuous improvements* enabled. Even the smallest organizations must have a well conceived plan for *organizational learning*. Further, people will find a new confidence founded in knowledge which will·strengthen their commitment for continuous performance improvement and stimulate organizational learning.

5.6.2 Learning through People's Performance Evaluation

> *Everything can be improved!*

Evaluation of energy and environmental performance that comes through the interpretation of performance data is at the same time also evaluation of people's performance. Energy and environmental performance is *people's performance*.

We have established EEMS as a process oriented performance management system with established targets and performance metrics. All processes are operated or supervised by people, hence process performance is a direct indicator of people's performance.

The energy and environmental targets agreed upon for each ECC represent specific performance objectives that people have accepted in order to take a full responsibility for their achievements. An energy and environmental management system focused on people provides an effective technical framework for *people's performance evaluation* in addition to the usual criteria that human resource departments apply. Such a process based people's performance evaluation can be an additional source of organizational learning.

5.7 Continuity and Communication

EEMS has to be fully implemented and has to evolve from a process of change to an everyday routine, in order to contribute to lasting internal changes and learning processes. Lasting changes will be supported by the visibility of:

- energy and environmental performance improvements;
- reduced resources consumption;
- operational efficiency gains; and
- resulting cost savings.

The most important criteria of success perceived *ex post* by all are cost reduction, improved product quality, and better working conditions. Therefore, it is important to keep the employees regularly *informed and updated* on the progress of the energy and environmental management program. The information and data on energy and environmental situation and results achieved have to be made visible, tangible and understandable, because energy and environmental issues are often perceived as an abstract, technical and complex.

Therefore, regular communication and public display of the results at all ECCs are important factors for assuring *continuity* of effort. Figure 5.8 and Figure 5.9 provide some examples of performance improvements related to more efficient steam use and CO_2 emission reductions. Both graphs show a part of base line period (March to December) prior to implementing proposed performance improvement measures and then *cumulative reductions* in steam use and CO_2 reductions onwards. These are so called CUSUM diagrams which enable reading out the achieved improvements at any moment in time.

Such positive results should be displayed and updated on the *notice boards* throughout the factory.

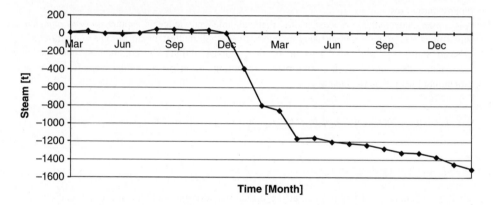

Figure 5.8   Example of Performance Improvements as Reductions of Steam Use

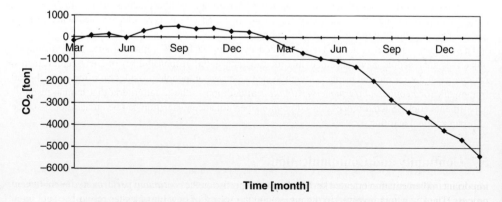

Figure 5.9   Example of Performance Improvements as CO_2 Emission Reductions

5.8 Integration of EEMS with Business Management System

Integrated energy and environmental management provides a powerful, effective and economic framework, both at organizational or project level, for:

- measurement, validation and verification of energy performance improvements and related or non-related emission reductions;
- consistent, transparent and credible quantitative monitoring and reporting on companies' GHG and other emissions;
- identifying and managing emission related liabilities, assets and risks;
- facilitating trade on emission allowances or credits;
- implementation of energy efficiency and environmental improvement projects, and tracking their results.

When people understand and master energy and environmental performance management processes, tools and techniques, data and information will become highly accurate and provide a good basis for sound *decision making* and knowledge creation.

For energy and environmental management not to be perceived as a topic *per se*, it has to be linked to other issues (productivity, quality, costs control), focusing on company's *multi-functional aspects of competitiveness*, and it has to be communicated and promoted as a core element of a broader concept of eco- and cost-efficiency. In addition to that, a higher motivational appeal, such as linking and concept broadening, opens up possibilities for synergies and further cost reductions, which are often hindered by the administrative and functional separation of organizations and the responsibilities within.

Special emphasis should be placed on promoting energy performance as a *quality indicator* for management and business performance because of the distinctive characteristics of energy in an industrial environment:

- It enters into every aspect of a business operation.
- It can be accurately and comparatively inexpensively measured at the point of use and at any given interval or frequency.
- An energy metering system can be designed to correspond with a production and cost control system, so it can provide accurate inputs for activity based costing.
- Energy consumption depends on production activities and this relationship can be established and monitored.
- Variation in energy use will reflect the varying performance of activities that use the measured amount of energy over a corresponding period, hence it can indicate on productivity in the measured area.
- Quality problems like rejects will also be detected by recording higher energy consumption per unit of good products finished over the corresponding period.

Therefore, *energy performance* may also be seen as an additional indicator for overall performance, because energy performance can be measured accurately and expressed clearly through standardized performance indicators. This fact is not often appreciated by management. When reported regularly, energy performance indicators will provide reliable information to management about other performance aspects in the same area – such as quality, productivity, maintenance, etc. In fact, an inadequate performance measurement system is often behind the disappointing results of various performance improvement initiatives.

The qualitative approach of EEMS development and operation, which is *focused on people*, generates important implementation oriented knowledge and expertise on the *continuous performance improvement process*. The key actors involved in the project are the source of new impulses for proposing new areas for action. They contribute to the internal stabilization of new routines, and communicate results and benefits inside and outside the company. Once continuous performance improvements are achieved, the various performance reports that the company is required to submit can be combined into one in order to reduce the overall workload and ensure consistency in communicating corporate performance to the general public and to the regulatory authorities.

Therefore, EEMS practice should not be seen as isolated from everyday company activities but needs to be *integrated* into the *process-oriented strategy* of overall company performance improvement. When EEMS is in full operation, with a *performance measurement system* working well, management often finds itself surprised by the new *insight* it achieves into the internal aspects of business operation.

It is only after the results are seen that EEMS starts to be appreciated as a powerful tool for improving overall business performance. That is why it would be good practice to integrate EEMS operation into the existing management system in order to enhance the effectiveness of *decision making* and performance improvement initiatives.

6

Energy and Environmental Management as a Driver for Integrated Performance Management

6.1 Introduction

Increasingly, the attainment of business objectives requires the deployment of multidisciplinary task forces within which the traditional lines of responsibility and hierarchy are redrawn. Very often, employees work simultaneously on several management initiatives or programs and participate in several work groups in order to put these initiatives into practice. All of these programs are ultimately intended to improve business operations and for that reason it is certainly useful to integrate energy and environmental management with other programs.

In fact, we have already achieved the first step in such integration by introducing energy and environmental management concurrently, whereas they are usually just loosely connected. It is a fact that modern environmental management system standards such as BS 7750, EMAS and ISO 14001 recognize the role of energy efficiency as just 'completing the picture' of integrated environmental management. However, in our approach we have extended the scope of 'energy' to all of the natural resources that a company uses, and so consequently the scope of energy management was extended to involve energy, water and raw materials. We have further argued that by optimizing the use of natural resources (energy, water and raw materials), we are also minimizing the environmental impacts from a company's operations. The fact is that any improvement of energy, water or material use efficiency contributes to the reduction of environmental pollution. If we used none of these, there would be no environmental impacts.

Therefore energy, water and raw material management comes first, to be followed by environmental management which then focuses on risk and compliance issues, and on the quantity and quality of 'end-of-pipe' releases and discharges.

Organizations which adopt this approach are taking the first step towards making resource management a central feature of their thinking, alongside critical issues such as operational effectiveness,

Applied Industrial Energy and Environmental Management   Zoran K. Morvay and Dušan D. Gvozdenac
© 2008 John Wiley & Sons, Ltd

waste management and minimization of emissions and discharges into the environment. Since this is an integrative chapter, we will draw here on several concepts which have been developed so far:

- performance policies and objectives;
- implementation plans for performance improvements;
- internal and external influencing factors;
- performance measurement;
- organizational structures and competencies;
- people's skills and communication;
- learning and knowledge capturing and creation;
- energy and environmental management system,

and extend them as required to propose an integrated performance management approach. In order to do that, we have to integrate the management of all internal and external factors that influence business performance (see Fig 6.1).

Figure 6.1 emphasizes that industrial operations are the focus of all considerations because that is where resources are used and where economic value is created. We were concerned with operational performance, which has several dimensions, but we were focused specifically on issues of energy and environmental management, and on some aspects of quality and productivity management. Since we

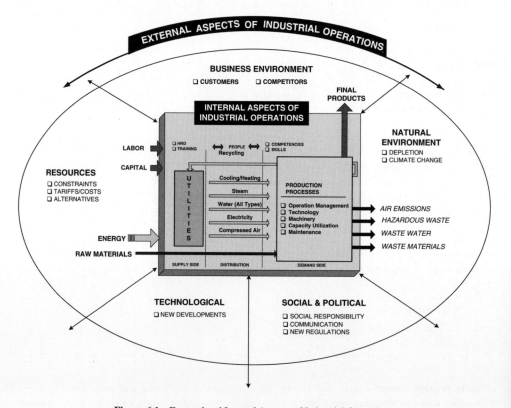

Figure 6.1   External and Internal Aspects of Industrial Operations

are venturing now into the fields of operations and strategic management, it would be useful to remind ourselves of the generally accepted definitions of these terms.

Strategic management is the process of specifying business objectives, developing policies and plans, allocating resources and making cross-functional decisions to implement these policies and plans in order to achieve the specified objectives. Strategic management aims to coordinate at least the polices, if not the actions, of the various functional departments in a company while attempting to achieve given performance targets.

The term '*operations*' refers to the production of goods and services, and generally represents the set of value-added activities or processes that transforms inputs into outputs. *Operations management* is concerned with the performance of producing goods and services, and involves the responsibility of ensuring that operations are efficient and effective. It also involves the management of resources including effective planning, scheduling and use of resources in manufacturing while focusing on quality, productivity and customer satisfaction. *Operational performance* management focuses on the efficient use of human and material resources by applying systematic and methodical procedures in order to improve business results and performance across a company.

The main aspects or dimensions of operational performance are related to (Fig. 6.1):

- customer satisfaction;
- operational productivity:

  - materials,
  - machines,
  - labor,
  - time,

- quality;
- energy performance;
- environmental performance;
- costs effectiveness.

The considerations so far were focused on energy and environmental performance management inclusive of water and raw materials and identification of internal factors that directly or indirectly influence these aspects of operational performance. Before we move on to the issues of strategic management and relevant external influencing factors, we should consider what other internal issues are important for operational performance thus warranting management's attention and requiring a system in place to control them. We will extend the considerations by looking into overall *operational costs* as a truly universal performance category that explicitly or implicitly reflects all of the other aspects of operational performance. Ultimately, everything comes down to the definition of appropriate performance indicators and the adequacy of the applied performance measurement.

6.2 Integrated Performance Management in Operations

Effective energy and environmental management requires that the integrated management process extends horizontally across the company, including production, utilities and sales. It must also extend vertically through the company's supply chain to include acquisition of energy and raw materials, and to link projected sales to actual manufacturing plans. A proficient EEMS is a tool that supports the improvement of skills, rewards correct actions, builds integrity of competent workforce and *creates and captures knowledge!* An EEMS is concerned with resources (energy, raw materials and water) and environmental impacts management that are operational issues. These issues are at the same time main cost categories.

The EEMS is also concerned with the efficiency of use of resources, which means money and therefore is a business management issue.

In fact, if we look again at our definition of an Energy and Environmental Management System (EEMS) as:

---

a specialized body of knowledge with an underlying organizational structure which integrates interrelated elements, such as:

- *people with skills and assigned responsibilities;*
- *declared policies with clear objectives and targets;*
- *defined procedures and practices for implementation;*
- *established measurement system for performance monitoring;*
- *action plan for continuous improvements;*
- *reporting system for checking progress and communicating results.*

These elements are applied with the main aim of reducing operational costs by improving energy performance, minimizing waste and reducing the environmental impact of a company's operations.

---

we could safely agree that the same definition may hold for any management system!

This definition broadens the objectives of an EEMS by declaring that the aim is reduction of *operational costs* (not only energy and environmental costs). Indeed, it is rather a small step forward to include operational costs optimization within the EEMS objectives once a proper organizational structure, and particularly a *performance measurement system*, are in place. What interests managers most is the cost of making a product or doing a specific activity. To provide accurate and timely information for operational management decisions, the use of all resources will need to be measured simultaneously during the course of production, and the gathered information needs to be related to the measured production's or activity's outputs. When such a performance measurement system is in place, it can also effectively support *activity based costing* which traces the costs of a product based on its actual demand for all resources during the production process.

Effective energy and environmental management thus becomes a backbone for improving operating efficiency and achieving *cost effectiveness* through the improved performance in using ALL resources. Improved cost management is at the heart of the *business value* that EEMS can deliver, therefore it provides competitive advantage and drives future financial performance. Therefore an energy and environmental management system, with established organizational structures, performance measurement, learning and continuous improvement procedures; can be a driver for integrated performance management in operations!

The only additional work that is required is to extend the scope of the performance measurement system so that it can provide reliable data on all relevant cost categories at the point of use of particular resources (Fig. 6.2). The '*cost centers*' approach is the basis for integrated performance management. These cost centers could also be established based on the same principles as the energy cost centers that already include raw material costs anyway. In most cases, whether we speak of *energy cost centers* or just *cost centers*, these will be the same parts of the plant, because they are established following the logic of production process. The main difference comes from the fact that *integrated performance management* requires monitoring the performance of all significant operational *cost items* – raw materials, water, energy, labor, maintenance, training, regulatory compliance, quality – incurred by a cost center.

The example in Box 6.1 illustrates a real case of integrated performance management, where the poor quality of products induced frequent customer complaints, which when investigated revealed a number of shortcomings in operations. These shortcomings were subsequently fixed, quality and customer service was improved and customer satisfaction was achieved.

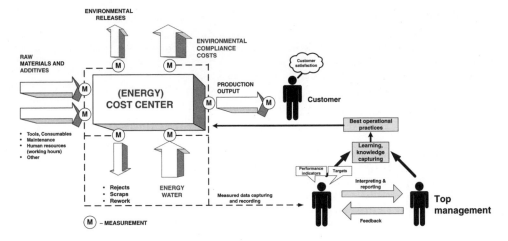

Figure 6.2   Cost center – a core of integrated performance measurement

Box 6.1: Textile factory – a case study: frequent customer complaints about quality of fabric and delivery times

A textile factory (in a subtropical climate area) with spinning, weaving and dyeing plants, was supplying fabrics of various designs to a large customer. The increasing number of complaints from the customer was recorded over a time period and was spotted by the EEMS operational team. The team has triggered an investigation into the causes of complaints. The complaints ranged from those about delayed deliveries, or unknown times for delivering specific shipments, to those about the quality of fabrics delivered. The following issues were examined:

(i)   Why the customer could not be given an accurate delivery schedule?
(ii) What could be the causes for delayed deliveries?
(iii) How it could have happened that company's own quality control (QC) have passed substandard quality products to the customer?
(iv) What was the cause for the substandard quality in the first instance?

The textile factory was quite large, so when specific orders were received and the order fulfilment process triggered, customer services did not have the means to track the particular order through the production stages (spinning, weaving and dyeing), therefore they were not able to tell the customer when exactly the shipment would be delivered.

The delays were tracked back to the spinning department, and after further analyses, attributed to an air conditioning system that was not working properly, therefore causing production stoppages. The problem was diagnosed as the malfunction of an air washer, and responsibility for maintenance was with the company's utility department.

The company QC was situated in a space that was not air conditioned. The temperature and humidity in the area were occasionally (depending on the weather) quite high, thus creating a harsh working environment, and making it difficult for the QC staff to concentrate on their tasks. That was the reason why substandard products were passed on to the customer.

And finally, the cause of poor quality was tracked back to the dyeing plant and to a particular machine where the production department was responsible for maintenance. In order to solve the problem of customers' complaints, the following improvements were implemented:

1. The air conditioning system in the spinning department was fixed, and additional performance indicators were defined and included in the regular performance monitoring system, in order to prevent recurrences of the problem.

Box 6.1: Continued

> 2. *A bar code tracking system was introduced enabling the tracking of any order through the stages of production thus ensuring accurate estimates of delivery times.*
> 3. *A new air conditioning system was installed at the QC area.*
> 4. *The malfunctioning dyeing machine was replaced by a new one, more energy efficient and less environmentally polluting, and more technologically versatile.*

> *This case illustrates also how all aspects of operational performance are intrinsically integrated in any event, because the customer complaint led to a quality issue, to an energy issue, to a HR issue, to a maintenance issue and to a production planning and control issue.*
>
> *The problem solving in this case did not take a great deal of time because performance management was already integrated and appropriate organizational structures were in place that included responsible people form all areas in the company. Should that not have been the case, the usual finger pointing and blame game might have started, delaying the solution of the problem and leading potentially to losing that big customer.*
>
> *The case further illustrates how an EEMS can grow by on-the-job learning, how it can shift the focus on quality issues when required, how the scope of a performance measurement system can be easily extended to include additional performance indicators, and how customer satisfaction does indeed drives improvement across production and utility departments.*

Integrated performance management in operations evolves in that way from concurrent and loosely connected, almost independent management systems for various performance categories (Fig. 6.3a) into an integrated system, which avoids duplication in performance measurement, avoids data incompatibilities and non-synchronal data records and provides a complete, reliable and true picture of a company's operational performance at any moment in time (Fig. 6.3b), as well as an efficient framework for continuous performance improvement and related learning.

So far we have studied internal influencing factors relevant to energy, environmental and operational performance management. The EEMS, as described, provides a powerful tool for managing continuous improvement of business performance in the reality of day-to-day operations. The most important features, which in our experience make EEMS effective in delivering performance improvement, are:

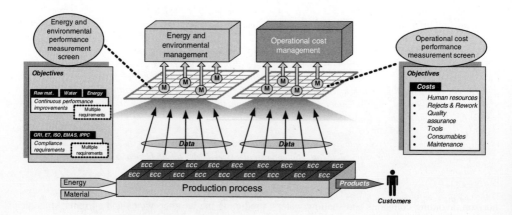

Figure 6.3a   Concurrent Performance Management in Operations

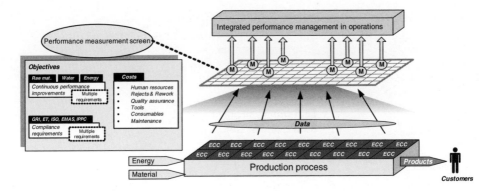

Figure 6.3b Integrated Performance Management in Operations

➤ Focus on people — because no tool, however powerful it may be, can do the work without people!

➤ Focus on performance measurement — because no performance objective could ever be reached if progress is not measured!

➤ Focus on implementation — because no concept can bring results if it is not applied successfully!

➤ Focus on learning — because learning and knowledge capturing are the only way to sustain improvement!

Reduction of scrap, rejects, energy, environmental compliance and waste treatment costs will directly improve a company's profit. Since energy is required for every activity, energy consumption will depend on the performance of these activities, and variations in energy use will be detected by EEMS, and the data interpretation process will lead to discovery of the causes of variations. Therefore it is prudent to integrate energy and environmental management with other aspects of operational management. Hence EEMS could be positioned as a tool for improvement of not only the technical, but also the financial and managerial performance of a business, the quality of products and customer satisfaction.

6.3 Strategic Aspects of Performance Management

Strategic management directs and coordinates the activities of the various functional areas of a company in order to achieve specified business objectives. It is an ongoing process that assesses the business and the industries in which the company is involved; assesses its competitors; and sets goals and strategies to meet the objectives as against changing circumstances, new technologies, new competitors, a new economic environment, or new social, financial, or political circumstances.

The key missing dimension for true integrated performance management is how to keep a focus on determining and maintaining a winning strategic direction. If we use the metaphor of a sailing boat, having perfect operational performance management in place is equivalent to keeping the boat best positioned against the wind with its sails perfectly trimmed, being the fastest in the pack. But to win a race, the most important issue is not to be fastest in a particular leg, but to know how to come first at the final destination. This requires, in addition to having a perfect team on board, looking ahead, scanning the horizon, avoiding collisions and shallow waters; predicting the weather, winds and currents; strategic planning and deciding on the best course over each leg and over the whole race, against the elements and against the competition. This is the skipper's task, i.e. the top management's task.

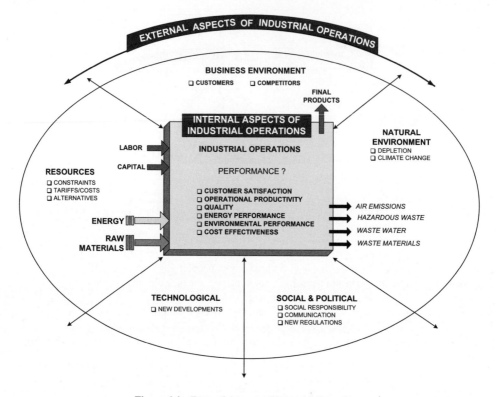

Figure 6.4   External Aspects of Industrial Operations

Of course we cannot get into the detail of strategic management issues because that would require another volume, but we can highlight the main aspects that are complementary to the systematic approach to operational performance management as elaborated here, starting with an examination of the external factors influencing performance on the short to medium term horizon (Fig. 6.4).

Significant changes in any of these areas will influence business performance, to which a strategic management response would be required. This response would be formulated through strategies, policies and objectives that need to be implemented in order to maintain a competitive business position, while satisfying customers, shareholders and other stakeholders.

However, due to the complexity of the modern business environment, these issues are not as straightforward as may be the case with internal factors affecting business performance. They will appear from different angles in a global market place, having an immediate or delayed impact and a direct or indirect influence, but requiring timely responses. If the responses are not timely, performance will deteriorate and the competitive position will be threaten, therefore constant alertness and vigilance with regard to changes in key external influencing factors is essential.

Looking more specifically at the technological area, we will outline a methodological approach for identifying and monitoring the external factors that influence business performance.

6.3.1 Identifying External Factors Influencing Performance

Rapid technological changes create a dynamic business environment that could influence companies' performance in the near or medium term, and sometimes even their survival. Some large companies

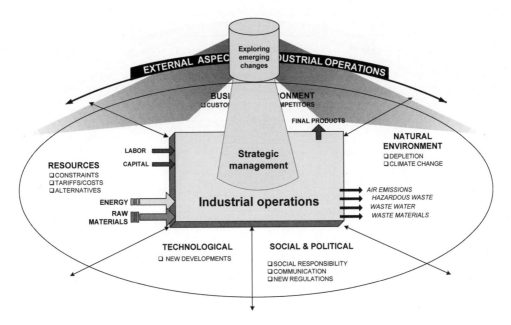

Figure 6.5   Searching actively for external factors influencing business performance

operating globally already have in place their own systems for the continuous scanning of the business environment and the identification of changes that would require preparing strategic responses.

Identifying these changes requires forward focused and lateral thinking which looks consistently at the alternative possible outcomes of technological changes which may confront the company's objectives and position. This forward focused thinking starts from the present and explores possible future developments and changes that can impact upon company performance (Fig. 6.5).

It begins with the identification and monitoring of trends of changes and emerging technological issues. Ensuring that we are considering all possible sources of change requires scanning the 'horizon' and beyond it (see Fig. 6.5), across related areas of the business environment. Even though our focus may be on technologies, we should not observe only upcoming technologies, but also developing trends in accepting new technologies, like for instance solar powered or electric cars, new political directions or regulations favoring renewable energy or putting an ever increasing price on carbon, that can affect a company's performance, thus threatening the company's market position, or opening new opportunities. These factors or the related changes thereof would come broadly from the areas summarized in Table 6.1.

Figure 6.5 shows that the impact of these factors, other trends and emerging changes also move inwards towards a company's operations, thus affecting critical decisions on operational issues and performance.

6.3.1.1 Identifying Technological Changes

> *'The telephone is a fantastic invention.
> I am sure that every city will get one!'*

This quote epitomizes the uncertainty of future outcomes with new technological developments at the time of their appearance, as opposed to the dramatic impact that they exert not only on businesses but also

Table 6.1   Dimensions and factors that influence strategic performance

	Strategic Performance Dimensions	Factors Influencing Performance
1	Markets, customers and competitors.	• market share; • profitability and other financial performance indicators; • competitors' business models; • customers' value perceptions; • new entrants; • customer satisfaction; • market opportunities; • pricing; • competitive positioning; • demand generation.
2	Technological advances.	• new materials; • new technologies (i.e. nanotechnology, biomedical, etc.); • new energy sources; • new knowledge; • product innovations.
3	Macro-economic, regulatory and political conditions.	• globalization; • international conflicts; • international treaties; • increasing price of carbon; • increasing prices of resources; • political will to push technological changes; • stricter regulations (on fuel efficiency, environmental compliance, etc.).
4	Social changes.	• aging societies; • life-long learning; • mobility; • social acceptance of new technologies; • sustainable and empowered communities.
5	Environmental changes.	• climate change; • global warming; • depletion of some resources.

on everyday life styles. Companies that could not identify significant technological changes in a timely manner were left struggling to maintain their market position, or in the worst cases were forced out of business. Figure 6.6 illustrates the fast pace of technological change and related developments over past decades. What lies ahead of us and what are the future changes that will impact upon our business, are questions that require constant searching for answers.

For the effective identification of important external influencing factors and signs of change, there is an array of so called *'foresight'* methods and techniques. Foresight – or *exploring the future* – is, first and foremost, a cross-functional *systematic approach* to considering alternative possible future developments and their outcomes and to planning adequate responses. The 'systematic approach' is equally important as an emphasis on 'exploring the future'. Changes occur as new properties and opportunities emerge from the complex systems of innovation, R&D, discoveries of new materials and technological advancement.

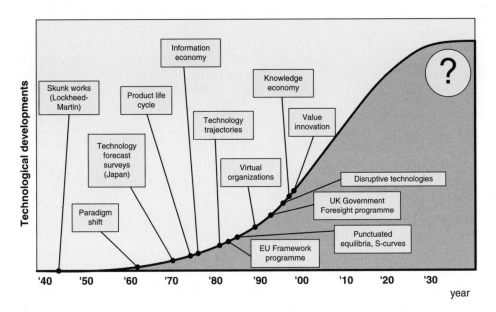

Figure 6.6   A timeline of developments in technology and management, Energy Foresight Network (2008) Institute of Studies for the Integration of Systems, Rome, Italy

Those changes cascade outward to affect businesses. New products, new policies, programs and regulations, or changed societal values can act as a sometimes unexpected enticement triggering the emergence of new dynamics within the business environment; and significant changes that have to be identified in a *timely manner*. The emphasis is on 'timely manner' because significant changes should be identified at an early stage, enabling a lead time for a company's preparation of strategic responses.

The key issues that any applied methodology needs to cover in order to enable the planning of the responses are:

(1) *identifying and monitoring change*;
(2) *analyzing the impacts of change*;
(3) *preparing a strategic response*;
(4) *defining adjusted policies, objectives and targets*;
(5) *implementing strategic responses and adjusted polices*.

Identifying and monitoring change involves the assessment of patterns of change in the past as well as collecting baseline data on current conditions against which future technological developments are monitored. This statement resembles what we have said and elaborated upon already in Part I Chapter 3, but in a different context. Such efforts provide a background against which we can look for ongoing cycles, trends, or emerging issues of change, such as technological innovation or societal value shifts or climatological aberrations.

Analyzing impacts of change means assessing both the effects that cascade from ongoing change throughout a company's macro environment and also evaluating what impact those effects have on its operations: how will day-to-day operations change? Who has been newly advantaged or disadvantaged by the advent of the change? What trade-offs might we face as a result of change?

Preparing a strategic response follows naturally from the extrapolation of trends and the consideration of the medium to long-term effects and impacts of emerging factors of change. Preparing a strategic response may involve scenario building and analysis that serves two purposes in technology foresight. First, scenarios of alternative future developments allow for exploring possibilities and uncertainties so that we may create contingency plans for the surprises the future might bring our way. Secondly, exploring alternative possible responses heightens the creativity and flexibility with which we imagine preferred future outcomes, because understanding potential alternatives helps us create richer visions.

Defining policy adjustments, adjusted objectives and targets involves creating models of the preferred future situations that we would like to create for the business. It is an act of visioning, which is the first step in response planning, setting objectives and providing leadership for achieving the targets. It requires a careful explicit articulation of our medium to long-term visions and missions and the values that contribute to them.

Implementing the desired polices and responses requires, first of all, a commitment to act. As already emphasized (Part I Chapter 1) this commitment must come firstly from the highest level of the company in order to champion actions which contribute to turning the vision into reality, and supporting the operational teams in finding the resources necessary to implement the plans which will result in the preferred outcomes.

These five activities are interrelated: data and actions flow from one to the next, and they are most effective when performed in concert, progressively and continuously. Once a significant change has been detected and a strategic response devised, the implementation plans must be effected and progress monitored, which in turn requires further monitoring of the changing emerging issues that have triggered the strategic response. Thus, these activities link back to create a full circle of foresight activities, an infinite loop.

As an organization begins the process of planning and implementing a response to an observed external change, it should monitor its own success in implementing the response actions and also keep an eye open for new external factors emerging on the horizon. Thus, the cycle continues with activities designed to identify, observe, and track changes, plan and implement responses and monitor success.

The disciplined practice of continuous performance monitoring and improvement, is precisely that which was introduced by implementing an EEMS, as described in Part I Chapter 5.

It is important to note that although strategic management, in a same manner as operational management, is a continuous process, the time scales may differ. While operational management almost as a rule would require daily data gathering and analysis in order to keep performance at its peak, strategic management, while it also needs reliable input data sets, would seldom require daily data gathering.

One of the main problems for strategic management, when managers have sought to drive strategy down and across their companies, was to transform strategies into practical metrics which could provide input for a meaningful analysis of the cause-effect relationships that, if understood, could provide valuable insight into operational decision making. This is the core problem of integrating strategic and operational management, which is only compounded by a lack of adequate data – both on the internal and external aspects of business operations.

At this point the performance measurement system comes into play again. As we have said earlier, only when based on hard and good data and information can the required strategic responses and operational management decisions be made. A well designed performance measurement system will ensure that a company's available performance data works for the benefit of specified goals and that the data provides information that is actually useful in achieving those goals. Integrated performance management, strategic and operational, needs reliably measured input in order to analyze the discerning patterns and meaning in the data, and to respond to the resulting information. Simply put, integrated performance management steers companies towards achieving their strategic goals by aligning strategic and operational objectives and by mobilizing and empowering people to achieve the agreed targets.

The backbone of an integrated performance management is an integrated performance measurement system. But, what is the *main barrier* to an integrated approach? In our experience, it is *organizational fragmentation!* This situation still persists in many companies where different business units and departments build walls between themselves, instead of creating and maintaining open and effective communication and coordination channels. Therefore, before proceeding further with integrated performance management, we will consider briefly the aspects of organizational fragmentation.

6.3.2 Organizational Fragmentation

The key for maintaining a company's competitive position is its overall performance in meeting customers' needs and maintaining or increasing its market share (Fig. 6.7). Organizations are usually designed to optimize each functional area, (Fig. 6.7a) with little reward for cross functional collaboration. In a typical organization, business units act independently of each other and make decisions based on the information that is relevant to them.

For instance, a procurement department may procure a fuel or a material that meets functional requirements and is cheaper at the moment, but the use of which will result in environmental impacts that would be more expensive to deal with in the future. Or, a sales department accepts small orders with differing product specifications that require the frequent stopping of production and retooling of machines, thus increasing energy consumption and the specific costs of production, consequently reducing or even diminishing the profits from such orders. Or, a production department prepares a production schedule that has unnecessarily high peaks in steam demand and as a result the utility department ends up with an oversized and expensive steam generation and distribution system.

Although a company may have a clear strategy related to its energy and environmental performance and overall profitability goals, all too often different functional areas rarely cooperate closely in meeting these strategic objectives. Besides the barriers between functional areas, there is often a divisive relationship between levels in the organizational hierarchy (Fig. 6.7b).

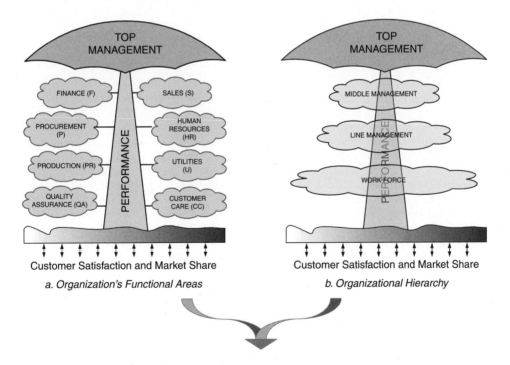

Figure 6.7   Functional areas, organizational hierarchy and organizational fragmentation

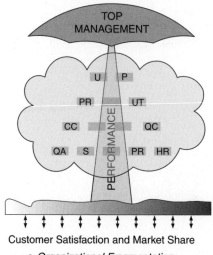

Customer Satisfaction and Market Share

c. Organizational Fragmentation

Figure 6.7   (*Continued*)

All of this can lead to organizational fragmentation (Fig. 6.7c) which creates a very difficult background for implementing any performance management system. If there is little or no encouragement for managers to make decisions that would benefit the organization as a whole and no sense of what is needed for mutual prosperity between employees it is hard to introduce a cross functional performance improvement program such as energy efficiency and environmental improvement or integrated performance management.

6.3.2.1 Data Fragmentation

Effective performance management and communication depends upon reliable and accurate information. Another consequence of organizational fragmentation is a fragmentation of data on an organization's performance (Fig. 6.8). Data is captured separately and kept separately at various business units according to their functional needs, while performance management normally requires simultaneous data inputs from several sections to be analyzed at the same time at regular intervals, in order to assess the energy and environmental or any other performance of the business as a whole.

Further, data fragmentation often leads to data redundancy. Various business units are collecting essentially the same data on their own and usually by different means, so that ultimately such data will differ and create more problems when analyzed by senior managers due to the inevitable inconsistencies. Such practices increase the cost of data gathering and processing and reduce the confidence that managers have in the data.

Organizational and data fragmentation increases monitoring costs and reduces a company's overall performance. For a company's performance data analysis to become a useful tool, however, it is essential that the company understands its strategic goals and objectives, i.e. that it knows the direction in which it wants the business to progress. The relevant performance indicators (PIs) need to be defined in order to assess the present state of the business and to indicate a future course of action. The emphasis is on 'relevant' because the value and importance of information must outweigh the cost of its acquisition and delivery. Identifying only the relevant performance data and determining how they are to be

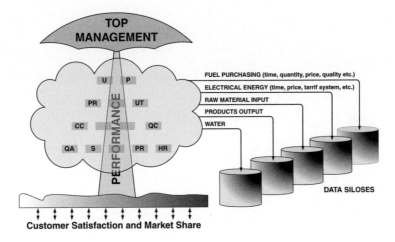

Figure 6.8   Data Fragmentation and Data Silos

measured enables companies to monitor performance across the board without getting lost in mountains of meaningless data; a scenario plaguing most companies today.

6.4 Integrated Performance Measurement System

Successful performance management depends upon the effective design and use of performance measurement. Organizations succeed when their business units and support functions work together to achieve a common goal. A well designed performance measurement system (PMsS), across business functions, is a cornerstone for the achievement of continuously improved performance at every level. The primary objective of an integrated performance measurement system (IPMsS) is to avoid the usual problems associated with data fragmentation, such as excessive cost and reduced reliability of gathering data on business performance.

Instead of gathering data and using it for a single purpose, a performance measurement system should be constructed so that data is gathered once from one source and than used in multiple places or for multiple purposes. This concept is illustrated in Fig. 6.9.

Therefore, designers of performance measurement systems need to take care to avoid data fragmentation and redundancy and secure a reliable and complete data base that will provide a platform for effective performance management.

Performance measurement – systematic data collection where accuracy is more important than precision – gives a comprehensive insight into activities and is to be utilized by employees and managers alike, in order to improve their performance. Performance measurement should be applied to every business process and function. There are no sustainable performance improvements without an adequate performance measurement system that should confirm, or otherwise, progress in meeting the objectives. The following questions can be used as a guide to determine performance measurements for each performance aspect and dimension that is to be monitored:

- What performance aspect or dimension has to be monitored?
- What has to be measured?
- Why it has to be measured?
- Where is the data source?
- How it will be measured?

- How can performance indicators be defined?
- Who will analyze and report on performance?
- To who are the reports are to be sent for comment and feedback?

With these questions answered, the performance measurement system can be developed by building upon the organizational structure already established by an EEMS, which is focused on energy cost centers (Fig. 6.2) and the people who work there. The above questions emphasize that the measurement is not just a technical issue but also necessarily involves people who will use data in order to manage the various performance aspects in the company.

For instance, quality management requires measuring customer satisfaction, rework and rejects. Production management requires measuring labor productivity, throughput quantities and cycle time. Environmental management requires measuring environmental impacts and regulatory compliance. Energy management requires measuring efficiency of resource use, including materials, energy and water at the point of use. Strategic performance management requires keeping an eye on competitors, new technological development, political and societal changes, and so on, Ultimately, everything comes down to the definition of appropriate performance indicators and the adequacy of the applied performance measurements.

An example of integrated performance measurement is provided by Table 6.2.

Table 6.2 is not and cannot be complete, because for any particular business it would have specific entries even under the same titles. But it demonstrates the approach to defining performance measurement which is the first step toward integrated performance management.

Performance measurement is where all the performance management systems ultimately meet each other. The IPMsS must be aligned with the performance improvement strategy, and if so it will influence behavior and encourage and enable implementation of the strategy. The essence of continuous improvement is to seek constantly ways in which products and processes can be made better. If additional improvement initiatives are to be endorsed, it would require upgrades of existing IPMsS, and not the introduction of a new one! As organizations grow in size, scope and complexity, integrated performance measurement systems become more and more important as effective providers of data and information on business performance.

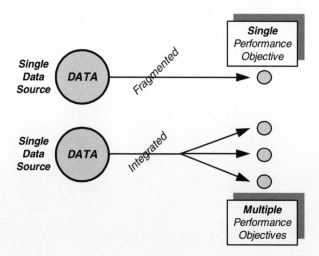

Figure 6.9   Fragmented and Integrated Performance Measurement

Table 6.2 A sample metrics for integrated performance measurement*

Performance aspect	What needs to be measured	Why	Where and how often	Performance indicator
I	\multicolumn{4}{c}{**Operational performance management**}			
Production output	• All types of products (P) or semi-products (SP) • Raw material (RM)	• To assess capacity utilization (C); • To be used for other performance indicators;	At all cost centers (CC), per day or per shift	$PI = \dfrac{RM}{P}$ $PI = \dfrac{RM}{SP}$
Energy efficiency	All energy types, including water (E)	• Increase efficiency at the point of use; • Reduce costs; • Reduce environmental impacts	At all cost centers (CC), per day or per shift	$PI = E$
Material productivity	• Raw material use (RM); • Additives (A); • Scrap (S); • Good products or semi-products	• Increase efficiency at the point of use; • Reduce costs; • Reduce environmental impacts	At all cost centers (CC), per day or per shift	$PI = \dfrac{A}{RM}$ $PI = \dfrac{S}{RM}$
Labor productivity	Good products or semi-products per shift (Sh)	• To improve skills; • To improve operational planning;	At all cost centers (CC), per day or per shift	$PI = \dfrac{P}{Sh}$ $PI = \dfrac{SP}{Sh}$
Machine productivity	• Machine output per shift (MO); • Number of cycles per shift (NC);	• To improve capacity utilization; • To avoid idling and holding;	At all cost centers (CC), per day or per shift	$PI = \dfrac{MO}{NC}$
Time productivity	Time required for completion of an operation (T);	• To improve skills; • To improve operational planning;	At all cost centers (CC), per day or per shift	$PI = T$
Quality	• Reworks (RW); • Rejects (RJ);	• Reduce costs; • Reduce environmental impacts;	At all cost centers (CC), per day or per shift	$PI = T$
Customer satisfaction	• Number of complains (NC); • Returned products (RP); • Warranty calls (WC);	• Reduce costs; • Reduce environmental impacts; • To improve skills; • To improve service;	Customer service centers, continuously	$PI = NC$ $PI = RP$ $PI = WC$

(continued overleaf)

Table 6.2 (*continued*)

	Performance aspect	What needs to be measured	Why	Where and how often	Performance indicator
	Environ-mental compliance	• Emissions to air (AE); • GHG emissions; • Waste water (WW); • Solid waste (SW) • Hazardous waste (HW);	• Waste minimization; • Cost reduction; • Environmental compliance; • Emission trading; • Social responsibility; • Green image; • Customers' goodwill;	At all cost centers (CC), per day or per shift	$PI = AE$ $PI = GHG$ $PI = SW$ $PI = HW$
	Cost effectiveness	All costs (TC) associated to an activity (C_i);	• Increase efficiency of resources use at the point of use; • Activity based costing; • Product costing;	At all cost centers (CC), per day or per shift	$PI = \dfrac{TC}{C_i}$
II	**Strategic performance management**				
	Margins in a product	• Cost of product (P); • Price of product (Pr);	• Relative importance of a product in products mix; • Shell we drop a product from the mix?	For all products, monthly	
	Overall profitability	• Total annual costs (TAC); • Total annual revenue (TAR);	• Value creation; • Shareholders' satisfaction; • Are we doing the right things? • Are we doing the things in the right way?	For the business, quarterly	
	Market share	• Annual sales of a product in a market (ASPM); • Our annual revenue from the product (ARPM); • Number of new entrants (NE);	• Are we among top three? • Where do we want to be? • Is there a threat from new entrants? • How we compare across the markets?	In a market or a jurisdiction, annually	
	Innovations	• Number of new products per year (NP); • % of TAR from NP;	• Are we innovative at all? • How do we compare with competitors?	For the business, annually	

Table 6.2 (*continued*)

New technological advancements	• Investments in R&D in new fields (R&D); • New ventures in these fields (NV); • Change in market capitalization of NV; • Annual sales of new products from NVs (ASNP);	• Is a new technology emerging? • Are new products being accepted? • How this affects our products line? • Is there a new threat? • Is there a new opportunity?	Globally and in particular markets where the company operates, annually and exploratory	
Political changes	• New international treaties signed • New local legislation enacted;	• Does it affect our supply chain? • Does it affect our technology? • Does it affect our products? • Does it affect our customers?	Globally and in particular jurisdictions where the company operates, annually and exploratory	

*At the level of each cost centers responsibilities for performance must be assigned, together with performance targets and reporting lines and frequencies.

Establishment of such a performance measurement system will initially require substantial resources in order to develop the knowledge and skill base needed to implement and use the performance measurements properly. But once such a system is established successfully, it provides a basis for systematic performance management across business functions enabling performance improvement in terms of quality, customer satisfaction, cost reduction, ultimately improving the profitability of the business.

A critical factor for the effectiveness of an integrated performance measurement system, as with the EEMS, is a company's ability to learn, to adjust and to capture the implicit knowledge that exists within the company. However, what is even more important for strategic management is to identify the pieces of relevant knowledge from the business environment that would shed light on the medium-term issues affecting business performance and which would support the preparation of adequate response strategies and polices.

6.4.1 Knowledge Discovery

Learning and knowledge capturing as a purposeful act was described in Part I Chapter 5 in the context of EEMS and the establishment of best operational practice through continuous performance improvement. The process was inward looking and focused on the company's operations. That is the main difference between that kind of learning and learning in the context of strategic management, where it should be outward looking and focused on the company's external environment.

This process starts with an awareness of the limits of the company's existing knowledge, of what we do already know and what we do not know. Discovering new knowledge that is relevant and important to the business is a process of purposeful exploration of how the changes in the company's environment may impact upon its business performance.

Again, a simple and an effective tool is a question! While interpreting performance data the main question was: 'Why?; but in this context it should be 'What if?'. Such a question triggers exploration of the possible outcomes of changes that may emerge and shape the process of strategic thinking. Exploring emerging changes in the external environment brings discoveries of patterns, develops an understanding of the cause-consequence relationship, and ultimately generates knowledge of how to enhance the capabilities and improve the competencies of the business and how to harness new opportunities and provide increased value for the customers.

The whole purpose of this process is to identify early fundamental changes in the external business environment and to search for effective responses to these changes. Management must react to external threats, no matter what the cause, in a timely manner and for that they must have the correct and timely information. The discovery of knowledge process provides useful information for timely decision making.

6.5 Integrated Performance Management

Performance measurement will not improve performance by itself. We still need to interpret performance data, to implement corrective actions and more than anything: we need people to change the way that they do things in order to achieve lasting performance improvement. This is again the same as with energy and environmental management, or indeed as with any performance management system.

But many companies do not have the experience, tools or established processes to address complex performance problems or to assess with confidence whether some new performance improvement project will deliver acceptable results. Performance problems can stifle growth, increase costs, reduce productivity, cause customer dissatisfaction and ultimately make a company less competitive.

An energy and environmental performance management system (EEMS) through its measuring, monitoring, controlling and efficient communication structures, as described here, can provide a foundation for the continuous performance improvement of production processes, which is a precondition for competitiveness. Based on these structures and foundations an integrated performance measurement system should be developed enabling combined strategic and operational management aiming at technical mastery, managerial excellence and business profitability and improvement in competitiveness .

An integrated performance management system (IPMS, Fig. 6.10.) provides an *infrastructure* that supports all management needs and enables the continuous evaluation of business performance. It reduces

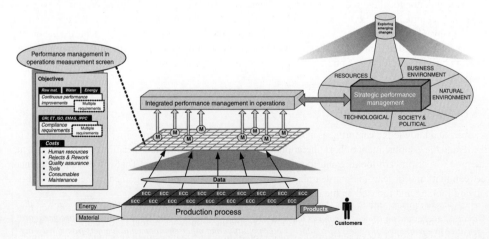

Figure 6.10   Integrated performance management system

redundancy in monitoring and reporting work and related equipment and ensures the seamless integration of data and information across business functions.

Let us emphasize once again what this infrastructure consists of:

1. Organizational provisions such as:

   - the conspicuous commitment of top management secured;
   - adequate policies in place,
   - established cost centers,
   - assigned responsibilities and reporting lines,
   - training needs assessment carried out,
   - motivation and awareness program designed.

2. Current position established and performance improvements targets set.
3. Strategic and operational performance indicators defined.
4. Integrated performance measurement system installed.
5. Implementation action plan approved.
6. Training carried out.
7. People empowered to make decisions and provide initiatives.
8. Pilot run of integrated performance management system (IPMS) kicked off.
9. Internal communication and coordination lines established.
10. Road wide open for continuous performance improvements, organizational learning and discovery of knowledge.

Therefore, integrating critical performance management elements such as a performance measurement system, performance indicators, organizational structures and a culture of continuous improvement, trained and empowered people at all levels and at all business functions and purposeful learning and knowledge discovery procedures, into a comprehensive solution in order to provide a framework for daily monitoring and improvement of business performance makes common sense.

Our case study of a textile factory, which was introduced in Box 6.1 with considerations of operational performance improvement, continues in Box 6.2 with the strategic management response to the threats from the company's technological, business, social and political environment.

Box 6.2: A textile factory – the case continue (from Box 6.1)

> *Additional issues considered at the same time as the initial customer complaints (see Box 6.1) on quality and delivery, were reflecting more on strategic business concerns. This very case evolved in South East Asia during the second half of the 1990s, when it became apparent that in the textile industry new low-labor cost competitors were arriving. The company recognized that in the medium term it would not be able to compete based only on cheap labor, because cheaper labor was already on offer. Therefore they realized that it was necessary to differentiate their products by increasing the quality and value of services to the customers, in order to remain competitive!*
>
> *Further, during the economic boom of 1980s, environmental issues were non-issues and nobody paid too much attention to clean production. It was the same with energy. The main focus was on production growth and energy efficiency was not an issue for top management.*
>
> *However, this situation started to change. At first, new regulations on energy conservation were introduced, which forced the company to act. This was the trigger that brought us to work with the company and as the result an energy management system was implemented by the company. Then, local communities became more sensitive to environmental issues, particularly water pollution and other environmental impacts from the dyeing plant.*

Box 6.2: Continued

> *The customers were expecting better service, because of stronger competition among textile manufacturers. Additionally, some customers started enquiring about the company's environmental management practices, in accordance with their own environmental policy of buying from 'clean' producers.*
>
> *The company's technology was gradually becoming obsolete as compared to new machines that had higher outputs, lower energy consumption, especially for dyeing machines, and lower water consumption, therefore reducing environmental impacts.*
>
> *All of these strategic issues and considerations were assessed at the same time as specific operational issues, and were the main reason that the company decided to undertake significant investment in the bar code tracking system and new dyeing machines, as pointed out already. They also decided to extend the already existing energy management system, in keeping with environmental issues and to go for ISO 14000 certification.*
>
> *This is an example of an integrated consideration of strategic and operational issues and the formulation and implementation of responses to strategic threats and immediate operational problems that have strengthened the company's competitive position as against emerging competition, against societal changes and against customers' increasing demands for quality and environmentally responsible manufacturing.*

The example in Box 6.2 shows how integrated performance management transforms conceptual and logical models into a practical working solution that optimizes the use of technical and human resources and enables the capturing the knowledge. Technical excellence achieved through learning by doing and related knowledge creation and innovation, enhanced capability for communication, product quality, work safety and an environmentally sound production, together with low energy costs, were the results achieved in that case study by an integrated approach to improvement of the company's performance.

Additionally, demand for reliable information about the social and environmental performance of companies is increasing. Intensified public concern and mandatory and voluntary reporting obligations, as already described, create additional data requirements for a company's social responsibilities and business and environmental performance (Fig. 6.11). Civil society, customers and investors increasingly expect businesses to demonstrate that they behave in a responsible way. More and more companies

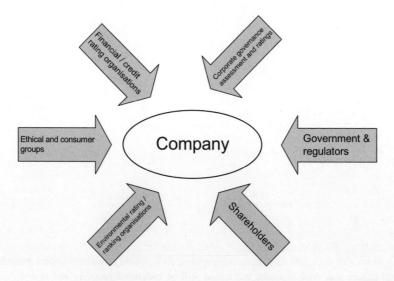

Figure 6.11 Organizations and groups generating the reporting requirements of a company

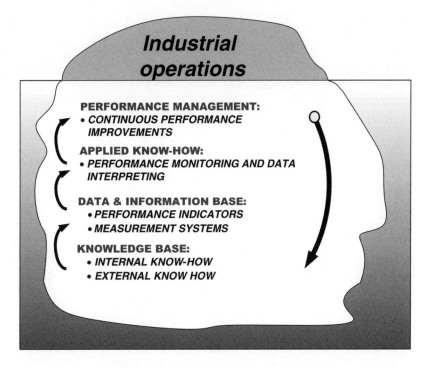

Figure 6.12   Performance Management 'Ice Berg'

recognize that sound performance measurements and an integrated performance management system can help them to manage the social issues effectively.

However, reporting in itself is just the tip of the performance management 'ice berg' (Fig. 6.12). The main challenge is to know how to design and apply a comprehensive performance management system that will deliver continuous performance improvement, thus realizing a company's strategic objectives, satisfying regulatory requirements and addressing public concerns about environmental issues.

At the bottom of the iceberg, there will always be an effective knowledge base, consisting of internal and, if necessary, external know-how that supports everyday performance management. New knowledge is being created continuously by learning through experience, i.e. organizational learning and the knowledge base is being updated and upgraded.

If we combine the issues of organizational fragmentation, operational and strategic management, organizational learning and business ethics, we acquire another view of integrated performance management. This is shown by Figure 6.13.

An integrated performance management system focuses on optimizing the productivity and profitability of a business as an integrated unit, including the internal and external business aspects and environment, and it requires adequate *business strategies* (Fig. 6.14). These strategies rely on people who ultimately deliver performance improvement and customer satisfaction and who create value and knowledge, thus contributing to organizational learning.

Deployment of such a strategy will integrate management processes in order to ensure that planning and decisions are done by examining the same databases, assumptions and methodology that include continuous improvement techniques, avoiding gaps but also avoiding overlapping and redundancy.

This will ensure that ALL resource utilization will be measured properly and that the supply and demand side of the business will be managed and coordinated properly. Inefficient utilization of resources

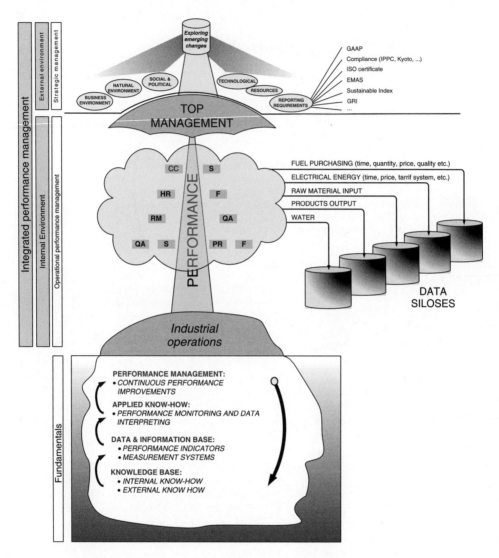

Figure 6.13 Knowledge base, internal and external aspects of integrated performance management

drives costs up and customer service down. Continual improvement of interrelated processes, individual skills and associated management systems is a permanent organizational objective that requires the continuous measurement and evaluation of results and satisfaction. Continuous improvement leads to quality enhancement and cost efficiency, thus improving productivity and improving overall business performance.

The focus is on results – improving competitiveness in the marketplace and improving the overall business performance of the company. The goals of the IPMS are the long-term growth of the business, timely adjustments to a changing world and global marketplace, purposeful learning and discovery of knowledge and continuous '*profit-ability*' improvement.

Figure 6.14   A Strategic Approach to Integrated Performance Management

6.6 Conclusion

> *Knowledgeable and Confident People!*

Energy is the life's blood of any business. Every business activity, be it a manufacturing process or an administrative task, requires energy in some form and amount. Therefore energy flows to every corner of a business. Similarly, wherever a business activity takes place, an impact on the environment may occur, significant or not. Hence energy and environmental issues, by their nature, cross the boundaries of business units and functional areas within an organization, therefore energy and environmental performance is everybody's concern. At the same time, people work in every business unit and every functional area; they use resources, operate machines, craft the products and provide services. Ensuring attention and care with regard to performance issues from the entire workforce in a company is the main challenge for achieving the good energy and environmental performance of a business as a whole.

These efforts should be extended to other business performance categories and integrated into the overall performance management process. Even if the most rudimentary integration is achieved, it can lead to significant reduction in costs, as well as significant increases in the ability of top management to 'see through' the organizational hierarchy and across the business functions and achieve a more accurate insight into the performance of specific activities. When an EEMS starts to produce results, top management becomes enthusiastic about the enhanced control of operations and the clarity of purpose and responsibility throughout the organization.

Therefore an EEMS is more than just an energy and environmental management system. It is people management, organizational behavior and culture management. It is a way of life for companies that want to achieve excellence. These companies would seek to establish their own unique corporate cultures and identities in order to survive and win in a global market place that brings new competition from all over the world, all of the time. Success will be derived from a deep organizational understanding of the underlying business processes, activities, and operations on the one hand and IPMS philosophy as proposed here, on the other.

At the core of companies' success and survival are knowledge and knowledgeable people able to use it, create it and capture it! In this global management environment, innovative thinking, innovative approaches and more effective management principles are needed. Part I of this book is dedicated to that need.

6.7 Bibliography

Cortada, J.W. (ed.) (1998) Rise of Knowledge Worker, Butterworth Neineman.

David, F. (1989) Strategic Management, Columbus Merrill Publishing Company.

Elliot, D. (1997) Energy Society and Environment, Routledge.

Hatten, K.J., Rosenthal, S.R. (2001) Reaching for the knowledge edge, AMACOM.

Holland Smith, D. J., Kemp, N.M. (2003) Text Analysis for S&T Intelligence, Innovation and Creativity: Final Customer Report. QinetiQ Customer Report.

Huzynski, A., Buchanan, D. (1991) Organizational Behavior, Prentince Hall.

Lamb, R.B. (1984) Competitive Strategic Management, Englewood Cliffs, NJ: Prentice-Hall.

Neely A. (1998) Measuring Business Performance, The Economist Books.

Oxelheim, L., Wihlborg, C. (1997) Managing in turbulent world economy, John Wiley & Sons.

Porter, M.E. (1980) Competitive Strategy – Techniques for Analyzing Industries and Competitors, The Free Press.

Swedish technology foresight, http://www.tekniskframsyn.nu/.

Part II
Engineering Aspects of Industrial Energy Management

1

Introduction to Industrial Energy Systems

1.1 Introduction

Industrial energy consumption accounts for roughly one third of the total energy consumption in the world. The constant growth of the price of energy and the increase of its share in the price of a unit product demands greater attention from the factory's management with regard to controlling energy consumption. This involves increased investment but it also produces significant benefits. A number of processes for energy transformation are used in industry, as well as processes in which energy is used directly as a raw material for the production of new added value. Production processes present a demand side for a factory energy system – therefore, production processes set the requirements for energy quantity and quality.

The basic principles for optimizing energy performance are to continuously monitor energy flows and relate the measured amount of energy used by a process or activity to the measured output of this process or activity (Fig. 1.1). As we have said earlier, wherever and whenever energy performance improvements are achieved, environmental performance will be improved simultaneously.

Industrial energy systems, also commonly called 'utilities', provide the energy needed for the conversion of raw materials and the fabrication of final products. Industrial energy systems transform fuels and power into a variety of energy utilities such as steam, direct heat, compressed air, chilled water, hot fluids and gases and power for compressors, fans, pumps, conveyors and other machine-driven equipment (Fig 1.1). Some industrial facilities have on-site generation of electricity or co-generation of electricity and heat for processes.

All manufacturing processes rely on energy supply. In energy intensive basic industries, such as the chemical industry, petroleum refining, iron and steelmaking and pulp and paper-making, energy systems are the backbone of the manufacturing process and are crucial for profitability and competitiveness. For these industries, changes in the efficiency and environmental performance of critical energy systems can impact significantly on the cost of production. With the growth of energy prices even within industries with lower energy intensity indicators, the significance of energy is becoming increasingly important.

Figure 1.2 illustrates the typical energy system of a factory. The total quantity of final energy entering into the factory is 100 units within a specified period of time (most often a year). The relationship between fossil fuels and energy produced in power plants in this example is 80:20. The factory is usually supplied by electrical energy from an outside power system but the outside supply of hot water or steam for heating and other technological processes is also possible. The ratio presented here is typical for the food and beverage industry. In petroleum refining industry, for example, this ratio is 96:4, and in the heavy machinery industry it is 55:45. For developed countries, this ratio for the whole of industry is around

Applied Industrial Energy and Environmental Management   Zoran K. Morvay and Dušan D. Gvozdenac
© 2008 John Wiley & Sons, Ltd

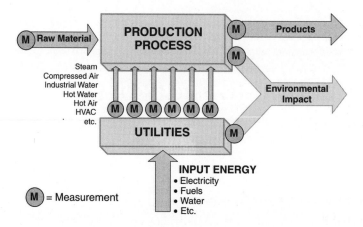

Figure 1.1   Basic Relationship of Energy and Production

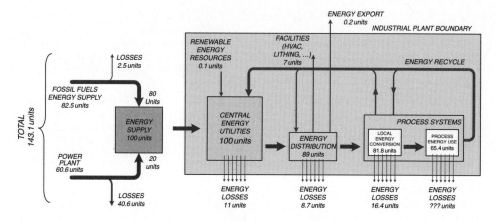

Figure 1.2   Typical Energy Flows in a Factory

85:15. Today, there is a trend in the world to increase one's own production of electricity within the industrial plant, which improves the cost effectiveness of used fossil fuels significantly.

In the example given in Figure 1.2, the efficiency of the power supply is 33 %, that is, for the production of 20 units of required energy, it is necessary to utilize 60.6 units of energy in the power plant. This figure varies and it mostly depends on the structure of the national electrical energy system and the fuels used in power plants. However, we are interested primarily in energy transformations which occur within the boundaries of the industrial plant. The incoming 100 units have their price and their structure depends on the type of production process, the available sources of energy, current regulations and the like.

Basic energy transformations are carried out in central energy utilities and imported energy is adjusted there for the production process which has to be performed. For example, the voltage of bought electricity is reduced (usually 0.4 kV), the pressure of natural gas is adjusted before entering the boilers, etc. Necessary forms of energy flows leave central energy utilities and are distributed to the production process where part of the energy is further transformed, the main part being used for the creation of new values – products.

Energy losses are unavoidable but often measurable. The first task of energy management is to identify and then minimize the amount of losses in the internal energy supply chain.

When energy transformations are carried out and when energy is supplied to the location of the end use in production, the delivered amount is reduced significantly in comparison to the that which has entered into the factory. In the example shown in Figure 1.2, out of the total 100 units only 65.4 units reach the production so it can be concluded that the efficiency of the energy system of the factory presented in Figure 1.2 is only 65.4 %. This does not take into consideration losses which occur at final utilization of energy in the production processes. However, in order to achieve maximum overall efficiency of energy use in an industry, it is necessary to account for the efficiency of energy use in production processes as well.

1.2 Industrial Energy Systems Analysis

In practice, unfortunately, the analysis of industrial energy systems is often reduced to checking the efficiency of main utilities, such as boilers, chillers or air compressors and the environmental impacts thereof. It is obvious that a simplified form of analysis can yield only partial results and cannot provide full benefits and inputs to a comprehensive energy and environmental management program. Often, the role and impacts of the human factor in energy and environmental performance are completely neglected.

Therefore, to avoid a piecemeal approach, the required scope for consideration of the industrial energy systems is given in Figure 1.3. It is obvious and common sense, that this covers all aspects of energy supply including distribution, metering and monitoring; control arrangements, end-use and recycling if there is any.

On top of this are people and everyday housekeeping, maintenance and operational practices. The energy system analysis aims at identifying opportunities to reduce energy costs. It starts from an understanding of the relationship between the supply and demand side of processes in the company's individual plants and departments and recognizing issues that are relevant for energy consumption.

A *systematic approach to industrial energy systems analysis* requires the study and understanding of the interactions among the main factors influencing energy performance, i.e.

- production;
- utilities;
- people; and
- technology.

It is focused equally on the physical aspects of energy supply and consumption and on the human aspects related to the operation and maintenance of facilities. Generally speaking, the scope of energy system analysis includes examination of the following:

- energy conversion (in boilers, transformers, compressors, furnaces, etc.);
- energy distribution (of electricity, steam, condensate, compressed air, water, hot oil, etc.);
- energy end-users' efficiency (equipment and buildings);
- waste minimization and reuse, recovery or recycling (energy, water);
- production planning, operation, maintenance and housekeeping procedures;
- management aspects (information flow, data analysis, feedback, education of employees, their motivation, etc.).

Therefore, the emphasis is not only on the singular aspects of the efficiency of a particular energy utility generation and distribution but also on the integration of the energy needs of a particular production

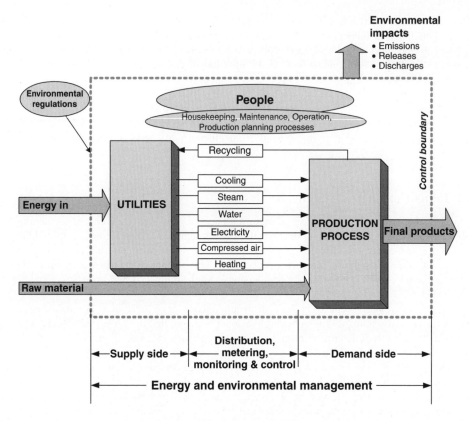

Figure 1.3   Extent of Industrial Energy Systems

process, the supply side options available to satisfy these needs, and management techniques, which tie together all the resources *(people, energy, technology)* in a cost effective way.

In this book we are going to apply this systematic approach to the consideration of the following industrial energy systems:

- Industrial Steam Systems;
- Electrical Power Systems;
- Compressed Air Systems;
- Refrigeration Systems;
- Industrial Co-generation.

With the exception of industrial co-generation, all of the other systems are very common in industry. Industrial co-generation is considered separately as it can substantially increase the efficiency of energy transformations not only at the factory level but also at the national economy level if it is used on a large scale. Recognizing that fact, many countries have started to promote co-generation in an organized manner and have elaborated mechanisms for the promotion of co-generation in industry and other sectors of the economy.

Electrical systems are unavoidable in industry and industrial steam systems are often present in many industries. Compressed air and refrigeration systems are both large consumers of electrical energy. Each of these systems contains a number of electric motors which drive pumps and fans. There are also a large number of heat exchanges of various types which are driven by more or less complex process requirements. All of these individual elements, as well as the overall system design, impact directly on energy efficiency. The individual elements of industrial energy systems are discussed separately in Part III – Toolboxes.

2

Industrial Steam System

2.1 System Description

Steam systems generate and distribute thermal energy in the form of steam, which is then used for a variety of process applications. On average, over 40 % of all the fuel burned in industry is consumed in order to raise steam. Steam is used to heat raw materials and treat semi-finished products. It is also a source of power for equipment, as well as for heat and electricity generation. Many companies can improve steam system performance through the installation of more efficient steam equipment and processes. The entire steam system must be considered in order to optimize energy use and achieve cost savings. A typical steam system (Fig. 2.1) includes the following subsystems:

(a) boilers;
(b) steam distribution system (including control valves, steam traps, insulation, etc.);
(c) steam end-users (including control systems, steam traps, insulation, etc.);
(d) condensation return system (including piping, storage tanks, insulation, pumps, etc.);
(e) metering, monitoring and control systems.

The basic characteristics of each of the above-mentioned subsystems and their energy performances will be analyzed separately. For each subsystem, the mass and energy balances that define its energy efficiency need to be established. The measurement points for steam system monitoring and control are indicated in Figure 2.1. These are the minimum measurement requirements for practical operational performance monitoring and analysis of the steam system as a whole. All of these measuring points often do not exist within companies, but even if they do exist, it is advisable to check the calibration records and assure oneself about the accuracy of the instrumentation.

2.1.1 Boilers

A boiler is a device that converts the chemical energy of fuel into useful heat output. Typical heat outputs are steam (saturated or superheated), hot water or thermal fluids like mineral oil.

There are many different types of boiler but all of them can be classified in two basic groups:

(a) *Water tube boiler* where water is contained in tubes and flames and hot combustion gases pass around them,
(b) *Fire-tube or shell boiler* where combustion gases pass through a furnace tube and after that enter tube bundles immersed in water within the shell.

Applied Industrial Energy and Environmental Management   Zoran K. Morvay and Dušan D. Gvozdenac
© 2008 John Wiley & Sons, Ltd

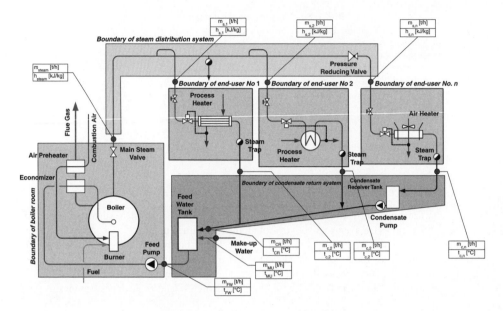

Figure 2.1   Overall Steam System Definition

Most boilers have a furnace chamber where the majority of the heat is transferred directly from the flame by radiation. After the furnace chamber the flue gas goes through passages where heat is transferred primarily by convection. Two-thirds of heat transfer takes place in the furnace and the remaining third in the flue gas passages.

Most of the boilers in industry are used for the production of saturated low and medium pressure steam. If operated correctly, all types of modern boilers are more or less equally efficient at converting fuel into steam, hot water or hot oil. Table 2.1 shows the expected thermal efficiencies obtainable for different boiler types, based on the Gross Calorific Value (GCV) of the fuel.

More detail on various boiler types and designs can be found in the literature (see References).

2.1.2 Steam Distribution System

This subsystem delivers produces steam for end-users. Steam flows depend on the needs of end-users. Steam condensate occurs due to the inevitable cooling of steam in pipelines and it is necessary to remove the condensate from the pipeline at certain points. For that purpose, steam traps are used, not only in the steam distribution system but also at the end-users. The types and characteristics of steam traps can be found in the available literature or directly from the manufacturers.

The insulation of steam distribution system is very important for the good performance of this subsystem and for the high efficiency of the whole steam system. Part III Toolbox 10 and Part III Toolbox 13 Software 8 deal with industrial insulation problems with all the necessary detail.

2.1.3 Steam End-users

There are a variety of steam end-users in industry and each of them should be the subject of energy management in respect of monitoring and estimating their individual effectiveness and energy efficiency. We are not going to enter into a detailed analysis of various end-users. However, for the proper operation

Table 2.1 Boiler Efficiency According to Boiler Type (based on GCV). Reproduced from: Good Practice Guide No. 30 (1993) Energy Efficient Operation of Industrial Boiler Plant, Energy Efficiency Office

Boiler Type	Efficiency, [%]
Condensing	88–92
High Efficiency Modular	80–82
Shell Boiler – Hot Water	78–80
Shell Boiler – Steam	75–77
Reverse Flame	72–75
Cast Iron Sectional	68–71
Steam Generator	75–78
Water Tube with Economizer	75–78

of any end-user, it is necessary to deliver the required steam flow at the corresponding pressure. This parameter is established by the manufacturer. The key issue in the efficient handling of any end-user is to respect the manufacturer's instructions for operation, to use it at the designed capacity, and to set up and implement a proper maintenance plan.

2.1.4 Condensate Return System

One of the oldest measures for increasing energy efficiency in industrial steam systems is condensate recovery. In the past, this measure did not draw much attention as energy was relatively cheap and condensate recovery systems comparatively expensive. In some processes, steam is used directly in the process and it is then not possible to consider condensate recovery. It is the same when the steam may be polluted by hazardous substances from the process and when for safety reasons condensate is discarded. However, even in such cases it is possible and desirable to utilize the condensate energy value by means of heat exchangers.

2.2 System Performance Definition

The *overall system energy efficiency* can be defined as the ratio of the energy delivered to and used by all end-users, and the energy supplied to the boiler by fuel.

$$\eta_{SS} = \frac{\text{Energy Used by End-Users}}{\text{Energy Supplied to System by Fuel}} = \frac{E_{EU}}{E_{FUEL}} \tag{2.1}$$

Energy efficiency defined in this way is in fact an energy performance indicator of the observed steam system (Fig. 2.1). Performance indicators are discussed in details in Part I Chapter 2.

Energy delivered to end-users can be defined as follows:

$$E_{EU} = \sum_{n=1}^{N} \left(m_{s,n} \cdot h_{s,n} - m_{c,n} \cdot 4.21 \cdot t_{c,n} \right) \quad (n = 1, 2, 3, \ldots N) \tag{2.2}$$

Where:

$m_{s,n}$ = flow rate of in-coming steam to end-user No. n, [t/h];

$h_{s,n}$ = enthalpy of in-coming steam, [kJ/kg] (enthalpy of water at 0.1 °C and 0.006113 bar (triple pint of water) is zero).

$m_{c,n}$ = flow rate of out-coming condensate from end-user No. n, [t/h];

$t_{c,n}$ = temperature of out-coming condensate, [°C];

4.21 = isobaric specific heat of water, [kJ/(kg °C)] (This value will be used for all calculations in this chapter. It is the average value of water specific heat for the temperature range which appears in industrial steam systems).

The mass flow rate of steam is equal or greater than the mass flow rate of out-coming condensate. The condensate flow rate can be zero if steam is used in the process and in that case it has to be fully compensated by make-up water.

The chemical energy of fuel used for steam production is as follows:

$$E_{FUEL} = M_{FUEL} \cdot GCV \qquad (2.3)$$

In this equation, M_{FUEL} is the flow rate of fuel and GCV is the Gross Calorific Value of fuel used for running the boiler. A more precise definition of fuel energy is discussed in Part III Toolbox 5.

Finally, we can rewrite the equation for the calculation of steam system energy efficiency (Eq. 2.1) as follows:

$$\eta_{SS} = \frac{\sum_{n=1}^{N} \left(m_{s,n} \cdot h_{s,n} - m_{c,n} \cdot 4.21 \cdot t_{c,n} \right)}{M_{FUEL} \cdot GCV} \qquad (2.4)$$

Similarly, it is possible to define the energy efficiency of the subsystem shown in Figure 2.1. In this way, the following efficiencies (performance indicators) are obtained:

- **For BOILER**

$$\eta_B = \frac{m_{steam} \times (h_{steam} - 4.21 \cdot t_{FW})}{M_{FUEL} \cdot GCV} \qquad (2.5)$$

- **For STEAM DISTRIBUTION SYSTEM**

$$\eta_{SD} = \frac{\sum_{n=1}^{N} m_{S,n} \cdot h_{S,n}}{m_{steam} \cdot h_{steam}} \qquad (2.6)$$

- **For CONDENSATE RETURN SYSTEM**

$$\eta_{CR} = \frac{m_{CR} \cdot 4.21 \cdot t_{CR}}{m_{MU} \cdot 4.21 \cdot t_{MU} + \sum_{n=1}^{N} m_{c,n} \cdot 4.21 \cdot t_{c,n}} \qquad (2.7)$$

where:

m_{steam} = mass flow rate of steam, [t/h];

h_{steam} = steam enthalpy, [kJ/kg];

m_{CR} = mass flow rate of condensate, [t/h];

t_{CR} = temperature of condensate, [°C];

m_{MU} = mass flow rate of make-up water, [t/h];

t_{MU} = temperature of make-up water, [°C].

2.2.1 The Cost of Steam

It is well known that the introduction of planned technical and operational improvements depends ultimately on economic factors. This requires the determination of the costs and benefits of proposed changes.

The main factor for this evaluation is the price of fuel and consequent cost of steam. The cost of ton of steam is also an important performance indicator for the steam system. The best way for determining it is to measure continuously steam generation and fuel consumption. If these data are available, the cost per ton of steam is as follows:

$$CS = \frac{M_F(\tau) \cdot FC}{m_S(\tau)} \ [US\$/t] \tag{2.8}$$

where:
CS   = cost per ton of steam, [US\$/t]
$M_F(\tau)$ = flow rate of fuel, [l/h, kg/h, nm³/h, etc.]
FC   = unit cost of fuel, [US\$/l, US\$/kg, US\$/nm³ etc.]
$m_S(\tau)$ = steam flow rate delivered to the process, [t/h]

Typically, the cost of fuel is obtained from fuel invoices. They represent the real cost of fuel as they also include transportation, taxes, etc. Another important parameter of fuel is its properties. Most fuels maintain consistent properties and are supplied based on certain specifications. In general, coal and biomass can have the widest range of properties. Other common fuels are delivered with certain tolerances. More detail concerning fuel readers can be found in Part III Toolbox 5.

Electricity costs in a boiler house are small compared to fuel cost, and they are mostly incurred by fans and pumps. The costs for boiler feed-water treatment also contribute to the cost of steam. The calculation of a unit energy price for various fuels will be demonstrated by the example of a vegetable oil factory (Table 2.2). The fuel prices are calculated on the basis of monthly bills and total measured monthly consumption.

Table 2.2   Example of Average Fuel Prices (year 2002)

Natural Gas		
Net calorific value (NCV)	MJ/nm³	35.71
Gross calorific value (GCV)	MJ/nm³	39.74
Density (15 °C; 1.013 bar)	kg/m³	0.692
Unit price	US\$/nm³	0.1923
Unit energy price (based on GCV)	US\$/kWh	**0.0174**
Heavy Fuel Oil		
Net calorific value (NCV)	MJ/kg	40.03
Gross calorific value (GCV)	MJ/kg	42.59
Density	kg/l	0.95
Unit price	US\$/l	0.2076
Unit energy price (based on GCV)	US\$/kWh	**0.0185**
Brown Coal (dried)		
Net calorific value (NCV)	MJ/kg	31.20
Gross calorific value (GCV)	MJ/kg	32.11

(continued overleaf)

Table 2.2 (*continued*)

Brown Coal (dried)		
Density	kg/m^3	–
Price	US$/kg	0.0698
Price (based on GCV)	US$/kWh	**0.0078**
Sunflower Shells		
Net calorific value (NCV)	MJ/kg	15.91
Gross calorific value (GCV)	MJ/kg	16.93
Density	kg/m^3	–
Price	US$/kg	0.0252
Price (based on GCV)	US$/kWh	**0.0054**
Silo Waste		
Net calorific value (NCV)	MJ/kg	10.00
Gross calorific value (GCV)	MJ/kg	10.64
Density	kg/m^3	
Price	US$ /kg	0.0100
Price (based on GCV)	US$/kWh	**0.0034**

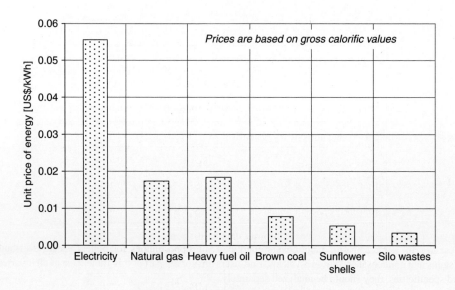

Figure 2.2   Example of Unit Energy Prices

Unit energy prices per kWh for all used energy carriers in that factory and electricity price are presented in Figure 2.2.

The annual primary heat energy consumed in this factory is 147 348 MWh. The majority of this energy is used for an annual steam generation of 63 206 t. An additional small amount of natural gas (only 2.6 %) is used for driers. Using the above-mentioned unit prices and the known participation of each of energy carriers in steam generation, the total cost of energy for steam generation is 1 906 566 US$. The efficiency of energy transformation is incorporated implicitly in generated steam. Maintenance, labor and other associated costs of the steam generation are not included in this amount. Now, the unit price of steam is, for a given example, as follows:

$$C_{Steam} = \frac{1\,906\,566}{63\,206} = 30.2\,[US\$/t] \qquad (2.9)$$

The value of steam unit cost is a practical performance indicator. Other performance indicators can be boiler efficiency, specific steam consumption per unit of product, and others, which will be described later.

2.3 Principles of Performance Analysis

The energy performance of a steam system has to be evaluated by analyzing all of the system components. Figure 2.3 suggests a systematic approach to performance analyses of a steam system. There are some activities within this procedure that are common to all system components, but one of them must be stressed. This is on-site measurement, which has to be carefully prepared, conducted and reported (see Part I Chapter 4).

We suggest the following steps in the steam system analysis:

Step 1: Determination or estimation of fuel cost and other costs related to the operation of the steam system and analysis of the production process organization according to steam consumption.

Step 2: Boiler operation investigation. This analysis has to be focused on evaluating the fuel to steam conversion efficiency of the boiler. It has to be based on existing data from daily log sheets and additional measurements of boiler operation.

Step 3: Measurement, calculation or estimation of the steam consumption of end-users and investigation of the operational practices of each of them.

Step 4: Investigation of energy losses throughout the distribution system.

Step 5: Investigation of condensate return.

The proposed approach can be changed and adapted in the ways most suitable for the particular steam system. The analyses can be performed simultaneously for all system components or starting with any one of them, but all conclusions, recommendations and implementations have to be reconciled in order to assess performance of the entire steam system.

2.3.1 Performance Improvement Opportunities

At the outset of preparations for steam systems performance analyses, it is useful to have in mind what are the potential opportunities for performance improvement. Such an awareness can guide the analysis and serve as a check list for the physical inspection of facilities. The most common opportunities are presented in Table 2.3. This table covers energy saving opportunities for boilers, steam distribution systems and condensate return subsystems. End-users are not the part of this Table. Due to their varieties and specificities, they should be analyzed separately.

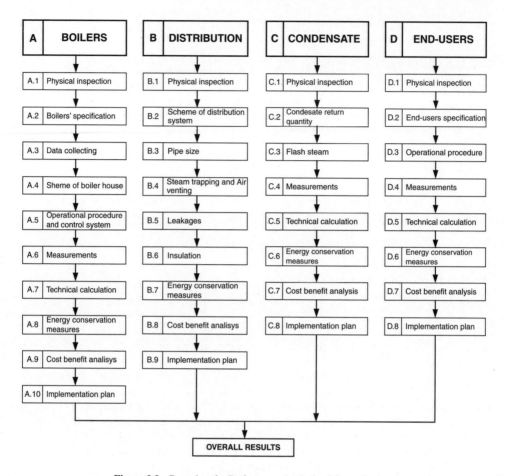

Figure 2.3 Procedure for Performance Analysis of Steam System

Table 2.3 gives the approximate values of the possible effects of performance improvement. Certainly, the mentioned percentages do not add up and they present only individual potentials for energy savings. For example, water treatment improvement will automatically reduce boiler-blow-down loss. Most of above-mentioned techniques or methods for improving energy performance refer to boilers (Table 2.3). This is common sense since the basic energy transformation is performed in boilers.

2.3.2 Example 1: Steam System Energy Performance Analysis

Let us consider the following real example. A group of end-users consumes a total of 5.34 t/h saturated steam of 6.2 bar. The medium temperature of condensate after the performed production process is 85 °C. All values are measured within the monitoring program. It is estimated that the total loss of steam of all end-users is around 10 %. When these values are given, we can calculate total heat consumed as follows:

$$E_{EU} = m_{s,EU} \cdot \frac{1000}{3600} \cdot h_{steam} - m_{c,EU} \cdot \frac{1000}{3600} \cdot 4.21 \cdot t_{c,EU} = 3631.01 \ [kJ/s] \ or \ [kW] \qquad (2.10)$$

Table 2.3 Common Performance Improvement Opportunities of Industrial Steam Systems *(Implementation of all cited techniques and measures does not lead to cumulative energy savings)*

	Technique or Method	Description	Potential for energy Efficiency Improvement relevant to Current Consumption
1.	Operation procedure and maintenance of boilers	When more than one boiler is in operation, sequencing control systems have to be analyzed. Maintenance must be fully in compliant with the equipment manufacturer's recommendations. Optimization of de-aerator vent rate will minimizes loss of steam. Monitoring system has to cover all relevant performance indicators (fuel steam and feed water flow rates, load profile of each boiler, steam pressure, water quality. etc.).	$\leq 5\%$
2.	Water treatment and boiler water conditioning	Reducing the amount of total dissolved solids in the boiler water allows less blow-down and therefore less energy loss.	$\leq 2\%$
3.	Total dissolved solids (TDS) control and boiler blow-down	Automatic control is desirable and in the long run it* can protect the boiler from undesirable water side scaling.	$\leq 2\%$
4.	Blow-down heat recovery	Transfer of available heat energy in blow-down stream back into the system reduces energy loss.	$\leq 4\%$
5.	Flash steam recovery	Recovery of available low pressure steam energy for preheating the feed-water.	$\leq 2\%$
6.	Boiler and burner management systems, digital combustion controls and oxygen trim	This measure is of particular importance for boilers which operate with variable loads. Application of this measure is often directly connected to the next one.	$\leq 5\%$
7.	Variable speed drives (VSDs) for combustion of air fans	This technique can reduce electricity consumption of boiler.	–
8.	Economizers	Recovery of available heat from flue gases and its transfer back into the system by preheating feed-water	$\leq 5\%$
9.	Combustion air preheating	Recovery of available heat from flue gases and its transfer back into the system by preheating combustion air. This measure can be implemented together with the previous one.	$\leq 2\%$

(continued overleaf)

Table 2.3 (*continued*)

	Technique or Method	Description	Potential for energy Efficiency Improvement relevant to Current Consumption
10.	Steam distribution system improvement	Steam system piping, valves, fittings, and vessels have to be well insulated. Implementation of effective steam-trap maintenance program. Isolate steam from unused lines. Repair steam leaks. Minimize vented steam.	≤ 10 %
11.	Condensate return increasing	Recovery of thermal energy in the condensate and reduction of the amount of make-up water added to the system, saving energy and chemicals treatment. High-pressure condensate can be used to make low-pressure steam and in that way the available energy in the returning condensate can be recovered too.	≤ 10 %
12.	Steam pressure	Adjusting the pressure of steam with minimum required for end-users.	≤ 2 %

This is the real energy consumption, which in combination with production volume determines the performance indicator that is the basis for performance monitoring, as has been explained in Part I Chapter 2.

In the observed example, heavy fuel oil (HFO) is used and its consumption in the same period of time as calculating heat energy consumption is 0.1145 kg/s. Its GCV is 42.0 MJ/kg. Now, it is:

$$\eta_{ss} = \frac{3613.01}{0.1145 \cdot 42\,000} = 0.7513 \quad (75.13\,\%) \tag{2.11}$$

The obtained results indicate that out of the total available primary energy of fuel, 75.13 % has been transformed into the required heat energy in order to conduct the production process. The remaining part up to 100 %, i.e. 24.87 is losses. These losses are the subject of energy management.

The boiler energy efficiency is established on the basis of an independent test which specifies that it is 82.93 %. Using this value and Eq. (2.5), we can determine the production of saturated steam delivered to the system:

$$m_{steam} = \frac{\eta_B \cdot M_{FUEL} \cdot GCV}{h_{steam} - 4.21 \cdot t_{FW}} = \frac{0.8293 \cdot 0.1145 \cdot 42\,000}{2759.2 - 4.21 \cdot 47.21} \cdot \frac{3600}{1000} = 5.61\,[t/h] \tag{2.12}$$

Now, it is also possible to determine the efficiency of the steam distribution system as (Eq. 2.6):

$$\eta_{SD} = \frac{3613.01}{5.61 \cdot \dfrac{1000}{3600} \cdot 2757.8} = 0.9519\,(95.19\,\%) \tag{2.13}$$

If the feed water flow is not measured directly, it can be estimated on the basis of the known flow of blow-down and on the basis of the estimate of steam distribution system's loss. These mass (and heat) losses are usually expressed in terms of percentages of produced saturated steam (m_{steam}). In our case, these values are estimated on the basis of the examination of installations and on the basis of boiler and make-up water quality. Therefore, the following has been estimated:

$x = 5\%$ – fraction of the feed water which is lost as blow-down
$y = 5\%$ – fraction of the total steam which is lost due to leakages and
defective steam traps in the steam distribution system.

Based on the above values and known steam production flow rate (Eq. 2.10), the mass flow rate of feed water can be calculated:

$$m_{FW} = (1 + x + y) \cdot m_{steam} = \left(1 + \frac{5}{100} + \frac{5}{100}\right) \cdot 5.61 = 6.17 \,[t/h] \tag{2.14}$$

After the end-users, steam condensates enters into the condensate return subsystem (see Fig. 2.1). The measured condensate temperature at the inlet of this subsystem is 82 °C. The condensate flow is 4.81 t/h ($= 0.1 \cdot 5.34$) but some 50 % of this condensate is not returned because end-users are far from the boiler house and this was used as an explanation for the reason why condensate is simply discharged into the sewerage.

Knowing that the temperature of make up water is 25 °C, and by using Eq. (2.7) the efficiency of condensate return subsystem is:

$$\eta_{CR} = \frac{0.5 \cdot 4.81 \cdot 4.21 \cdot 82}{3.76 \cdot 4.21 \cdot 25 + 4.81 \cdot 4.21 \cdot 85} = 0.3920 \quad (39.20\%) \tag{2.15}$$

The low efficiency of this subsystem is a direct consequence of a large quantity of condensate which is thrown away. Consequentially, the required supply of make-up water of a relatively low temperature is also large.

The feed water temperature is determined on the basis of a simple energy balance of inlet flow and outlet flow of feed water (see Fig. 2.1). Therefore, the following equation can be obtained:

$$t_{FW} = \frac{m_{CR}}{m_{FW}} \cdot t_{CR} + \left(1 - \frac{m_{CR}}{m_{FW}}\right) \cdot t_{MU} = \frac{2.40}{6.17} \cdot 82 + \left(1 - \frac{2.40}{6.17}\right) \cdot 25 = 47.21 \,[°C] \tag{2.16}$$

Now, all of the individual energy performance indicators have been obtained for the separate systems shown in Figure 2.1 and for the overall energy performance indicator of the whole steam system.

Table 2.4 provides a complete computation of these indicators in column 4. The values in rows 1 to 7 are the parameters that influence performance and which could and should be controlled. The values in rows 8 to 11 and 15 are constants for this example.

Column 5 provides the computation of the improved operational parameters of the system. Therefore, the loss of steam and condensate is reduced at the end-users from 10 % to 5 % by fixing some steam leakages and several steam traps which were defective and leaking live steam. By insulating several non-insulated sections, it will be possible to increase this temperature from a given 82 to 83 °C. The percentage of condensate return can be increased from the current 50 % to 75 %. The boiler efficiency can be increased from 82.93 % to 86 %, by an adjustment of O_2 content and by cleaning heat exchange surfaces. The quality of boiler water is not satisfactory and this has caused fouling of the water side in the heat exchange surfaces of the boiler. Improvement of the make-up water quality will remove the consequences of loss caused by fouling on the water side and it will also reduce blow-down water loss from 5 % to 3 %. As in the case of steam and condensate loss, similar defects have been noticed in the operation of the steam distribution subsystem and it is estimated that simple measures may lead to the reduction of steam loss from 5 % to 3 %.

Table 2.4   Calculation of Energy Performance Indicators (Example)

	Description	Unit	CURRENT	NEW
	2	*3*	*4*	*5*
	INFLUENCING FACTORS			
1	Steam and condensate loss by end-users	%	10	5
2	$t_{C,EU}$	°C	85	85
3	$t_{CR} < t_{C,EU}$	°C	82	83
4	Percentage of condensate return	%	50	75
5	η_{Boiler} **(given)**	%	**82.93**	**86**
6	X (blow-down)	%	5	3
7	Y (steam loss in distribution system)	%	5	3
	END-USERS			
8	$m_{s,EU}$	t/h	5.34	5.34
9	p_{steam}	bar	6.2	6.2
10	t_{steam}	°C	160.14	160.14
11	h_{steam}	kJ/kg	2757.8	2757.8
12	$m_{c,EU}$	t/h	4.81	5.07
13	ΔQ_{EU}	kJ/s	3613.01	3586.47
	MAKE-UP WATER			
14	t_{MU}	°C	25	25
	BOILER			
15	GCV	MJ/kg	43.9	43.9
16	M_{fuel}	kg/s	0.1145	0.1054
17	p_{steam}	bar	6.4	6.4
18	t_{steam}	°C	161.39	161.39
19	h_{steam}	kJ/kg	2759.2	2759.2
20	m_{steam}	t/h	5.58	5.50

Table 2.4 (*continued*)

	Description	Unit	CURRENT	NEW
	2	*3*	*4*	*5*
	CONDENSATE RETURN			
21	m_{FW}	t/h	6.17	5.83
22	m_{CR}	t/h	2.40	3.80
23	m_{MU}	t/h	3.76	2.03
24	t_{FW}	oC	47.71	62.85
25	η_{CR}	%	39.20	62.85
	STEAM DISTRIBUTION			
26	η_{SD}	%	95.19	97.04
	STEAM SYSTEM			
27	η_{SS}	%	75.13	81.29

Repeated computations for changed parameters provide for (column 5) an increased degree of efficiency of the steam system from 75.13 % to 81.29 %, which means that relatively simple measures can significantly increase the performance of the steam system and its overall energy efficiency.

2.4 Analysis of Boiler Performance

Boiler performance relates to its ability to transfer heat from fuel to water and steam (or some other fluid) while meeting operating specifications. Boiler performance includes all aspects of the operation. Performance specifications include operating capacity and the factors for adjusting that capacity, steam pressure, boiler water quality, boiler temperatures, boiler drafts and draft losses, flue gas analysis, fuel analysis, and fuel burned. Boiler efficiency, as the most important energy performance indicator, is presented as a percentage ratio of heat absorbed in the boiler water and heat supplied to the boiler by fuel. A boiler is essentially a heat exchanger in which the thermal energy released by combustion process is transferred from the products of combustion to the circulated fluid. An energy balance applied to a boiler within the dotted line shown in Figure 2.4 gives the following equation:

$$M_F \cdot GCV + M_F \cdot h_F + M_A \cdot h_A + (1 + x) \cdot m \cdot h_1 = m \cdot h_2 + M_{FG} \cdot h_{FG} + x \cdot m \cdot h_{BD} + Q_c \quad (2.17)$$

where:
M_F = flow rate of fuel, [kg/s]
GCV = gross calorific value, [kJ/kg]
h_F = enthalpy of fuel, [kJ/kg]
h_A = enthalpy of combustion air, [kJ/kg]
M_{FG} = mass flow rate of flue gas, [kg/s]
h_{FG} = enthalpy of flue gas, [kJ/kg]
m = flow rate of steam, [kg/s]

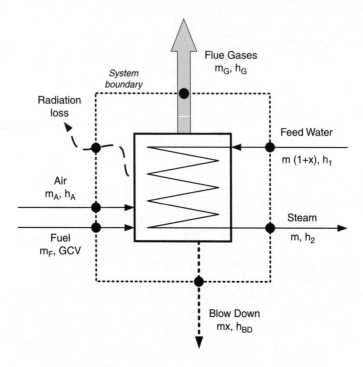

Figure 2.4   Relevant parameters for boiler performance analysis

h_1 = feed water enthalpy, [kJ/kg]
h_2 = steam enthalpy, [kJ/kg]
x   = fraction of the feed water flow which is lost as blow-down, [-]
h_{BD} = enthalpy of the water at working pressure of the boiler, [kJ/kg]
Q_C = radiation loss, [kJ/s]

The left-hand side of the equation represents the sum of all incoming energy streams and the right-hand side of the equation represents the sum of leaving energy streams. The increase in the enthalpy of the circulated fluid to and from the process represents the *useful output*. Of course, when the circulated fluid is hot water or hot oil, instead of steam, the enthalpy of hot water or hot oil is used. There is no blow-down for such boilers (x = 0).

Let us consider now the balance (Eq. 2.17) applied to the boiler from Example 1. The additionally required data from this example are:

O_2 = 7 % – oxygen content in flue gas,
t_{FG} = 245 °C – temperature of flue gas,
t_F = 10 °C – temperature of fuel,
t_A = 14 °C – temperature of combustion air,

and it produces the following specific case:

$$4809.0 + 2.1 + 63.4 + 325.0 = 4279.9 + 974.9 + 53.0 + 65.0 \qquad (2.18)$$

Equation (2.18) refers to the actual operation of the analyzed system (column 4 in Table 2.4) before any improvements. It is evident that the largest single item in the equation is the chemical energy of fuel

(4809.0 kJ/s). The heat energy of fuel (2.1 kJ/s) is really small compared to the chemical energy of fuel and it can be neglected. Combustion air heat energy (63.4) depends strongly on its temperature but it only makes up a small percentage of chemical fuel energy. Feed water heat energy (325.0) also depends strongly on water temperature and its influence on boiler energy efficiency cannot be neglected.

On the left side of the balance equation (Eq. 2.18), the dominant item is the one which refers to the heat energy of steam, i.e., to useful output (4279.9). The second one (974.9) represents the heat energy released into the atmosphere by flue gas. As a rule, this is the largest single loss in boilers. The third item on the left side represents the loss occurring in the blow-down process and may amount to a few per cent of the chemical energy of fuel. The last item on the left side of the equation represents loss occurring due to transfer of heat from boiler surface into the environment.

Example 1 demonstrates the impact of improvements of certain parameters of the system by applying available technical measures aimed at increasing the system's efficiency. If the balance equation is written for these changed conditions, the following is obtained:

$$4411.7 + 1.9 + 43.5 + 416.4 = 4215.7 + 622.0 + 31.2 + 4.6 \tag{2.19}$$

The difference between individual items in Equations (2.18) and (2.19) is obvious, and it reflects the improvement in boiler energy performance or its efficiency.

There are two practical methods for calculation of boiler efficiency:

(a) direct or input/output method;
(b) indirect or heat loss method.

Both of these methods can be used, but it is generally accepted that the indirect or heat loss method provides the more accurate results in practice.

2.4.1 Direct (Input/Output) Method

The basic expression for the direct or input/output method is:

$$\text{Boiler Efficiency} = \eta_B = \frac{Q_l}{Q_F} = \frac{\text{Useful Heat}}{\text{Fuel Chemical Energy Input}} = \frac{m \cdot (h_2 - h_1)}{M_F \cdot \text{GCV}} \tag{2.20}$$

The enthalpy of fuel is usually negligible compared with the chemical energy of fuel and because of that it is left out in Eq. (2.17).

Useful heat is defined as the multiplication of *useful steam* (m) delivered to the process and the difference between enthalpy of this steam (h_2) and enthalpy of feed water (h_1) which enters into the boiler.

The GCV (Gross Calorific Value) is used in the equation which implies that water vapor in flue gases *is condensed out* (see Part III Toolbox 5). This occurs only in a condensing boiler. The NCV (Net Calorific Value) is sometimes used as well. When expressing boiler efficiency, it must be clear which calorific value is used. Efficiency using the NCV will give higher values. We will always use the GCV.

Boiler efficiency can then be defined as:

$$\eta_B = \frac{m \cdot (h_2 - h_1)}{M_F \cdot \text{GCV}} \text{ or } \eta_B = \frac{m \cdot (h_2 - h_1)}{M_F \cdot \text{NCV}} \tag{2.21}$$

The limitations of using the direct method are:

• accuracy of steam flow meters;
• accuracy of fuel weight (solid fuel);
• fuel calorific value (measurement accuracy);
• enthalpy of steam and water, which depends on relevant pressures and temperatures.

Let us discuss the possible error that occurs by using the direct method of boiler efficiency calculation. The value of (η_B) is deduced by direct measuring and other quantities related to it as given by Eq. (2.20). The formula for error calculation applied to Eq. (2.20), gives:

$$\Delta\eta_B = \pm\sqrt{\left(\frac{\partial\eta_B}{\partial m}\cdot\Delta m\right)^2 + \left(\frac{\partial\eta_B}{\partial h_2}\cdot\Delta h_2\right)^2 + \left(\frac{\partial\eta_B}{\partial h_1}\cdot\Delta h_1\right)^2 + \left(\frac{\partial\eta_B}{\partial M_F}\cdot\Delta M_F\right)^2 + \left(\frac{\partial\eta_B}{\partial GCV}\cdot\Delta GCV\right)^2}$$

(2.22)

where:

$$\frac{\partial\eta_B}{\partial m} = \frac{h_2 - h_1}{M_F\cdot GCV}; \quad \frac{\partial\eta_B}{\partial h_2} = \frac{m}{M_F\cdot GCV}; \quad \frac{\partial\eta_B}{\partial h_1} = -\frac{m}{M_F\cdot GCV}; \quad \frac{\partial\eta_B}{\partial M_F} = \frac{m\cdot(h_2 - h_1)}{M_F^2\cdot GCV} \quad \text{and}$$

$$\frac{\partial\eta_B}{\partial GCV} = \frac{m\cdot(h_2 - h_1)}{M_F\cdot GCV^2}$$

are a partial derivation of boiler efficiency equation per independent variables. Δm, Δh, ΔM_F and ΔGCV are errors of independent variables measurements.

Example 1 can be used for estimating the influence of measurement errors on boiler efficiency calculation by the *direct method*. The results of boiler efficiency are presented in Table 2.5. The measured steam flow rate is 5.61 t/h with an error of $\pm 2\%$. For industrial instruments this error is very low and in order to achieve such accuracy by practical measurements a great deal of experience and skill is necessary. The same error ($\pm 2\%$) is assumed for fuel flow rate and GCV measurements. These errors are common in industry, but the GCV error can even be greater. It depends strongly on the fuel type.

The enthalpy cannot be measured directly. It is common to measure the steam pressure for saturated steam and pressure and temperature for superheated steam. If the error of pressure measurement is again $\pm 2\%$, it will produce a very small error in saturated steam enthalpy determination. If the measured saturated steam pressure is 6.4 bar and if pressure measurement accuracy is $\pm 2\%$, consequently enthalpy error will be only 0.03 %. In our case, the enthalpy of saturated steam is 2759.8 kJ/kg. The feed water temperature

Table 2.5 Example of Boiler Efficiency Calculation by Using the Direct Method

Item	Unit	Value (1)	Value (2)
Steam flow rate, m	t/h	5.61 ($\pm 2\%$)	5.61 ($\pm 1\%$)
Gross Calorific Value, GCV	MJ/kg	42.0 ($\pm 2\%$)	42.0 ($\pm 1\%$)
Mass flow rate of fuel, M_F	kg/h	0.1145 ($\pm 2\%$)	0.1145 ($\pm 1\%$)
Saturated steam pressure, p_1	bar	6.4 ($\pm 2\%$)	6.4 ($\pm 2\%$)
Steam temperature, t_1	°C	161.4	161.4
Enthalpy of steam, h_1	kJ/kg	2759.8 ($\pm 0.03\%$)	2759.8 ($\pm 0.03\%$)
Feed water temperature, t_{FW}	°C	47.21 ($\pm 1\%$)	47.21 ($\pm 1\%$)
Water specific heat at feed water temperature, c_{pFW}	kJ/(kg °C)	4.21	4.21
Boiler efficiency (direct method)	%	82.93	82.93
Boiler efficiency error	%	$\pm$ **2.95**	$\pm$ **1.47**

in our case is 47.21 °C. The error of $\pm 2\%$ in our case means app. ± 1 °C, which is quite possible to achieve in industrial measurements.

Applying all these data in Eq. (2.22), the error in boiler efficiency calculation is as follows:

$$\Delta \eta_B = \pm 2.95\% \tag{2.23}$$

If we reduce the measurement errors of steam and fuel and steam flow rate, and GCV from 2 % to 1 %, the boiler efficiency error is reduced from 2.95 % to 1.47 % (see Table 2.5). This example demonstrates how the error can be estimated and how accurately the measurements have to be performed in order to obtain results useful for further analyses. The attached Part III Toolbox 13 Software 9: *Efficiency of Steam Boilers* enables boiler efficiency calculation by using the direct method.

2.4.2 Indirect (Heat Losses) Method

The indirect method calculates individual losses, add them up, and determines boiler efficiency by subtracting total losses from the total primary energy input.

The heat quantity Q_F contained in the fuel is introduced into the boiler and transformed into the following types of heat energy:

$$Q_1 = heat\ energy\ transferred\ to\ the\ heat\ receiver\ (useful\ heat)$$

and

$$\sum_{i=2}^{7} Q_i = 6\ types\ of\ heat\ losses$$

where:
  $Q_2 =$ heat loss of boiler flue gases;
  $Q_3 =$ heat loss occurring because of chemically incomplete combustion;
  $Q_4 =$ heat loss caused by mechanically incomplete combustion;
  $Q_5 =$ heat loss due radiations from boiler surface;
  $Q_6 =$ heat energy loss due to ash removal;
  $Q_7 =$ blow-down heat loss

The boiler efficiency is now defined as follows:

$$\eta_B = \frac{Q_1}{Q_F}\ (direct\ method) = 1 - \frac{\sum_{i=2}^{7} Q_i}{Q_F}\ (indirect\ method) \tag{2.24}$$

Individual specific energy losses are expressed as follows:

$$q_i[\%] = \frac{Q_i}{Q_F} \cdot 100\ (i = 2, 3, \ldots 7) \tag{2.25}$$

Let us consider the procedure of finding particular losses.

2.4.2.1 Heat Energy Loss due to Flue gas

The temperature of flue gas at the boiler outlet is higher than the ambient temperature and, therefore, the quantity of heat contained in this gas is the heat lost for the boiler. However, the overall physical heat of the flue gas cannot be considered as a loss because it also contains the physical heat of the cold combustion air. The real loss is then defined by the following equation:

$$Q_2 = M_{FG} \cdot h_{FG} - M_A \cdot h_A \tag{2.26}$$

Table 2.6 Flue Gas Oxygen Content Control Parameters

FUEL	Automatic Control of O_2		Positioning Control of O_2	
	Minimum [%]	Maximum [%]	Minimum [%]	Maximum [%]
Natural Gas	1.2	3.0	3.0	7.0
Diesel	2.0	3.0	3.0	7.0
Heavy Fuel Oil	2.5	3.5	3.5	8.0
Pulverized Coal	2.5	4.0	4.0	7.0
Stoker Coal	3.5	5.0	5.0	8.0

where enthalpies of flue gas and combustion air are per mass of dry gas. This loss is almost always the biggest loss. The major contributors are the flue gas temperature and excess air amount. The excess air component is in direct connection with the oxygen content of flue gas.

Table 2.6 shows the typical limits of flue gas oxygen content for industrial steam boilers. The positioning and automatic controls, as two main types of boilers controls, are analyzed. Positioning control does not use flue gas oxygen measurement. In steam boilers with such a control the steam pressure is typically measured and used as control variable. If the steam pressure decreases, the controller will increase the fuel flow rate to increase the steam generation and consequently to increase the steam pressure. The combustion air-flow rate will be adjusted to the fuel flow rate by another control system. In this system of control the fuel flow rate dictates the airflow using the previously established and fixed ratio of fuel and air. Periodically, the fuel and air ratio is adjusted for different boiler loads and based on the oxygen content in the flue gas. As this generally fixed ratio cannot meet all possible loads, the bigger values of flue gas oxygen are proposed for this type of control as compared to automatic control.

An automatic combustion control system means the system that self-adjusts burner/boiler operation (based on O_2 [%] measurement) in order to maximize energy efficiency. It must include at least the following capabilities: fuel/air ratio adjusted automatically, fuel flow metered/monitored and continuous monitoring of nitrogen oxides (NOx) and carbon monoxide.

Generally, the limits of flue gas oxygen content are defined by the periodical control of carbon monoxide (CO) content in the flue gas for different loads. CO content less than 10 ppm can be accepted within all ranges of load.

Besides flue gas oxygen content, which is highly recommended to be measured and controlled, flue gas temperature measurement is also very important. This measurement should be recorded at least daily and should be analyzed with respect to boiler load and ambient conditions. If there are some additional heat transfer surfaces which use flue gas for feed water preheating or air preheating, the sensor has to be installed just after the last heat transfer surface. In that way, the real flue gas loss can be determined.

To summarize, flue gas loss depends strongly on flue gas temperature which is affected mainly by the following factors: boiler load, boiler design and combustion-side and water-side heat transfer surfaces fouling.

The specific flue gas loss is:

$$q_2 = \frac{Q_2}{M_F \cdot GVC} \cdot 100 \ [\%] \qquad (2.27)$$

If the following is known from the flue gas analysis:

Pressure of the ambient air $= P_A$ [bar]
Temperature of air $= t_A$ [°C]
Relative humidity of air $= RH_A$ [%]
Pressure of the flue gas $= P_{FG}$ [bar]
Temperature of flue gas $= t_{FG}$ [°C]
Oxygen content in flue gas (by volume) $= v_{O2}$ [%]

then the flue gas loss can be calculated. The calculation procedure is elaborated upon in Part III Toolbox 5. The main outputs of this calculation will be:

- specific flue gas loss q_2;
- the air-fuel ratio;
- excess air;
- specific heat of dry flue gas;
- absolute humidity of flue gas;
- wet flue gas per kg of fuel;
- enthalpy of wet flue gas;
- heat energy loss with wet flue gas.

For the estimation of possible errors that can appear in the calculation of flue gas loss, the calculations in Table 2.7 can be used. The test is performed at a saturated steam package boiler fired by HFO (nominal capacity 6 t/h, 14 bar). The oxygen content is 6.5 % and the flue gas temperature is 225 °C. The combustion

Table 2.7   Example of Calculation of Flue Gas Loss

Name	Unit	Value	Measurement Error	Expected Error [%]
Content of oxygen, O_2	%	6.5	±0.8 %	
Flue gas temperature, t_{FG}	°C	225	0	
Gross Calorific Value, GCV	MJ/kg$_F$	42.59	0	
Flue gas loss, q_2	%	16.22		+3.23/ − 2.87
Content of oxygen, O_2	%	6.5	0 %	
Flue gas temperature, t_{FG}	°C	225	±5 °C	
Gross Calorific Value, GCV	MJ/kg$_F$	42.59	0 %	
Flue gas loss, q_2				+1.65/ − 1.59
Content of oxygen, O_2	%	6.5	%	
Flue gas temperature, t_{FG}	°C	225	0 °C	
Gross Calorific Value, GCV	MJ/kg$_F$	42.59	±2 %	
Flue gas loss, q_2				+1.95/ − 2.07

air temperature is 25 °C and its relative humidity is 55 %. The GCV of fuel is 42.59 MJ/kg, as specified by the supplier. The estimated accuracy of GCV is ± 2 %. The manufacturer of the instrument for emission measurement guarantees the accuracy of the instrument as ± 0.8 %. Repeating the flue gas loss calculation with the implemented prescribed possible errors in measurement one by one and comparing the obtained results with actually read out values shows the expected errors in percentages of read or taken values as given in Table 2.7. Part III Toolbox 13 Software 4: *Fuels, Combustion and Environmental Impacts* from the Toolbox was used for the calculations. By changing one by one all the relevant independent variables (oxygen content, flue gas temperature and GCV) we calculate deviation of boiler efficiency from its benchmark value.

One can say that errors are not very big, but for the proper interpretation of obtained results, it is necessary to specify the range of errors. For repeated measurements and the later comparison of data obtained in different periods, it is highly recommended to use always the same method of measurements and the same instrumentation. In that way, the influence of errors on final conclusions will be relatively reduced.

2.4.2.2 Heat Energy Loss due to Chemically Incomplete Combustion

Chemically incomplete combustion causes the appearance of CO, H_2 and C_nH_m in the flue gases. H_2 and C_nH_m appear only during very poor combustion. The main cause of this loss is the insufficient mixing of fuel and combustion air. It is relatively small and its influence on total boiler efficiency is also small. Most commonly, only the CO content in dry flue gases is measured and, as it is much higher than H_2 and C_nH_m, this loss can be estimated based only on this measurement.

If the CO (vol.) is known, the calculation of energy loss because of incomplete combustion can be done by using the following equation:

$$Q_3 \, [kW] = 22.4 \cdot 12\,644 \cdot \frac{v_{CO} \, [\%]}{100} \cdot B \, [kmol/kg_F] \cdot M_F \, [kg_F/s] \qquad (2.28)$$

This equation assumes energy release by supplementary combustion of carbon monoxide contained in the flue gas. The content of CO is commonly expressed in parts per million [ppm]. The following relation is valid: 1 [%] = 10 000 [ppm].

The specified energy loss is as follows:

$$q_3 [\%] = \frac{22.4 \cdot 12\,644 \cdot \dfrac{v_{CO} [\%]}{100} \cdot B}{GCV} \cdot 100 \qquad (2.29)$$

where:

v_{CO} [%]   = volumetric concentration of CO in dry flue gas
B [kmol/kg$_F$] = kmol of dry flue gas per 1 kg of fuel
GCV [kJ/kg$_F$] = gross calorific value of fuel
M_F [kg/s]   = mass flow rate of fuel

This loss is very small in a well-operated combustion process. The content of CO in combustion products is more relevant as an environmental impact than as overall heat loss. However, if the content of CO is large (> 1000 ppm), it indicates a poor combustion process and requires comprehensive analysis.

2.4.2.3 Heat Energy Loss due to Mechanically Incomplete Combustion

This loss assumes:

1. loss of part of fuel throughout the grade ($Q_{4'}$) (exists only for solid fuels);
2. combustible part of fuel accumulated in red-hot slag ($Q_{4'}$) (exists only for solid fuels); and
3. combustible parts of fuel discharged with flying ash in the flue gas ($Q_{4'''}$) (exists only for solid fuels and, sometimes, for liquid fuels).

In particular, coal and biomass fired boilers have an additional energy loss through solid refuse. Solid refuse takes the form of the mixture of minerals and carbon left in the ash pit. Depending on the firing technique and the skill of the operator, there is more or less carbon left in the refuse. The energy loss occurs in two ways. Firstly, the refuse is removed from the ash pit at a temperature between 100 to 250 °C, and secondly, the remaining carbon in the refuse has not been burned, but paid for. The more ash the fuel has the higher the danger of carbon left in the refuse.

The unburned fuel or combustibles content of ash is a generally negligible loss for other types of fuel but it can be significant in coal-fired boilers. This loss is not related to the partial combustion of fuel forming carbon monoxide in the flue gas, but to the amount of fuel remaining in unburned ash. The ash content in coal varies widely from less than 5 % to more than 20 % of the total mass of coal. Unburned carbon is the function of many factors, but most of them are related to combustion and fuel conditions.

This loss can be calculated by taking the ash after the fuel has been burned and measuring the amount of carbon in the laboratory, expressed as the percentage of incoming fuel energy as follows:

$$q_4[\%] = \frac{Q_4}{M_F \cdot GVC} \cdot 100 \tag{2.30}$$

Generally speaking, the unburned carbon loss is expected to be less than 0.5 % of the total fuel input energy.

2.4.2.4 Heat Energy Loss due to Radiation

This loss results from the radiation of outside boiler surfaces and is equal to the quantity of heat exchanged between the boiler jacket and the surrounding air. It is very hard to determine the precise quantity of this loss. Therefore, its value is usually shown in relevant references either in diagrams or in tables, dependant on the boiler capacity or boiler heat output. The relation of this loss as against the rated heat output for steam boilers is represented in Figure 2.5. As losses are calculated in relation to heat input, read-out or

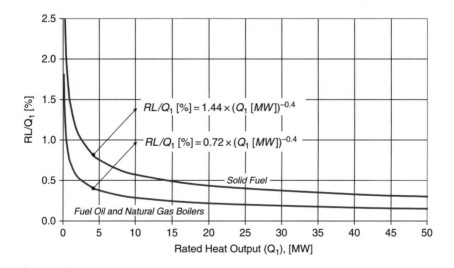

Figure 2.5 Heat Lost in the Surroundings. Reproduced from: EN 12953–11: Shell boilers, European Committee for Standardization, 2003

computed value should therefore be multiplied by the relation between rated heat output and heat input (Q_1/Q_F).

For boiler loads lower than the nominal one, the value of this loss (q_5) is calculated by using the following formula:

$$q_5 = \left(\frac{RL}{Q_1}\right) \cdot \frac{Q_1}{Q_F} \cdot \frac{m_n}{m} [\%] \qquad (2.31)$$

where m_n is nominal and m is actual steam flow rate.

2.4.2.5 Heat Energy Loss due to Slag

This loss depends directly on slag temperature. This implies only heat energy accumulated in slag without carbon remaining in it.

2.4.2.6 Heat Energy Loss due to Blow-down

Steam boilers regularly eject some water in order to prevent precipitated salts from forming sludge, which can foul heat exchange surfaces and can be carried over into the super-heater and steam mains. This process is known as boiler *blow-down*. It represents a loss of money concerning: (a) reduction of boiler efficiency due to the loss of energy caused by hot water blow-down; and (b) a fraction of the cost of chemical treating of the feed water before entering the boiler. The most efficient system of blow-down is one that is fully automatic with continuous measurement of totally dissolved solids or water conductivity.

Blow-down can be as high as 10 % or more of the steam flow rate. However, 5 % would be extraordinarily high for a system with high-quality water treatment systems. Recovering part of the thermal energy of the blow-down water by using it to heat the feed water can reduce this loss. The most effective way of doing this is to allow the flash steam formed from the blow-down to mix directly with the feed water. This also recovers some of the costs for water treatment.

Blow-down loss is as follows:

$$Q_7 = x \cdot m \cdot (h_{BD} - h_1) \qquad (2.32)$$

where:
  x = fraction of the feed-water flow which is lost as blow-down,
  h_1 = enthalpy of the make up water, and
  h_{BD} = enthalpy of the saturated water (not steam) at working pressure of the boiler.

Finding out the amount of blow-down is more or less guesswork. Only in installations that have in-line steam and feed-water meters is it simple because the blow-down is the difference between feed-water input and steam output. In all other cases, it is necessary to collect the blow-down water and measure its volume. Recommendations to reduce boiler blow-down should be made only if there is a complete picture of the present feed-water chemistry and after feed-water chemical analysis. As experience has shown, almost all boiler operators need help to adjust their feed-water chemistry and chemical suppliers rarely provide such assistance.

The best way to determine this loss is the measurement of the blow-down flow rate. When the flow rate is known, it is very simple to calculate the loss of heat energy and money because of lost water. However, if there is no metering system for blow-down water, the boiler blow-down requirement can be calculated from the following equation:

$$x = \frac{TDS_{FW}}{TDS_{BW} - TDS_{FW}} \cdot 100\,\% \qquad (2.33)$$

where:
  TDS_{FW} = Total Dissolved Solids in feed water (mg/l or ppm)
  TDS_{BW} = Total Dissolved Solids in boiler water (mg/l or ppm)

Very often, only the water conductivity is measured. In such a case, the TDS can be replaced by conductivity. This does not imply that TDS is equal to conductivity but that their relationship is constant for particular water.

Specific blow-down loss is calculated as follows:

$$q_7 = \frac{Q_7}{Q_F} \cdot 100 = \frac{x \cdot m \cdot (h_{BD} - h_1)}{M_F \cdot GCV} \cdot 100 \, [\%] \tag{2.34}$$

2.4.2.7 Summary of Indirect Boiler Efficiency Determination

The sum of above-mentioned seven heat flow rates must be equal to the heat quantity Q_F contained in the working fuel. This is shown by the following equation:

$$Q_F = Q_1 + Q_2 + Q_3 + Q_4 + Q_5 + Q_6 + Q_7 \tag{2.35}$$

If this equation is divided by Q_F and then multiplied by 100, it will read:

$$100 = q_1 + q_2 + q_3 + q_4 + q_5 + q_6 + q_7 \, [\%] \tag{2.36}$$

The value of $Q_1/Q_F \times 100 \, [\%]$, i.e., the ratio of the heat quantity used and the heat quantity introduced into the boiler is already defined as the boiler efficiency. However, the equation for boiler efficiency by using the indirect or heat loss method is:

$$\eta_B = 100 - (q_2 + q_3 + q_4 + q_5 + q_6 + q_7) \quad [\%] \tag{2.37}$$

Typical values of heat losses and some comments on them are presented in Table 2.8 (for solid fuels) and in Table 2.9 (for liquid and gaseous fuels). Blow-down loss is not reported in those tables as they belong only to steam boilers and the tables below also include hot water and oil boilers.

2.4.3 Preparing a Measurement Plan for Boiler Performance Analysis

Figure 2.6 shows the scheme for a boiler with an economizer, an air pre-heater and a system for blow-down heat recovery with marked measurement points. Only on the basis of concrete measurements, is it possible to quantify certain losses and consider the effects of measures for improving energy performance, when these measures have been applied.

The degree of measuring requirements is directly dependent on the planned performance improvement measures that are to be analyzed.

Table 2.8 Losses for Boilers Fired by Solid Fuel

Loss q_i	Layer	Dust	Comment
$q_2[\%]$	See Eq. 2.27	See Eq. 2.27	
$q_3[\%]$	≥ 0.5		Manual feeding, no induced fan, light volatile C_nH_m, good mixing, high excess air
	≤ 2.5		Manual feeding, induced fan, light volatile C_nH_m, good mixing, high excess air
		≥ 0.2	Fine dust, liquid slag, no induced fan, good mixing, enough high excess air
		≤ 0.6	Opposite than previous

(continued overleaf)

Table 2.8 *(continued)*

Loss q_i	Layer	Dust	Comment
$q_{4'}$ [%]	≥ 1.5	0.0	Without relative moving of the layer, steam production over 25 t/h, ranked solid fuel, no induced fan
	≤ 5.0	0.0	With relative moving of the layer, steam production lower than 10 t/h, unranked solid fuel, induced fan
$q_{4''}$ [%]	≥ 1.0	0.0	Without relative moving of the layer, steam production over 25 t/h, ranked solid fuel, no induced fan, low ash/GCV, dry slag
	≤ 4.0	0.0	Opposite than previous, but for steam production lower than 10 t/h and soft slag
$q_{4'''}$ [%]	≥ 0.5		Ranked fuel, low volatile, no induced fan, steam production over 100 t/h
	≤ 3.5		Opposite than previous, but for steam production lower than 50 t/h
		≥ 2.2	Fine dust, no induced fan, steam production lower than 100 t/h
		≤ 6.5	Rough dust, with induced fan, steam production lower than 50 t/h
q_5 [%]	See Fig. 2.6	See Fig. 2.6	
q_6 [%]	≥ 0.2		One chamber furnace, small ash/GCV, dry slag
	≤ 3.0		Two chambers furnace, big ash/GCV, liquid slag
		≥ 0.3	Small ash/GCV, induced fan, dry slag
		≤ 2.0	Big ash/GCV, no induced fan, liquid slag

Table 2.9 Losses for Boilers Fired by Liquid and Gaseous Fuel

Loss q_i	Liquid Fuel	Gaseous Fuel
q_2	See Eq. (2.27)	See Eq. (2.27)
q_3	0.1–0.6	0.0–0.5
q_4	0.5–1.0	0.0
q_5	See Fig. 2.6	See Fig. 2.6
q_6	0.0	0.0

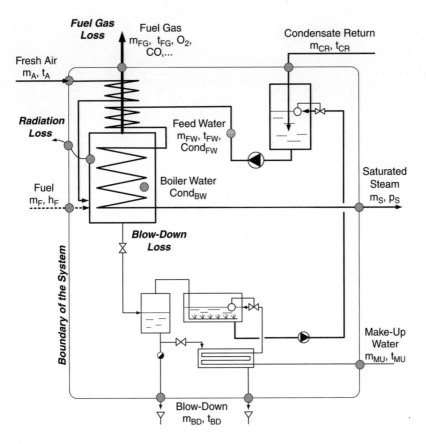

Figure 2.6 Boiler Scheme and Measuring Points

Values to be measured:

Fuel consumption:	**Make-up water (soft water):**
– Fuel flow rate, m_F [l/h, t/h, nm^3/h ....]	– Mass flow rate of make-up water, m_{CR} [t/h]
	– Temperature, t_{MU} [°C]
Fresh air:	
	Feed water:
– Temperature, t_{FA} [°C]	
– Relative humidity, RH_{FA} [%]	– Mass flow rate of feed water, m_{FW} [t/h]
	– Temperature, t_{FW} [°C]
Steam (saturated):	– Conductivity, $Cond_{FW}$, [μS/cm]
– Pressure, p_S [bar]	

(continued)

Flue gas:	Boiler water:
– Flue gas temperature, t_{FG} [°C] – Oxygen content, O_2 [%] – CO content, CO [ppm] – Other	– Conductivity, CondBW, [μS/cm] **Blow-down:** – Mass flow rate of blow-down water, m_{BD}, [t/h]
Condensate return:	**Average temperature of boiler surface** (Average of app. 10 measurements):
– Mass flow rate of condensate, m_{CR} [t/h] – Temperature, t_{CR} [°C]	– Temperature of boiler surface, t_{BS}, [°C]

Although it is not always necessary to make all of the above-mentioned measurements, they are all desirable. For example, by measuring feed water and make-up water flow rates, it is possible to determine the condensate return flow rate from the mass balance. This means that the separate measurement of the condensate flow rate is not necessarily required in order to establish energy balance.

2.5 Factors Influencing Boiler Performance

This section will analyze in more detail some of the factors affecting boiler energy efficiency with the aim of indicating specific opportunities for performance improvement.

2.5.1 Boiler Load

As boiler load increases, flue gas temperature generally rises. The heat transfer surface of a boiler is fixed and, in the case of steam generation, the temperature on the water side is practically constant. By increasing the load (steam generation), the fixed heat transfer surface allows less heat transfer per unit mass of combustion products and consequently the flue gas temperature rises. It seems that a boiler has to be run under a lower load in order to reach higher efficiency. However, as the boiler load decreases, the flue gas oxygen has to be increased to maintain proper combustion. Generally speaking, the boiler load will affect the flue gas temperature, but this effect is essentially down to design characteristics and there is no room for a significant influence on this parameter for the given boiler.

The boiler efficiency and flue gas temperature versus load is presented in Figure 2.7 and Figure 2.8, respectively. The data are the results of measurements performed on seven package boilers for saturated steam generation. The fuel is heavy fuel oil. The nominal steam production is in the range from 5.5 to 14 t/h and the steam pressure for all boilers is 13 bar.

The control of all tested boilers is of the positioning type. The tests are performed for a full load of 100 % and partial loads of 60 % and 30 %. These tests prove that significant efficiency reduction appears with the reduction of load and that the flue gas temperature will also be reduced with the decreased load.

2.5.2 Boiler Design

Boiler design defines overall boiler efficiency. The heat transfer area, combustion gas paths, etc. for the given boiler can hardly be changed and is not recommended without having a highly qualified service

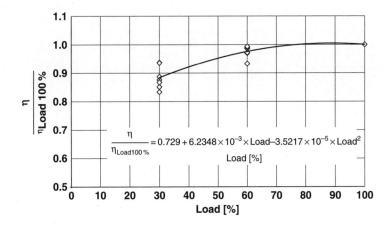

Figure 2.7   Boiler Efficiency versus Load

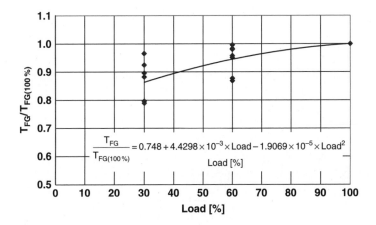

Figure 2.8   Flue Gas Temperature versus Load

company. But installing additional heat recovery equipment can effect improvements. The most common heat recovery equipment is a feed-water economizer and a combustion air pre-heater. The job of both of them is to extract energy from the flue gas.

Practical limits exist and have to be analyzed carefully. The practical minimum of flue gas temperature is dictated by the corrosiveness of the flue gas. The sulfur content of the fuel is responsible for the occurrence of sulfuric acid (H_2SO_4), which is corrosive for many boiler components. The sulfur combustion product is sulfur dioxide (SO_2), which in contact with water (H_2O) forms the sulfuric acid. The problem arises when condensate occurs and the condensate is in a direct relation to the flue gas temperature. This is the reason why the flue gas temperature has to be maintained at a temperature greater than the dew point of sulfuric acid.

Some fuels do not contain sulfur, and in this case, the dew point of water vapor will be the flue gas temperature limit. The existing carbon dioxide (CO_2) in the reaction with water forms the carbonic acid

(H_2CO_3) which can cause carbonic acid corrosion. If the sulfur content is high, generally more than 1 %, special materials will be required for heat exchangers.

2.5.3 Fouling of Heat Transfer Surfaces

If the heat transfer surfaces of boiler become fouled, heat transfer intensity will be weaker and the efficiency of this process will be lowered. Unfortunately, fouling can occur on both heat transfer surfaces and gas and water sides. But, besides reduction, the heat transfer intensity and the accumulation of material on boiler surfaces can also cause overheating and corrosion.

2.5.3.1 Gas-side Fouling

Natural gas generally does not produce significant gas-side deposits if the burner is functioning properly. Soft black soot, which can be easily removed by brushing, is the main characteristic of deposits and appears during the fuel oil combustion process. Low grade fuel oil sometimes contains large quantities of alkaline sulfates and vanadium oxide which can cause serious deposits on the gas-side. Because of their low fusion temperatures, the majority of deposits appear on convective heat transfer surfaces. Solid fuels produce deposits containing ash based slag and soot. As the boiler temperatures are high, the process of sintering or melting can appear.

Anyhow, each type of fouling on the heat transfer surface reduces the heat flux and in that way the boiler's efficiency as well.

Roughly, it can be estimated that every 4.5 °C increase of flue gas temperature appears as the fouling of heat transfer surfaces and causes a loss of combustion efficiency of 1 %.

2.5.3.2 Water-side Fouling

Boilers are delivered clean without sludge and scale. Consequently, sludge and scale is a classic management and operational problem that has very little to do with boiler design. These two deposits, which appear on water-side of boiler heat transfer surfaces, can be defined as follows:

- *Sludge* deposits are precipitated elsewhere and transported to the metal surface by flowing water. Sludge is the accumulation of solids that precipitate in the bulk boiler water or enter the boiler as suspended solids.
- *Scale* is crystallized directly onto tube surfaces. Salts that have limited solubility in the boiler water form scale. These salts reach the deposit site in a soluble form and precipitate when concentrated by evaporation.

Sludge and scale prevention is one of the most important tasks of a boiler operator in controlling flue gas losses.

Chemical deposits or scaling on the water-side can significantly reduce efficiency. A more serious side effect is that these deposits are very good insulators since they reduce the heat transfer and increase the metal temperature. Thus, this can cause premature tube failure. Even a thin layer of scale causes an evident increase in tube temperature and decrease in its expected life.

Mineral impurities in boiler water can also cause other operating problems, for example foaming and consequent moisture carry over into the steam line.

The influence of waterside boiler scaling is illustrated in the following example. Figure 2.9a shows the temperature gradient through a clean tube wall and also the film effect on the gas-side and water-side. The heat transfer coefficient on the water-side is 5000 W/(m^2 K) and on the gas-side it is 460 W/(m^2 K).

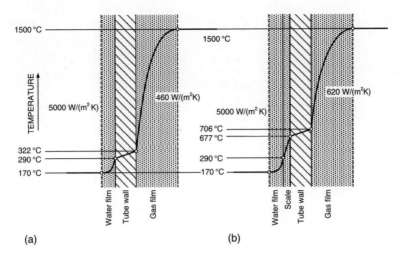

Figure 2.9 Effect of Water-Side Scale in the Boiler

Thermal conductivity of wall material is 45 W/(m K). The outside diameter of the tube is 50 mm and wall thickness is 2.5 mm. For these data, the heat flow rate per 1 meter of tube is:

$$\frac{Q}{L} = q_L = K_L \cdot \Delta t = \frac{t_g - t_w}{\dfrac{1}{d_o \cdot \pi \cdot h_1} + \dfrac{\ln(d_o/d_i)}{2 \cdot \pi \cdot \lambda_w} + \dfrac{1}{d_i \cdot \pi \cdot d_i}} =$$

$$= \frac{1500 - 170}{\dfrac{1}{0.045 \cdot \pi \cdot 5000} + \dfrac{\ln(50/45)}{2 \cdot \pi \cdot 45} + \dfrac{1}{0.05 \cdot \pi \cdot 460}} = 64 \cdot 1330 = 85\,110\,[\text{W/m}] \tag{2.38}$$

The temperature of water-side surface is:

$$t_{ws} = 170 + \frac{85110}{0.45 \cdot \pi \cdot 5000} = 290.4 \approx 290\,°C \tag{2.39}$$

and the temperature of gas-side surface is:

$$t_{gs} = 290 + 85110 \cdot \frac{\ln(50/45)}{2 \cdot \pi \cdot 45} = 321.7 \approx 322\,°C \tag{2.40}$$

As the temperature of metal rises, the allowable stress of material for that temperature decreases until it cannot carry the same pressure. If the scale is formed over the water-side of the heat transfer surface and if the thickness of the scale is only 0.5 mm (thermal conductivity of scale is 0.7 W/(m K)), the heat flow rate will be:

$$q'_L = \frac{1500 - 170}{\dfrac{1}{0,045 \times \pi \times 5000} + \dfrac{\ln(45/44)}{2 \times \pi \times 0.7} + \dfrac{\ln(50/45)}{2 \times \pi \times 45} + \dfrac{1}{0.05 \times \pi \times 620}} = \tag{2.41}$$

$$= 58.15 \times 1330 = 77\,340\,[\text{W/m}]$$

The calculated heat transfer coefficient of gas for new temperature is 620 W/(m² K) (Fig. 2.9b).

The temperatures of the scale surface at water-side, walls at water-side and gas-side are:

$$t_{\text{ScaleWater}} = 170 + \frac{77340}{0.044 \cdot \pi \cdot 5000} = 281.9 \approx 282\,°C \qquad (2.42)$$

$$t_{\text{Water-sideWall}} = 282 + 77340 \cdot \frac{\ln(45/50)}{2 \cdot \pi \cdot 0.7} = 677.1 \approx 677\,°C \qquad (2.43)$$

$$t_{\text{Gas-sideWall}} = 677 + 77340 \cdot \frac{\ln(50/45)}{2 \cdot \pi \cdot 45} = 705.9 \approx 706\,°C \qquad (2.44)$$

In the case of a clean tube, the temperature of the gas-side wall surface is 322 °C. After scale formation, the same temperature will rise to 706 °C, which may be near the point of failure for the grade of used steel. At the same time, the heat flow rate is reduced from 85.11 kW/m to 77.34 kW/m or will be lower by 9.1 %.

2.5.4 Boiler Operation Controls

Control of energy and mass flows is the source of significant performance improvement opportunities. The most common boiler control systems which are applied to industrial boilers can be classified in two groups:

- *On/Off controls* – assumes simple On/Off control and High/Low/Off, which has a high and low fire *On* condition and the *Off* condition.
- *Modulating controls* – assume two basic classes: *Positioning* and *Metering*.

The simplest and the least costly control is the On/Off control. This type of control is limited to small boilers. The control is initiated by a steam pressure (or hot water temperature) switch. This switch has two limits (maximum and minimum) which can be adjusted. In that way, frequent switching *on* and *off* will be eliminated. If the pressure drops and reaches the minimum pressure switch setting, the fuel valve is opened (or the fuel pump started) along with the combustion air fan motor. The fire is ignited, usually with continuous pilot flame. The full firing rate is established and the pressure (or temperature) rises until the maximum setting of pressure (or temperature) is reached. When the upper limit of pressure (or temperature) is reached, the fire is stopped.

Such a control system may maintain steam pressure or hot water temperature within acceptable limits, but there is no combustion control. While firing, the combustion efficiency is the result of mechanical burner adjustment.

This simplest control is presented in Figure 2.10. The measurement of parameters at the output gives information which has to be used to regulate input parameters in order to achieve optimum output conditions. The main measurement is the steam pressure at the output. Steam pressure will vary depending on the steam flow rate which is dictated by the process of using steam. If the flow rate of steam increases, caused by starting or increasing steam consumption by some of machines, the pressure of the steam will decrease. Following the scheme of regulation presented in Figure 2.10, which assumes that the On/Off regulation of burner is applied, the regulator will switch on the burner to produce more steam and to increase the pressure.

The boiler water level sensor has two main functions:

- To start and stop the feed water pump automatically in order to keep the water level within the range prescribed by the boiler manufacturer. If the level is at the minimum, the level regulator will start the feed water pump and, when the maximum level is reached, the regulator will stop the pump. This control is independent and it is only indirectly in connection with the pressure of the steam or steam flow rate.

- For safety reasons, there are two more limiters in the level regulator, which have to protect the boiler from too low and too high a level of water. The range of these switches is wider than the previous one, and they are connected to the fire regulator directly in order to switch off the combustion process if the water level is extremely low or high.

The effects of On/Off control actions are represented in Figure 2.11. When the burner is *on*, the excess air is subjected to variation in the fuel (pressure and temperature and consequently changing of gravity and viscosity; GCV variation and mechanical adjustment tolerance). The excess air is also subjected to variation of changes in combustion air parameters. However, it is important to mention that each time the burner is *off*, cold air passes through the boiler carrying heat up the stack unless the flue gas damper is closed. If the process permits a wide range of pressure control (and because of that the amount of switching on the burner will be reduced), the cooling down of water and the boiler structure will occur. A significant amount of heat energy (fuel) has to be added.

The starting and shutdown periods when the boiler is in transition will cause some smoking as a result of excess fuel that is sent to the boiler when ignition is not complete.

A boiler that is constantly cycled will have a higher probability of soot build-up compared to one which is left at a steady firing rate. The cycling boiler will not be allowed to reach and maintain an even combustion temperature. The cooler regions in the boiler will exhibit excessive soot build-ups and tend to further promote incomplete combustion. Figure 2.12 illustrates the effects of firing period changes on designed boiler efficiency (continuous running).

Modulating burner control will alter the firing rate to match the boiler load over the whole turn-down ratio. Full modulation means that the boiler keeps firing, and fuel and air are carefully matched over the whole firing range to maximize efficiency and minimize materials' thermal stresses.

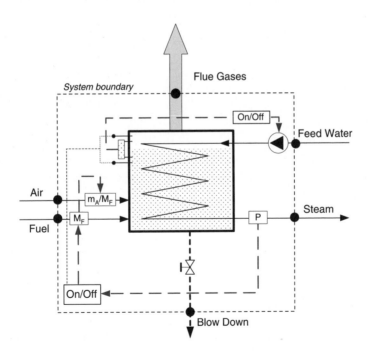

Figure 2.10   On/Off Regulation of Steam Boiler

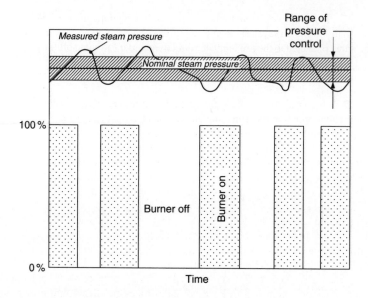

Figure 2.11   On/Off Control Action

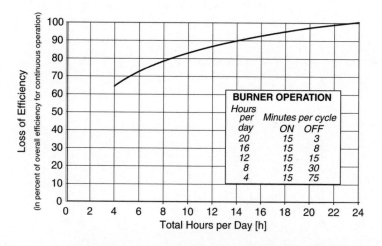

Figure 2.12   Burner Operation versus Boiler Efficiency

In matching burner and control system to the boiler, the following factors have to be taken into account:

- the maximum output of the plant;
- whether the load is steady or changeable;
- the fuel which is used.

2.5.5 Water Quality

Operational problems related to the chemical composition of water can be classified in the following four categories:

- problems with the formation of scale;
- problems with corrosion of steel materials for boilers (mild steel);
- problems with the fragility of steel materials for boilers (mild steel);
- problems with the carry-over (foaming and priming).

To prevent these troubles, the control of water characteristics (water treatment) is necessary. The relation between dissolved matters in water and the above-mentioned main problems are classified as follows in the Table 2.10.

Table 2.10   Relation between Water Characteristics and Possible Problems

Problem		Scale		Corrosion			Caustic Deterio-Ration	Carry over
Water	Where problem occurs → Water Characteristic ↓	Water feeding system	Boiler internal	Water feeding system	Boiler internal	Water returning	Boiler internal	Where steam is used
Feed Water	pH			♣	♣			
	Hardness	♦	♦♦					
	Dissolved Oxygen			♦	♦♦	♦		
	Oil Fat	.	♦					♦♦
Boiler Water	pH				♣♣		♦	♦
	Alkalinity							♦
	Total Dissolved Solids		♦		♦			♦♦
	Silica		♦					♦♦
Condensate	Free Carbon					♦♦		

Legend:

♣   When pH gets out of suitable range, corrosion becomes more severe.
♣♣ Implies especially severe corrosion
♦   The higher pH or of concentrated substance, the more severe problem
♦♦ Implies especially severe problem

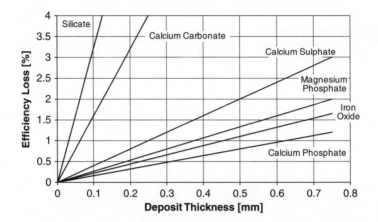

Figure 2.13   The Effect of Scale on Boiler Efficiency. Reproduced from: Good Practice Guide No. 30: Energy efficient operation of industrial boiler plant, Energy Efficiency Office, 1993

The effect of the formation of scale on loss of efficiency is represented in Figure 2.13. It is obvious that even a very thin layer of scale causes high efficiency loss.

Some of the most common physical and chemical characteristics of water are:

- *Turbidity:* The presence of colloidal solids gives liquid a cloudy appearance which is aesthetically unattractive and may be harmful. Turbidity in water may be due to clay and silt particles, discharges of sewage or industrial wastes, or to the presence of large numbers of microorganisms.
- *Dissolved solids:* This refers to any minerals, salts, metals, cations or anions dissolved in water. These include anything present in water other than pure water (H_2O) molecules and suspended solids. Suspended solids are any particles/substances that are neither dissolved nor settled in water. In general, total dissolved solids concentration is the sum of cations (positively charged) and anions (negatively charged) ions in water. Total Dissolved Solids (TDS) are the total amount of mobile charged ions, including minerals, salts or metals dissolved in a given volume of water, expressed in units of mg per unit volume of water (mg/l), also referred to as parts per million (ppm). TDS is directly related to the purity of water and the quality of water purification systems and affects everything that consumes, lives in, or uses water, whether organic or inorganic, whether for better or for worse.
- *Electrical conductivity:* Pure H_2O has virtually zero conductivity. The conductivity of a solution depends on the quantity of dissolved salts present. TDS and electrical conductivity are not the same thing but TDS can be calculated by converting electrical conductivity with the factor (K) of 0.5 to 1.0 times electrical conductivity:

$$K = \frac{\text{Conductivity [S/m]}}{\text{TDS [mg/l]}} \tag{2.45}$$

With known approximate value of K for particular water, the measurement of conductivity provides rapid indication of TDS content.
- *pH:* The intensity of the acidity or alkalinity of a sample is measured on the pH scale which actually measures the concentration of present hydrogen ions. This is expressed in a scale from 0 to 14 with 7 as neutrality, below 7 being acid and above 7 being alkaline.
- *Hardness:* This is the property of water which prevents lather formation with soap and produces scale in hot water systems. It is mainly due to metallic ions Ca++ and Mg++ although Fe++ and Sr++ are also responsible. The metals are usually associated with HCO_3-, SO_4-, $Cl-$, and NO_3-. There is

no health hazard, but the economic disadvantages of hard water include increased soap consumption and higher fuel costs. Hardness is expressed in terms of $CaCO_3$ and is divided into two forms:

- carbonate hardness: metals associated with HCO_3-
- non-carbonate hardness: metals associated with $SO_4 =$, $Cl-$, NO_3-

Boiler water chemistry is complicated and the key to success is testing feed-water as often as possible. The issue of chemical treatment is not only important for reducing fuel costs, but it is even more important for avoiding costly repairs and maintenance costs due to the danger of pipe cracks and corrosion. If there are no strictly recommended values of *boiler water* characteristics prescribed by the boiler manufacturer, the data presented in Table 2.11 can be used. These data are applicable for industrial boilers.

The quality of needed make-up water depends mostly on boiler design, operational pressure and temperature. The degree of required treatment also depends on boiler capacity, quality of available condensate return and raw-water analysis.

Table 2.11 Standard Boiler and Water Values. Reproduced from: Good Practice Guide No. 30: Energy efficient operation of industrial boiler plant, Energy Efficiency Office, 1993

Type of Boiler		Fire Tube Boilers		Water Tube Boilers				
Max. working pressure [bar]		–		Less than 10	10–20		20–30	
General treatment method		Raw Water	Soft Water	Soft Water	Soft Water	De-Mineralized Water	Soft Water	De-Mineralized Water
pH at 25 °C		> 7	> 7	> 7	> 7	> 7	> 7	> 7
Hardness $CaCO_3$	ppm	< 40	< 2	< 2	< 2	0	< 2	0
Oil and fat	ppm	0	0	0	0	0	0	0
Dissolved oxygen O_2	ppm	Lower as possible		< 0.5	< 0.5	< 0.5	< 0.1	< 0.1
General treatment method		Alkali treatment		Alkali treatment				
pH (25 °c)		11.0–11.8		11.0–11.8	11.0–11.5		11.5–11.0	
M alkalinity $CaCO_3$	ppm	500–800		500–800	< 600	< 300	< 300	< 150
P alkalinity $CaCO_3$	ppm	300–600		300–600	300–400	< 200	< 300	< 100

(continued overleaf)

Table 2.11 (*continued*)

Type of Boiler		Fire Tube Boilers			Water Tube Boilers			
Total dissolved solids	ppm	< 2500		< 2500	< 2000	< 700	< 1000	< 500
Chlorine ion Cl	ppm	< 400		< 400	< 300	–		–
Phosphoric acid ion $PO^3/_4$	ppm	30–50		30–50	20–40	10–30	20–40	10–30
Silica SiO_2	ppm	< 300		< 300	< 200	< 50	< 150	< 40
Hydrazine N_2H_4	ppm	0.05–0.3		0.05–0.3	0.05–0.3		0.05–0.3	
Sulfurous acid ion $SO^2/_3$	ppm	10–20		Oct-20	10–20		10–20	
Conductivity	μS/cm	< 3500		< 3500	< 2800	< 1000	< 1500	< 750

REMARKS:

Total Iron, Fe [ppm]	If Fe content in the feed water is more than 0.5, the Iron arrester has to be applied	Chlorine Ion, Cl⁻ [ppm]	Keep total solid matters under 20 %. It is indirect control of TDS.
Hydrazine, N_2H_4 [ppm]	It appears in the case hydrazine is used as de-oxidizer. It will not be used with sodium sulfur.	M alkalinity, $CaCO_3$ [ppm]	Keep total solid matters in the range of 20–30 %. It is indirect control of TDS.
Sulfurous acid ion, SO2⁻/3⁻ [ppm]	It appears in the case sodium sulfur is used as de-oxidizer. It shall not be used with hydrazine.	P alkalinity, $CaCO_3$ [ppm]	Keep silica over 1.7 times. Scale fixing prevention of silica.
Conductivity, [μS/cm]	Correctly speaking, totally dissolved solids have to be measured. Conductivity is used as substitute to measure matter impurity. It is used for indirect control of TDS.		

2.6 Opportunities for Boiler Performance Improvement

It has to be noted that, in practice, maximizing the steam system's energy efficiency does not start at the boiler house but with end-users. Poor control or heat use at this point is potentially the biggest cause of low steam system efficiency. The next reason can be steam distribution and condensate systems. After solving energy efficiency problems in these areas, attention should be paid to the boiler house. But, as has already been mentioned, it is possible to take action simultaneously or by using your own scheme. Anyhow, the scheme has to cover whole steam system and the steps in the implementation of the plan must be logical and without elimination of each other.

Concerning boiler energy efficiency improvement, we will consider the following energy conservation measures:

(a) combustion efficiency improvement;
(b) load scheduling;
(c) heat recovery techniques:

    (i) heat recovery from flue gas;
    (ii) combustion air pre-heating system;
    (iii) heat recovery from boiler blow-down;

(d) water treatment improvement;
(e) control system improvement.

2.6.1 Combustion Efficiency Improvement

Stable combustion requires an adequate mixture of fuel and oxygen. In practice, combustion conditions are never ideal and more air must be supplied to complete burning than in the case of theoretical or ideal combustion processes. The amount of air above the theoretical requirement is referred to as excess air. If insufficient air is supplied to the burner, unburned fuel and smoke and carbon monoxide appear in the flue gas. These may result in

- fouling of the heat transfer surface;
- pollution of the environment;
- decreasing in combustion efficiency;
- instability in the combustion process.

The main factors affecting flue gas loss are the flue gas temperature and its O_2 content. Typically, 15 %–20 % of fuel chemical energy is lost by flue gas. A gradual increase of flue gas temperature and consequent decrease of combustion efficiency indicates that

- minor adjustments or repairs in the control linkages are necessary;
- fuel valve or air damper have to be checked;
- the replacement of a worn burner tip or control may have to be done;
- the cleaning of heat transfer surfaces has to be performed.

To achieve proper combustion, the fuel must be well-mixed with combustion air. Excess air is a crucial variable affecting combustion efficiency and every attempt should be made to provide the lowest excess air level possible for an effective and efficient combustion process. Flue gas temperature is a good measure for unrecoverable heat discharged into the atmosphere. The higher the flue gas temperature, the greater the heat loss carried by dry flue gas and water vapor contained in the gas. The effect of excess air in flue gas loss is presented in Figure 2.14. The graph shows the variation of oxygen and carbon dioxide contents in the flue gas. The value zero on excess air axis represents theoretical combustion conditions. The curves in the diagram depend on fuel and have to be calculated for each particular fuel.[1] The flue gas loss depends, as has been mentioned, on its temperature and excess air. It also depends on combustion air temperature. Such a diagram provides a good opportunity to analyze the parameters influencing the

---

[1] Part III Toolbox 13 Software 4: *Fuel, Combustion and Environmental Impacts* can be used for drawing such a diagram for any fuel. More information on different fuels can be found in Part III Toolbox 5.

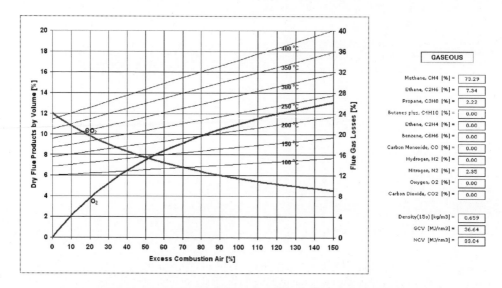

Figure 2.14 O_2 and CO_2 Contents and Flue Gas Loss versus Excess Air

combustion process in the particular boiler, but some measurements have to be performed in order to enable an investigation of the true situation.

The reasons for a high excess air, oxygen content in the flue gas or a high flue gas temperature can be very different and because of that any change in the operation of the burner has to be analyzed carefully. At the beginning, benchmark values have to be established. The best way is to accept the equipment manufacturer's recommendations.

A spot check of combustion conditions using the flue gas measurements of oxygen, CO, smoke and temperature is an effective method for assessing performance. These measurements have to be taken when the boiler operates under steady-state conditions and compared to those obtained during previous boiler adjustment tests and to benchmark values. These data help to identify any deterioration in combustion efficiency, as well as its possible causes. In many plants, efficiency-related data are checked on a daily basis, sometimes during each work shift.

Changes in the level of excess oxygen may be caused by problems in the air or fuel supply systems or may reflect changes in fuel properties. Other possibilities include deterioration in furnace baffling or setting leakage, excessive tube fouling, air heater plugging, or changes in atmospheric conditions. If excess oxygen has remained constant, but smoke or CO have increased, fuel specifications may have changed, or there may be a problem with burner operation.

While spot checks of combustion conditions are valuable for identifying efficiency degradation, long-term performance monitoring, including documentation in the boiler log, is a more effective means for ensuring peak efficiency. The objective of such a monitoring program is to document deviations from desired performance levels as the function of time.

The real case of fine-tuning of a 9.3 MW boiler is presented in Figures 2.15 and 2.16. The fuel is natural gas.

Boiler efficiency variation versus excess air is shown in Figure 2.16. The boiler efficiency is calculated based only on losses q_2 and q_3. It can be seen that very flat maximum efficiency exists in the range of excess air from 1.09 to 1.13. In this range, the CO in the flue gas will be practically zero, which is the most important criterion in fine-tuning the burner (minimum O_2 with lowest possible CO in the flue gas). Before fine-tuning the O_2, boiler efficiency was 83.34 % and after adjusting the burner, it is 84.41 %.

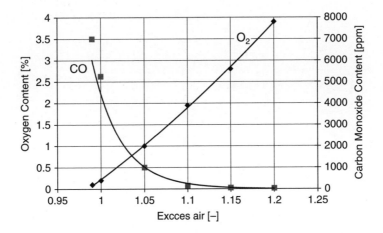

Figure 2.15 O_2 and CO versus Excess Air

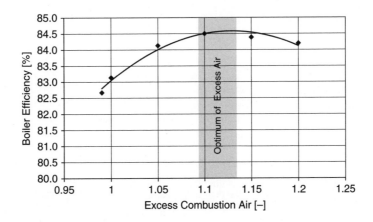

Figure 2.16   Boiler Efficiency (only q_2 and q_3 are taken) versus Excess Air

2.6.2 Load Scheduling

It has already been mentioned that the greatest efficiency of a boiler can be expected at full load. If more than one boiler is in the boiler house, the best way to reach the highest boiler house energy efficiency is to optimize the sequences of boilers being turned on and off bearing in mind their capacities and efficiencies versus loads and instantaneous steam consumption in the processes.

2.6.3 Waste Heat Recovery Techniques

2.6.3.1 Heat Recovery from Flue Gas

The temperature of flue gas leaving typical industrial boilers is generally in the range from 180 to 300 °C. In order to increase overall boiler efficiency, the waste heat of flue gas is used to increase the feed water temperature. A simple schematic diagram of a boiler economizer arrangement is presented in Figure 2.17.

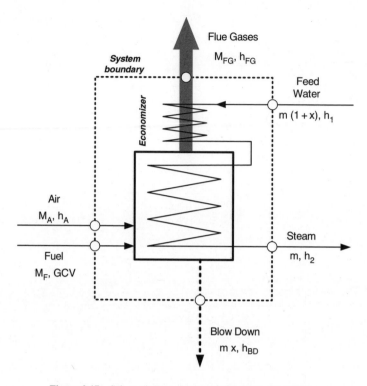

Figure 2.17   Schematic Presentation of Economizer Installation

By using Part III Toolbox 13 Software 7: *Efficiency of Steam Boilers*, the effect of feed water temperature increase on boiler efficiency can be calculated.

The potential of energy saving depends on the type of boiler and the fuel used. However, generally speaking, it can be said that boiler efficiency improvement will be up to 5 % (Table 2.3).

Economizers are usually constructed of steel or cast iron, which is more resistant to acid corrosion. The low temperature on heat transfer surfaces can occur during start up and shut down or running the boiler on a low load. However, in the case of clean fuels with minimum or without sulfur content, it is possible to get flue gas temperature below water dew point without corrosion problems. A condensing economizer provides the opportunity to increase boiler efficiency much more than mentioned above as the huge content of condensing water energy can be used for feed water or make-up water preheating.

2.6.3.2 Combustion Air Pre-Heating System

In order to improve boiler efficiency by 1 %, the combustion air temperature must be raised for approximately 40 °C. Most existing natural gas and oil burners used in industry are not designed for higher air temperatures. This means that the combustion air temperature, which is typically less than 50 °C can be increased to up to 50 °C maximum.

2.6.3.3 Heat Recovery from Boiler Blow-down

To keep the concentration of total dissolved solids (TDS) below the recommended level (see Table 2.11), water at the steam temperature must be blown down from the boiler and replaced with cooler low TDS feed water. This process removes some of the suspended and dissolved solids in the boiler water. Typically between 1 % and 5 % of the energy input to the boiler is lost in blow-down. Before discharge into drains, it is also necessary to cool blow-down water to below 40 °C (to comply with permit conditions). This is achieved by diluting it in the blow-down vessel with cold water.

In manual blow-down systems, a representative sample of the boiler water is tested. The correct TDS level is maintained by the operator controlling how often and how long the manual boiler bottom blow-down valve is opened. This form of control is basic and can result in wide fluctuations in measured levels of TDS. Provided the boiler can withstand these variations, this is generally not a problem as long as the severity and duration of variations are within the limits accepted by the boiler manufacturer and approved by the water treatment specialist.

Operation at TDS levels significantly below recommended levels wastes fuel due to the excessive amounts of boiler water discharged into drains. It is therefore preferable to provide for automatic control of TDS levels. Automatic control devices react upon sensors and monitoring and control equipment to regulate blow-down by using an automatically controlled boiler blow-down valve.

Table 2.12 indicates the percentage of fuel savings from a reduction of blow-down.

Both energy and water treatment efficiency can be improved by recovering flash steam and residual heat from blow-down. Up to 80 % of the heat in the boiler blow-down water can be recovered. Heat recovery can be effected using flash steam recovery vessels, heat exchangers or both. Such measures save energy by increasing the temperature of feed-water to the boiler and reducing the amount of fuel consumed in the boiler. Figure 2.18 shows a typical system with recovery of both flash steam and residual heat from blow-down water.

2.6.4 Boiler Water Treatment Improvement

The quality of water should and must be maintained in conformity with the instructions of the boiler manufacturers. We have already emphasized that not only energy efficiency but also safety of boiler operations are jeopardized by water quality which is not maintained within the prescribed limits. 'Excessively good' water quality, on the other hand, increases the cost of produced steam.

2.6.5 Boiler Control System Improvement

There is no doubt that a good control system provides for the high energy performance of a boiler and the entire steam system. In addition to compulsory standard control loops which ensure safe boiler operation, it is recommended that one establishes the monitoring of all factors relevant for boiler performance. Thus, each of the previously proposed measures for increasing energy efficiency implies the definition

Table 2.12 Percentage of Fuel Saving by Reduction of Blow-Down. Reproduced from: Good Practice Guide No. 30: Energy efficient operation of industrial boiler plant, Energy Efficiency Office, 1993

Boiler Pressure [barg]	Percentage of fuel saving for 1 % reduction in blow-down
7	0.19
10	0.21
17	0.25
25	0.28

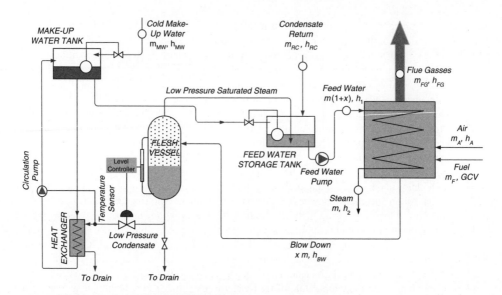

Figure 2.18   Heat Recovery from Blow-Down

of an adequate performance indicator that needs to be monitored in order to maintain maximum boiler efficiency.

2.7 Software for Boiler Performance Analysis

The software which can be used for analyses of boiler performance and quantification of the described performance improvement opportunities is in the Toolbox and will be presented here shortly. The data and formulae specified in Part III Toolbox 5 are used for the preparation of Part III Toolbox 13 Software 4: *Fuels, Combustion and Environmental Impacts*. The analysis of a steam system requires knowledge of the thermodynamic parameters of steam. Therefore, Part III Toolbox 13 Software 5 *Thermodynamic Properties of Water and Steam* is also provided. When only saturated steam is in question, it is possible to use the polynomial fittings equations obtained by using the method of least squares. These equations are provided in Part III Toolbox 4.

Specific software is attached for analyzing boiler losses (Part III Toolbox 13 Software 7: *Efficiency of Steam Boilers*). This software requires data input and some measurements to define the operation of the analyzed boiler. In addition to that, it enables the estimation of the boiler's efficiency improvement effects arising from the application of mentioned performance improvement measures.

Part III Toolbox 13 Software 9 *Heat Exchanger Operating Point Determination* supports analyses of applying economizers as a performance improvement opportunity, but it can be used for other purposes as well.

2.7.1 Software 4: FUELS, COMBUSTION AND ENVIRONMENTAL IMPACTS

This software is used for the computation of relevant parameters during combustions in two cases:

- when oxygen content in combustion products is zero;
- when combustion products contain oxygen.

In order to make this computation, it is necessary to input fuel composition (Fig. 2.19). Fuel composition is entered by mass percentage for solid and liquid fuels and by volume for gas fuels. If the fuel composition is not known, it is possible to select some of provided fuels. These fuels are classified in four groups. The

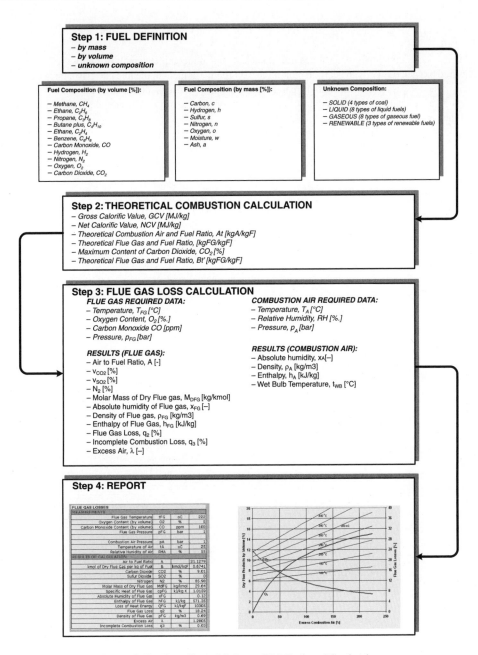

Figure 2.19 Flow Chart of Software IV-4: Fuels and Combustion

first group includes solid fuels, the second includes liquid fuels and the third group includes gas fuels. The fourth group includes biomass fuels (Step 1 in Fig. 2.19).

After entering the fuel composition or selecting some of provided fuels, the software computes the theoretical parameters of combustion (Theoretical AIR and FUEL Ratio [kg_{Air}/kg_{Fuel}]; Theoretical FLUE GAS and FUEL Ratio [$kmol_{Flue\ Gas}/kg_{Fuel}$]; Maximum Content of Carbon Dioxide in Flue Gas [%] and

Theoretical FLUE GAS and FUEL Ratio [$_{kgFlue\ Gas}$/kg$_{Fuel}$]) and Gross and Net Calorific Values (Step 2 in Fig. 2.19).

After this computation, it determines flue gas loss (q_2) and incomplete combustion loss (q_3) based on measured oxygen (O_2 [%]) and carbon monoxide (CO [ppm]) content in the flue gas. Also, it is necessary to know the flue gas temperature and the parameters of combustion air. Additionally, calculations provide data such as: enthalpy, density and absolute humidity of flue gas, loss of heat energy per kilogram of fired fuel and etc. (Step 3 in Fig 2.19).

For the selected fuel, it is also possible to obtain a diagram of oxygen and carbon dioxide concentration [%] in flue gas versus excess combustion air. The same diagram may also include flue gas loss [%] for the corresponding flue gas temperature. The results can be reported in a tabular and graphical format and exported as an Excel data file (Step 4 in Fig. 2.19).

2.7.2 Software 5: THERMODYNAMIC PROPERTIES OF WATER AND STEAM

The software is based on standard equations and constants.[2] It can be used in the range defined in Figure 2.20 (gray scale rectangle). In the superheated region of steam, any combination t & v; t & p; t & h; t & s; v & p; v & h; v & s; p & h; p & s and s & h can be selected and the rest of parameters can be calculated. The thermo-physical properties of saturated steam can be calculated too for any temperature within the range defined in Figure 2.20.

2.7.3 Software 7: EFFICIENCY OF STEAM BOILERS

A flow chart for this software is presented in Figure 2.21. There are six basic steps of efficiency analysis. The first defines fuel which is used in the same way as described in Software 4: *Fuels, Combustion and Environment*. After the fuel has been selected, the user enters the boiler's name plate data. Based on pre determined data, the software computes the so called best operation practice parameters of the boiler. These values are later used for comparisons with current operational performance.

In step 3, data regarding current operation are entered. Depending on the available data, the method of energy efficiency calculation (Direct or Indirect) can be selected. Data input requirements assume

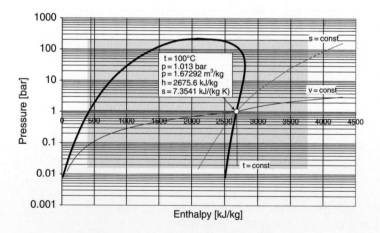

Figure 2.20   Pressure – Enthalpy Diagram of Water and Steam

---

[2] Reynolds, W.C.: *Thermodynamic Properties in SI* (Graphs, tables and computational equations for 40 substances), Department of Mechanical Engineering, Stanford University, 1979.

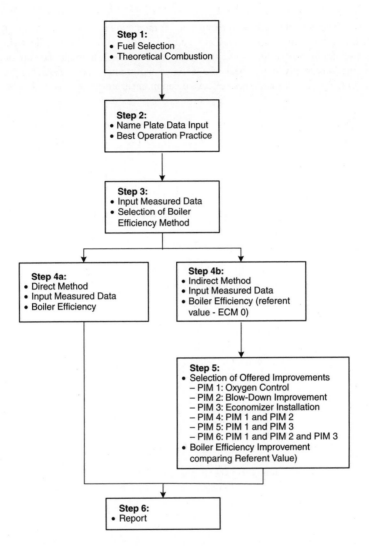

Figure 2.21   Flow Chart of the Software 7: Efficiency of Industrial Steam Boiler

that certain measurements have been carried out or that average values are used for some representative period. These values should be carefully prepared and analyzed.

If the direct method is selected, the software computes only the boiler efficiency and prepares a report in step 6.

In the indirect method, besides the computation of losses q_2, q_3, q_5 and q_7 for the current operational regime, the software provides opportunities for the quantification of effects arising from the following performance improvement measures:

- improved oxygen control;
- blow-down improvement;
- economizer installation.

It is possible to select any of the above measures or a combination thereof provided oxygen control is always used. The effect of each improvement is computed in relation to previously achieved improvement. Therefore, the use of all three measures gives a total improvement in relation to the current operation.

CO_2 and C emissions are given in a special frame for each of the analyzed cases. Therefore, it enables the assessment of the positive environment effects of certain measures.

2.7.4 Software 9: THE HEAT EXCHANGER'S OPERATING POINT DETERMINATION

The calculation procedure for this software is presented in detail in Part III Toolbox 13 Software 11. This is a flexible program for determining the parameters of the heat exchanger. It fully defines all of the relevant parameters required for the selection and design of heat exchangers. In the case of industrial boiler analyses, this program may be used for defining the economizer and heat exchanger for heat recovering of blow-down water.

The flow chart for the software is presented in Figure 2.22.

2.7.5 Example 2: Boiler Efficiency Improvement

2.7.5.1 Facility Description

The scheme of a case study boiler house is presented in Figure 2.23. The main characteristics of a boiler used in the factory are reported in Table 2.13. The boiler uses natural gas and its nominal capacity is 14.7 t/h of 11 barg saturated steam (name plate data). In real conditions, this capacity was impossible to achieve. This is frequently the case in practice.

Some simple computations were made and the results compared to the values specified by the manufacturer. The data presented in Table 2.13 are given as delivered by the suppliers of the boiler. As saturated steam pressure and feed water temperature are known, the heat output is equal to:

$$Q_{1,nom} = \frac{14.7}{3.6} \cdot (2784.3 - 4.21 \cdot 75) = 10079.9 \,[kJ/s]\,[kW] \tag{2.46}$$

Natural gas nominal consumption is also known, therefore it is also possible to compute heat input. The heat input proportional to fuel burned will include heat in fuel (chemical heat) and heat of combustion air as follows:

$$Q_{F(G\ or\ N)} = M_F \cdot (GCV\ or\ NCV + A_s \cdot h_A) \tag{2.47}$$

Where M_F is flow rate of fuel and GCV and NCV are gross or net calorific values of fuel. h_A is enthalpy of combustion air and A_s [kg/kg$_F$] is stochiometric air/fuel ratio.

By using given data, the following results are obtained:

$$Q_{F,GCV} = \frac{900 \cdot 0.9381}{3600} \cdot (38.72 \cdot 1000 + 12.88 \cdot 80.75) = 9324.7 \,[kJ/s]\,[kW] \tag{2.48}$$

$$Q_{F,NCV} = \frac{900 \cdot 0.9381}{3600} \cdot (34.91 \cdot 1000 + 12.88 \cdot 80.75) = 8431.2 \,[kJ/s]\,[kW] \tag{2.49}$$

The enthalpy of air is calculated for $p_A = 1.013$ bar, $t_A = 35\,°C$ and $RH_A = 50\,\%$. In this case, we use air to fuel ratio for stoichiometric combustion. The difference compared to the real ratio is negligible for this analysis. The fuel parameters are taken for the natural gas used in this case study (Table 2.15).

It is obvious that $Q_{1,nom}$ cannot be greater than any of Q_F. The reasons for great errors are wrong and imprecise data specifications. However, the manufacturer's specified data were used for establishing the so called best practice operation procedure as referent data for comparison of the boiler's real practice operation.

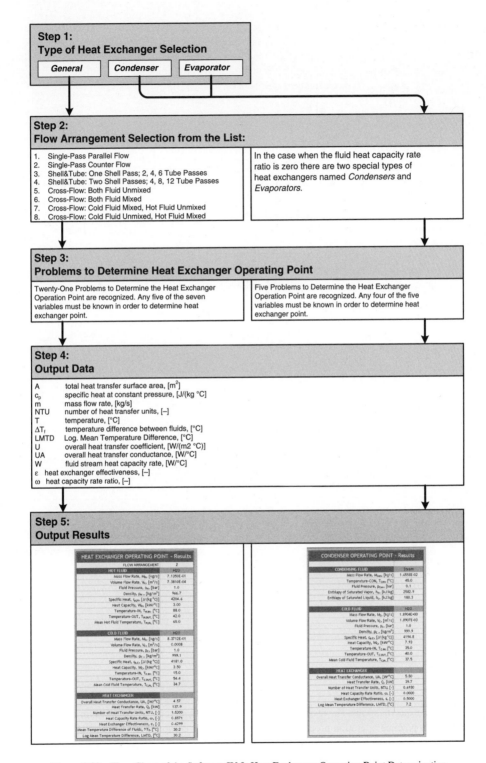

Step 1:
Type of Heat Exchanger Selection

General   Condenser   Evaporator

Step 2:
Flow Arrangement Selection from the List:

1. Single-Pass Parallel Flow
2. Single-Pass Counter Flow
3. Shell&Tube: One Shell Pass; 2, 4, 6 Tube Passes
4. Shell&Tube: Two Shell Passes; 4, 8, 12 Tube Passes
5. Cross-Flow: Both Fluid Unmixed
6. Cross-Flow: Both Fluid Mixed
7. Cross-Flow: Cold Fluid Mixed, Hot Fluid Unmixed
8. Cross-Flow: Cold Fluid Unmixed, Hot Fluid Mixed

In the case when the fluid heat capacity rate ratio is zero there are two special types of heat exchangers named *Condensers* and *Evaporators*.

Step 3:
Problems to Determine Heat Exchanger Operating Point

Twenty-One Problems to Determine the Heat Exchanger Operation Point are recognized. Any five of the seven variables must be known in order to determine heat exchanger point.

Five Problems to Determine the Heat Exchanger Operation Point are recognized. Any four of the five variables must be known in order to determine heat exchanger point.

Step 4:
Output Data

A   total heat transfer surface area, [m^2]
c_p specific heat at constant pressure, [J/(kg °C)]
m   mass flow rate, [kg/s]
NTU   number of heat transfer units, [–]
T   temperature, [°C]
ΔT_f temperature difference between fluids, [°C]
LMTD   Log. Mean Temperature Difference, [°C]
U   overall heat transfer coefficient, [W/(m2 °C)]
UA   overall heat transfer conductance, [W/°C]
W   fluid stream heat capacity rate, [W/°C]
ε   heat exchanger effectiveness, [–]
ω   heat capacity rate ratio, [–]

Step 5:
Output Results

HEAT EXCHANGER OPERATING POINT - Results	
FLOW ARRANGEMENT	2
HOT FLUID	H2O
Mass Flow Rate, M_H, [kg/s]	7.1350E-01
Volume Flow Rate, V_H, [m^3/s]	7.3810E-04
Fluid Pressure, p_H, [bar]	1.0
Density, ρ_H, [kg/m^3]	966.7
Specific Heat, c_{pH}, [J/(kg °C)]	4204.6
Heat Capacity, W_H, [kW/°C]	3.00
Temperature-IN, $T_{cH,in}$, [°C]	88.0
Temperature-OUT, $T_{H,out}$, [°C]	42.0
Mean Hot Fluid Temperature, $T_{H,m}$, [°C]	65.0
COLD FLUID	H2O
Mass Flow Rate, M_C, [kg/s]	8.3712E-01
Volume Flow Rate, V_C, [m^3/s]	0.0008
Fluid Pressure, p_C, [bar]	1.0
Density, ρ_C, [kg/m^3]	999.1
Specific Heat, c_{pC}, [J/(kg °C)]	4181.0
Heat Capacity, W_C, [kW/°C]	3.50
Temperature-IN, $T_{C,in}$, [°C]	15.0
Temperature-OUT, $T_{C,out}$, [°C]	54.4
Mean Cold Fluid Temperature, $T_{C,m}$, [°C]	34.7
HEAT EXCHANGER	
Overall Heat Transfer Conductance, UA, [W/°C]	4.57
Heat Transfer Rate, Q, [kW]	137.9
Number of Heat Transfer Units, [-]	1.5200
Heat Capacity Rate Ratio, ω, [-]	0.8571
Heat Exchanger Effectiveness, ε, [-]	0.4299
Mean Temperature Difference of Fluids, ΔT_f, [°C]	30.2
Log Mean Temperature Difference, LMTD, [°C]	30.2

CONDENSER OPERATING POINT - Results	
CONDENSING FLUID	Steam
Mass Flow Rate, M_{cond}, [kg/s]	1.6558E-02
Temperature-COM, T_{com}, [°C]	46.0
Fluid Pressure, p_{cond}, [bar]	0.1
Enthalpy of Saturated Vapor, h_v, [kJ/kg]	2582.9
Enthalpy of Saturated Liquid, h_L, [kJ/kg]	188.3
COLD FLUID	H2O
Mass Flow Rate, M_C, [kg/s]	1.8904E+00
Volume Flow Rate, V_C, [m^3/s]	1.8907E-03
Fluid Pressure, p_C, [bar]	1.0
Density, ρ_C, [kg/m^3]	999.9
Specific Heat, c_{pC}, [J/(kg °C)]	4194.8
Heat Capacity, W_C, [kW/°C]	7.93
Temperature-IN, $T_{C,in}$, [°C]	35.0
Temperature-OUT, $T_{C,out}$, [°C]	40.0
Mean Cold Fluid Temperature, $T_{C,m}$, [°C]	37.5
HEAT EXCHANGER	
Overall Heat Transfer Conductance, UA, [W/°C]	5.50
Heat Transfer Rate, Q, [kW]	39.7
Number of Heat Transfer Units, NTU, [-]	0.6930
Heat Capacity Rate Ratio, ω, [-]	0.0000
Heat Exchanger Effectiveness, ε, [-]	0.5000
Log Mean Temperature Difference, LMTD, [°C]	7.2

Figure 2.22   Flow Chart of the Software IV-9: Heat Exchanger Operating Point Determination

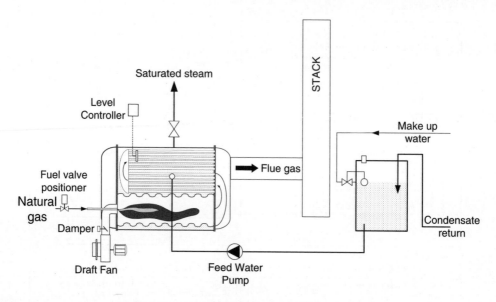

Figure 2.23 Scheme of boiler from a case study of example 2

Table 2.13 Boiler Specification

Title	Unit	Values
Type of boiler		Package
Number of passes		3 pass fire tube
Maximum allowable working pressure	barg	14
Nominal steam production	t/h	14.7
Working pressure	barg	11
Steam temperature	°C	185
Type of fuel – oil Burner		Natural gas $Q_{max} = 11.5\,MW$ $P_{max} = 250\,mbar$ $P_{min} = 150\,mbar$
Number of burners		1 set
Rated fuel consumption (Natural gas)		900 nm^3/h
Heat output		9000 kW
Air temperature	°C	35 °C
Feed water temperature	°C	75 °C

Table 2.13   (*continued*)

Title	Unit	Values
Electric power	kW	18.5 kW (2 sets)
– Feed water pump – Draft fan		30 kW (12 415 m³/h) (70 mbar)
Dimensions (diameter x length)	mm	3000 × 5800 (mm)
Manufacturer of boiler		–
Manufacturer of burner		– (Rotating cap type)
Age	year	12

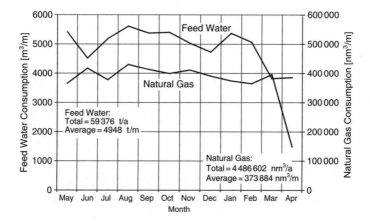

Figure 2.24   Monthly Steam Generation and Natural Gas Consumption

2.7.5.2 Steam Generation and Fuel Consumption

Monthly consumption of feed water and natural gas for a period of 12 months is presented in Figure 2.24. Both of them are measured regularly. However, two independent teams performed the measurements. Normally, these two curves should have similar trends. This figure shows that when comparing March and April this is not the case. The natural gas consumption dropped dramatically in April, compared to March, but the feed water (in direct relation with steam production) was practically the same for these two months. The explanation is found in the fact that natural gas readings were made at different times in the month and the feed water (steam) consumption was fitted to the every first day in the month. The NG reading in April was late and the gas consumption made in April was attached to March consumption.

This is why measured and recorded feed water consumption is taken as referent values for further calculations. Of course, a noted problem was resolved and daily readings of both values were taken at the same time of day.

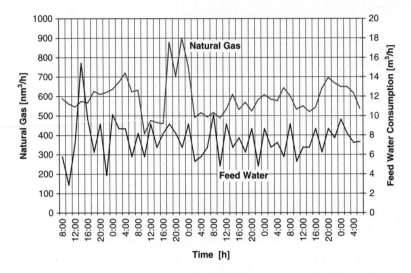

Figure 2.25   NG and Feed Water Measurement Results (during Detail Energy Audit)

Additional measurements of feed water and natural gas flow rates provided the results as presented in Figure 2.25.

The number of operating hours during the observed 12 months was 7 200 hours.

The average obtained values are:

$$m_{NG} = 593.9 \pm 95.69 \, \frac{nm^3}{h} \tag{2.50}$$

$$m_{FW} = 7.52 \pm 2.03 \, \frac{m^3}{h} \tag{2.51}$$

$$t_{FW} = 76.1 \pm 6.43 \, °C \tag{2.52}$$

With annual working hours of 7 200 and measured values of natural gas and feed water, the average hourly consumptions are:

$$M_{NG} = \frac{4\,486\,602}{7200} = 623.1 \, \frac{nm^3}{h} \tag{2.53}$$

$$M_{FW} = \frac{59\,376}{7\,200} = 8.25 \, \frac{m^3}{h} \tag{2.54}$$

2.7.5.3 Measurements, Technical Calculations and Analysis

The results of measurements performed during a detailed energy audit DEA are presented in Table 2.14.

The conductivity of boiler water is 1366 μS/cm with standard deviation of 116.5 μS/cm and feed water conductivity is 9.3 μS/cm with standard deviation of 7.6 μS/cm. For this analysis, daily conductivity measurements for the seven days before the stated DEA measurements are used. Acceptable values for boiler water are from 1000 to 2500 μS/cm and for feed water it has to be less than 100 μS/cm (the manufacturer's recommendation). This means that the quality of water is better than necessary. Although this is good for the boiler, the steam cost is unjustifiably increased.

The condensate return is estimated based on temperatures of feed, condensate and make-up waters. The average temperatures are 76.1, 83.0 and 25.0 °C, respectively.

Table 2.14   Measurement results (DEA)

Name		Unit	Low load	High load
Ambience	Temperature	°C		34.6
	Relative humidity	%		74
Fuel	Temperature	°C		–
	Flow	m^3/h		593.9
Feed water	Temperature	°C		76.1
	Flow	m^3/h		7.52
Feed water tank	Surface temp.	°C		37
Blow-down tank	Surface temp.	°C		–
Steam	Pressure	barg		10.5
	Flow (main)	t/h		–
Flue gas	Flue gas temperature	°C		207
	Oxygen content	%		6.0
	CO_2	%		8.5
	SO_2	ppm		1
	NO_2	ppm		0
	CO	ppm		11
Conductivity of feed water	C_{FW}	µS/cm		9.3
Conductivity of boiler water	C_{BW}	µS/cm		1366

Simple heat balance gives the following:

$$CR = \frac{76.1 - 25.0}{83.0 - 25.0} \cdot 100 = 88.1\,\% \tag{2.55}$$

This result indicates that approximately 88 % of condensate is returned to the boiler.

2.7.5.4 Best Operation Practice

The best practice operation parameters (the manufacturer's recommendations for a new or freshly cleaned boiler) are as follows:

- Load: Full
- Flue gas oxygen content   1.2 %
- Temperature of flue gas   186 °C
- Feed water conductivity   100 µS/cm
- Boiler water conductivity   2500 µS/cm
- Natural gas consumption   900 nm^3/h

Table 2.15   Fuel Parameters

FUEL (Content by volume):	Unit	Value
CH_4	%	74.28
C_2H_6	%	5.71
C_3H_8	%	1.45
C_4H_{10}	%	0.91
C_2H_4	%	0.00
C_6H_6	%	0.00
CO	%	0.00
H_2	%	0.00
N_2	%	2.27
O_2	%	0.00
CO_2	%	15.38
Density (15 °C, 1.01325 bar)	kg/m^3	0.9381
GCV	MJ/nm^3	36.32
NCV	MJ/nm^3	32.75
GCV	MJ/kg	38.72
NCV	MJ/kg	34.91

The composition of fuel is provided by Table 2.15. The results of the calorific values calculations are also represented.

Assuming that the boiler runs full load (10.75 t/h steam generation) and that flue gas oxygen content (1.2 %), flue gas temperature (186 °C), boiler and feed water conductivities (2500 and 100), feed water temperature (75 °C), combustion air temperature (35 °C) are as specified by the manufacturer, Part III Toolbox 13 Software 7: *Efficiency Steam Boilers* provides the results as presented in Table 2.16. The calculated fuel consumption is 900 nm^3/h or 844.3 kg/h. For this particular boiler, this is the best efficiency of fuel energy transformation.

We can conclude that for the specified maximum fuel consumption and for the given parameters of saturated steam and feed water and for the available natural gas, the efficiency of the boiler is 81.19 % and the steam production is 10.75 [t/h]. This steam production is less than that specified by the manufacturer (14.7 t/h). It should be noted that conductivity of boiler water was 2500 [μS/cm] and of feed water 100 [μS/cm]. By comparing the data measured during a detailed energy audit it can be seen that these data are much higher. Later, will use this fact to calculate the effects of blow-down improvement.

Table 2.16 Best Practice Operation

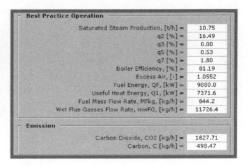

Name Plate Data (ultimate data)	
Rated Fuel Consumption, [nm3/h] =	900
Working Pressure (abs), [bar] =	12
Flue Gas Oxygen Content, [%] =	1.2
Feed Water Temperature, [oC] =	75
Flue Gas Temperature, [oC] =	186
Boiler Water Conductivity, [µS/cm] =	2500
Feed Water Conductivity [µS/cm] =	100
Make-Up Water Temperature [oC] =	25
Combustion Air Temperature [oC] =	35

Best Practice Operation	
Saturated Steam Production, [t/h] =	10.75
q2 [%] =	16.49
q3 [%] =	0.00
q5 [%] =	0.53
q7 [%] =	1.80
Boiler Efficiency, [%] =	81.19
Excess Air, [-] =	1.0552
Fuel Energy, QF, [kW] =	9080.0
Useful Heat Energy, Q1, [kW] =	7371.6
Fuel Mass Flow Rate, MFkg, [kg/h] =	844.2
Wet Flue Gasses Flow Rate, mwFG, [kg/h] =	11726.4

Emission	
Carbon Dioxide, CO2 [kg/h] =	1827.71
Carbon, C [kg/h] =	498.47

2.7.5.5 Efficiency Calculation Based on Measured Values

Boiler testing and daily data log analyzing proved that most of time the boiler operated with a higher oxygen content and consequently higher excess air. In spite of higher excess air and the reduced load of the boiler, the flue gas temperature was higher than that suggested by the best practice operation. The CO was low (only 11 ppm). During DEA, the specific measurements were performed and the results are presented in Table 2.17. The data were analyzed using the software described and the results are presented in Table 2.18. The calculated fuel consumption is 593.9 nm³/h, and the boiler efficiency is 79.48 %.

The main purpose of this software is to estimate the effects of feasible performance improvement measures (PIM). The first measure was to apply the automatic control of flue gas oxygen and to improve the blow-down system (PIM 4 = PIM 1 + PIM 2). The continuous conductivity measurements existed but blow-down water and generated flash steam were not used for the preheating of make-up water.

The calculation of the effects of the implementation these two measures is presented in Table 2.19. The total improvement of boiler efficiency is from 79.48 % to 80.85 %. It is interesting, as by applying only

Table 2.17 Input Measured Data

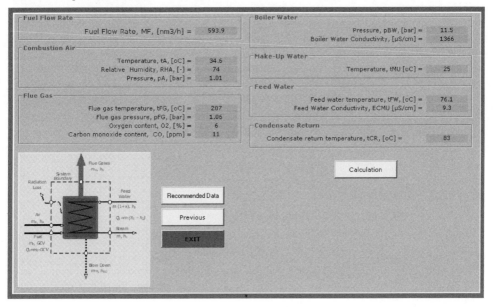

Table 2.18   Current Operation of the Boiler

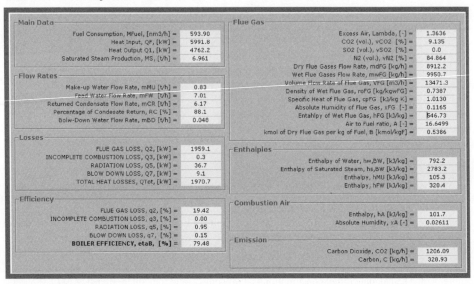

Table 2.19   Effects of Implementation of PIM 1 and PIM 2

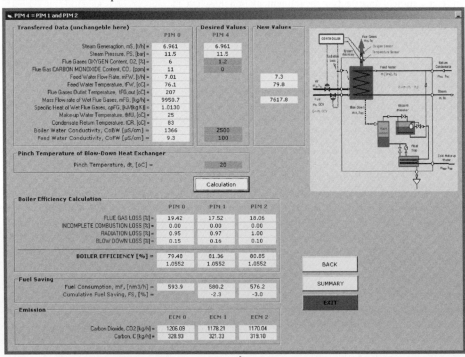

PIM 1 (oxygen trim), efficiency is increased to 81.36 %. The reason is to be found in an unnecessary high quality of feed water. By applying PIM 2, the quality of feed water is reduced and feed water conductivity is increased from the current 11 to 100 [μS/cm], as recommended by the manufacturer. The blow-down water flow rate will be increased but the cost of chemical water treatment will be reduced.

In addition to applying these measures (PIM 1 and PIM 2), it is possible and advisable to apply an economizer as well (PIM 3). The computation and the results in the case of applying this measure in addition to the previous two measures are presented in Table 2.20. The application of these measures enables an efficiency increase from 79.48 % to 83.23 %. There is no sulfur in flue gas so the flue gas temperature can be reduced by up to 120 °C. The feed water temperature will be increased from the current 76.1 to 104.6 °C.

If all of the performance improvement measures are applied boiler efficiency can be improved by 9.1 %.

For the economizer calculation, Part III Toolbox 13 Software 9: *Heat Exchanger Operating Point Determination* was used. For the cross flow heat exchanger (both fluids unmixed), flow arrangement No. 5 and task No. 4 were used (Table 2.21).

2.7.5.6 Summary of Performance Improvement Measures (PIMs)

2.7.5.6.1 (1): Automatic Control of Combustion Process and Blow-Down Heat Recovery

Following the manufacturer's recommendation of an oxygen content of 1.2 % (see also Table 2.6) and repeating the last calculation it is possible to find the difference in fuel consumption for the same steam production:

$$\Delta NG = 593.9 - 576.2 = 17.7 \, \text{nm}^3/\text{h}.$$

Table 2.20 Results of Calculation for PIM 6

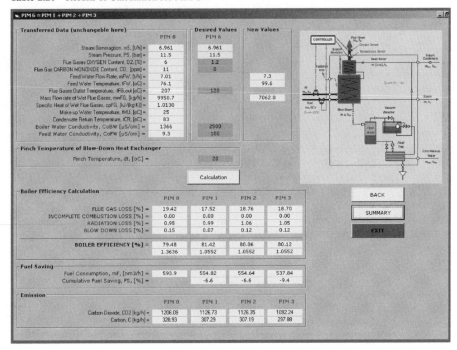

Table 2.21   Economizer Calculation

INPUT DATA

HOT FLUID			COLD FLUID	
		PROBLEM No. 4		
		Flow Arr. No. 5		
Wh [kW/K] =	2.8		Wc [kW/K] =	8.537
Th,in [oC] =	207	UA [kW/K] = Unknown	Tc,in [oC] =	76.1
Th,out [oC] =	120		Tc,out [oC] = Unknown	
(Th,out<Th,in)			(Tc,out>Tc,in)	

Calculation

OUTPUT DATA

HOT FLUID				COLD FLUID	
Wh [kW/K] =	2.80	UA [kW/K] =	3.75	Wc [kW/K] =	8.54
Th,in [oC] =	207.0	Q [kW] =	243.6	Tc,in [oC] =	76.1
Th,out [oC] =	120.0	NTU =	1.34	Tc,out [oC] =	104.6
Th,ave [oC] =	163.5	Omega =	0.3280	Tc,ave [oC] =	90.4
		Effectiveness =	0.6646		
		dTfluids [oC] =	65.0		
		LMTD [oC] =	69.1		

This is worth $0.2090 \cdot 7200 \cdot 17.7 = 26\,635$ \$ US/year. The NG price was 0.2090 \$US/nm^3.

The ratio of blow-down water and feed water is only 0.69 %. It is proposed to recover heat from blow-down by installing a flash vessel and heat exchanger in the way represented in Figure 2.19. The existing blow-down control system is connected to new controller.

2.7.5.6.2 PIM (2): Installation of Economizer
Using software and setting the flue gas temperature at 120 °C the additional heat energy recovery will be as follows:

$$\Delta Q = 243.6\,\text{kW} \tag{2.56}$$

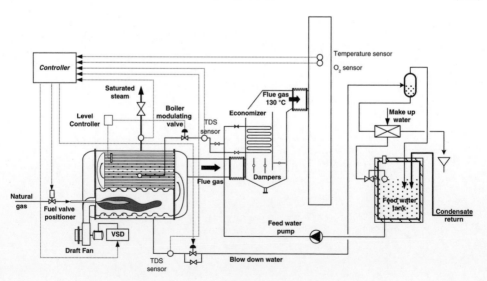

Figure 2.26   Automatic Control System and Economizer

The NG reduction is then:

$$\Delta NG = 576.2 - 539.5 = 36.7 \, \text{nm}^3/\text{h} \tag{2.57}$$

The expected NG savings are 276 480 nm^3/year or 57 784 \$US/year. The steam production is the same. For the boiler's full steam production of 10.75 t/h (maximum) and reduced flue gas temperature, the power of the economizer has to be specified at approximately 375 kW. As the pressure drop on the gas side of the economizer is only approximately 5 mbar, it is not necessary to replace the existing draft fan or install additional one.

The scheme of automatic control of the combustion process and blow-down is shown in Figure 2.26. The economizer is also represented in this figure. The oxygen sensor gives the signal to the controller (set value is 1.2 %) which adjusts the flow of air by a frequency regulator in order to keep the adequate fuel to air ratio. The pressure of the steam is controlled by the fuel flow rate.

2.7.5.6.3 Basis for Engineering Design and Cost Estimate

Project: **BOILER HOUSE**	**PIM (1):** **Automatic control of combustion process and blow-down**	
Investment specification:		*Energy cost reduction:*
Equipment:		**0.02664 m\$US**
− Oxygen sensor (zircon) − Temperature sensor − Motorized NG valve − Frequency regulator (VSD) − Feed water flow meter − Controller		*Simple pay back period:*
Installation: Training:		**0.66 years**
TOTAL Cost:	0.0175 m\$US	

Project: **BOILER HOUSE**	**PIM (2):** **Installation of economizer**	
Investment specification:		*Energy cost reduction:*
Economizer (turn key price):		**0.0578 m\$US**
− *Flue gas flow rate = 16000 kg/h;* − *Inlet temperature = 207 °C;* − *Outlet temperature = 120 °C;* − *Water flow rate = 14700 kg/h;* − *Water inlet temperature = 75 °C;* − *Heat flow rate = 375 kW)* (installation is included)		*Simple pay back period:* **0.88 years**
TOTAL Cost:	0.0510 m\$US	

2.8 Boiler Performance Monitoring

The purpose of performance monitoring is to compare actual performances with expected (targeted) values. Continuous analysis is based on measured values and operational corrections oriented towards the fulfillment of best performance practice. In the previous chapters (Part I Chapter 3), we have provided examples of how to process and interpret data. A framework for both energy and environmental monitoring is provided in Part I Chapter 4. We will elaborate here on the values that need to be monitored.

2.8.1 Values to be Monitored

All of the relevant parameters which should be measured in order to determine the performance indicators of boilers have already been discussed in the examples of performance improvement measures. Successful monitoring requires defining which data are needed and what data collection frequency is required for the anticipated measurements. There is no doubt that continuous measurement is the best although monitoring success does not depend only on that, but also on the manner of using measured values.

Figure 2.27 shows the basic measuring points and values which should be measured. The cells marked in grey represent the values which are not compulsory for measurement, but which are useful if obtained. For example, steam temperature (M2) should be measured if overheated steam is concerned. Or, if steam

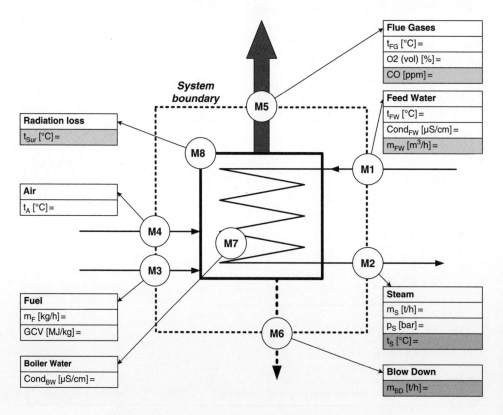

Figure 2.27 Measuring Points for Boiler Performance Monitoring

flow is measured with satisfactory accuracy, it is not necessary to measure feed water flow although this value is in fact measured very often.

Table 2.22 provides recommendations not only for the parameters which must be measured, but also for the frequencies of those measurements. In addition to that, the monitoring purposes for which these values are used are specified. Normally, for small capacity boilers, the requirements are not so rigorous.

Table 2.22   Relevant Parameters for Steam Boiler Monitoring

PARAMETER TO BE MONITORED	PURPOSE	Checking and Recording Frequency	
		< 20 t/h steam	20–100 t/h steam
OIL FIRED BOILERS			
Calorific value	Heat input calculation.	No checks. Only figures provided by the supplier can be used.	Density and sulfur content at each delivery.
Fuel content (distribution company certification)	Possible negative effects on heat transfer surfaces and flue gas emissions.	No checks. Only figures provided by the supplier can be used.	Control the fuel content annually.
Consumption	Heat input calculation.	Metering and daily recording.	Metering and recording continuously.
Viscosity	Burner operation control.	From time to time	At each delivery.
Flue gas temperature	Efficiency calculation	Metering and recording continuously.	Metering and recording continuously. Automatic control
Flue gas analysis	Efficiency Calculation (q_2 loss using O_2 content; q_3 loss using CO content)	Daily analysis. Automatic control is desirable.	Metering and recording continuously. Automatic control
GAS FIRED BOILERS			
Calorific value	Heat input calculation	No checks. Only figures provided by the supplier can be used.	No checks. Only figures from supplier can be used.
Fuel content	Possible negative effects on heat transfer surfaces and flue gas emissions.	No checks. Only figures provided by the supplier can be used.	Control the fuel content annually.

(*continued overleaf*)

Table 2.22   (*continued*)

PARAMETER TO BE MONITORED	PURPOSE	Checking and Recording Frequency	
		< 20 t/h steam	20–100 t/h steam
Consumption	Heat input calculation	Metering and daily recording.	Metering and recording continuously.
Flue gases temperature	Efficiency calculation (q_2).	Metering and recording continuously.	Metering and recording continuously. Automatic control
Flue gases analysis	Efficiency calculation (q_2 loss using O_2 content; q_3 loss using CO content)	Daily analysis. Automatic control is desirable.	Metering and recording continuously. Automatic control
COAL FIRED BOILERS			
Calorific value	Heat input calculation.	Only figures provided by the supplier can be used. Monthly measurement of moisture and ash.	Monthly measurement of calorific value
Fuel content (distribution company certification)	Possible negative effects on heat transfer surfaces and flue gas emissions.	No checks. Only figures provided by the supplier can be used.	For each delivery, coal content has to be measured.
Consumption	Heat input calculation.	Daily gauging.	Metering and recording continuously
Flue gases temperature	Efficiency calculation.	Metering and recording continuously	Metering and recording continuously.
Flue gases analysis	Efficiency calculation.	Daily O_2 and CO analysis.	Metering and recording continuously.
Slag analysis	Efficiency calculation.	Daily visual inspection.	Shift analysis.
WATER			
Feed water analysis	Blow-down rate.	Shift measurements	Metering and recording continuously
Feed water temperature	Efficiency calculation.	Metering and recording continuously	Metering and recording continuously

Table 2.22 (*continued*)

PARAMETER TO BE MONITORED	PURPOSE	Checking and Recording Frequency	
		< 20 t/h steam	20–100 t/h steam
Feed water flow rate	Steam production calculation if steam flow rate is not measured.	Metering and recording continuously	Metering and recording continuously
Make up water analysis	Condensate return calculation	Daily measurements.	Metering and recording continuously.
Boiler water analysis	Bow-down rate.	Shift measurements.	Metering and recording continuously.
Condensate return temperature	Calculation of feed water temperature.	Shift measurements.	Metering and recording continuously
STEAM			
Steam flow rate	Efficiency calculation, heat output.	Metering and recording continuously or feed-water flow rate metering.	Metering and recording continuously.
Steam pressure	Heat output calculation.	Metering and recording continuously.	Metering and recording continuously.

2.9 Steam Distribution and Condensate Return System

Example 1, as described earlier, shows that possible losses in the steam distribution and condensate return systems are considerable and thus their reduction can contribute significantly to steam system performance improvement. Let us consider in more detail these issues and estimate the potential for performance improvement of steam distribution and condensate return systems. These two systems will be analyzed together in this section.

2.9.1 Steam Distribution System Performance Analysis

In practice, a generally' accepted indicator for the assessment of the efficiency of steam distribution systems does not exist. Performance assessment is reduced to individual analyses of the state of insulation, leakages and functionality of steam traps. However, the most accurate manner of defining the performance of a steam distribution system is through the relation between the total delivered energy of steam to the end-users and the energy transformed into steam and delivered into the distribution system from the boiler house (see Fig. 2.1). In other words, the efficiency of a steam distribution system can be estimated as follows:

$$\eta_{SD} = \frac{\sum_{n=1}^{N} m_{S,n} \cdot h_{S,n}}{m_{Steam} \cdot h_{steam}} \tag{2.58}$$

Where:
η_{SD} = efficiency of steam distribution system, [-]
m_{Steam} = mass flow rate of steam entering the steam distribution system, [kg/s]
h_{Steam} = enthalpy of steam entering steam distribution system, [kJ/kg]
$m_{S,n}$ = mass flow rate of steam n-th (out of N) end-users, [kg/s]
$h_{S,n}$ = enthalpy of steam i-th (out of n) end-users, [kJ/kg]

The total saturated steam flow rate entering the distribution system is 10 t/h and pressure is 12 bar. An inevitable drop of steam pressure will occur in the system, as well as condensate of a small quantity of steam due to pressure drop and heat losses. Produced condensate is separated regularly at certain points in the system but part of the live steam can be lost because of uncontrolled leakage through unwanted holes in the system. When leaving the system at the boundary towards end-users, saturated steam is of lower pressure and flow than the steam which has entered the system. Figure 2.28 shows that the pressure drop has very little influence on efficiency. However, the decrease in the mass flow rate of steam can seriously jeopardize it. Thus, the loss of some 10 % of mass flow rate of steam (for the given example) which has occurred because of leakage and/or condensate can reduce efficiency of the system by around 10 %.

2.9.2 Influencing Factors on Steam Distribution System Performance

Typical steam distribution system losses are as follows:

- steam leaks;
- heat transfer loss through insulation;
- condensate loss; and
- flash steam loss.

These losses are easily observed and should be one of the priorities for energy management. By proper maintenance they can be reduced to a minimum.

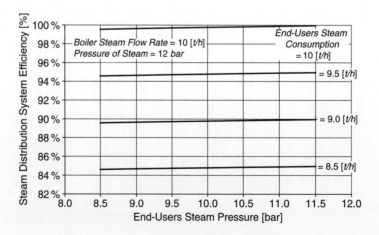

Figure 2.28   Steam Distribution System Efficiency versus End-Users Pressure

2.9.2.1 Steam Leaks

Reducing steam leaks is an important area for potential savings for industrial facilities. Two main types of failures result in steam leaks:

(1) steam trap failures; and
(2) valve, flange and pipe failures.

However, steam loss through a leak is difficult to determine. One of many approximate methods of determining heat loss from steam leak is based on the approximate size of the leak hole and is presented in Figure 2.29. This diagram can be used only as rough quantification of steam leaks. However, the use of this and similar diagrams or analytical estimates can produce sufficiently reliable data for estimating losses and defining strategies for the elimination of these losses.

2.9.2.2 Insulation

The determination of the amount of energy lost from non-insulated or poorly insulated equipment will provide the basis for determining the extent of an insulation improvement project. The main factors that affect the amount of energy lost from non-insulated or poorly insulated equipment are:

- process fluid temperature;
- ambient temperature;
- surface area exposed to heat transfer.

Tables and diagrams have been developed for typical arrangements such as horizontal and vertical pipe vessels, valves and flanges. These tables and diagrams allow for simple heat loss estimation. One example of the tabular presentation of heat losses, which can be useful for losses estimation, is presented in Table 2.23. The bare pipe surface temperature is app. 75 °C for condensate pipes and approximately 150 °C for steam pipes.

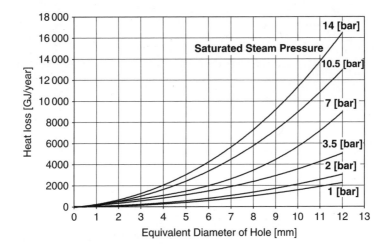

Figure 2.29 Heat Loss versus Leaks (Heat loss is calculated for the working pattern based on 8760 hours per year)

Table 2.23 Typical Steam Piping Heat Losses. Reproduced from: Improving Steam System Performance (a source-book for industry), DOE/GO-102002-1557, June 2002

Piping Item	Pipe Status	Unit	Heat Loss	
			Steam	Condensate
All pipes	*Well-insulated*	W/m	35–130	15–35
Flat surface	*Well-insulated*	W/m²	180	70
32 mm N.B. pipe	*Bare*	W/m	340	110
50 mm N.B. pipe	*Bare*	W/m	480	150
65 mm N.B. pipe	*Bare*	W/m	580	185
80 mm N.B pipe	*Bare*	W/m	660	210
100 mm N.B. pipe	*Bare*	W/m	820	260
125 mm N.B. pipe	*Bare*	W/m	960	310
150 mm N.B. pipe	*Bare*	W/m	1 150	370
200 mm N.B. pipe	*Bare*	W/m	1 500	–
Flat surface	*Bare*	W/m²	2 300	700

The heat loss of an un-insulated or insulated valve can be calculated using the practical rule that this heat loss can be approximated with 1.0 m of bare or insulated pipe of the same diameter. A similar case is that for an un-insulated flange. This heat loss can be approximated to that of 0.5 m of bare or insulated pipe.

2.10 Condensate Return System

The steam trap management program should investigate every steam trap and determine if condensate is captured or lost from the system. This has been pointed out earlier. This section serves to highlight the primary aspects associated with condensate recovery that include collecting condensate and returning it to the boiler.

A simple calculation indicates that energy in the condensate can be more than 10 % of the total steam energy content of a typical system. Figure 2.30 shows the heat remaining in the condensate at various condensate temperatures, for a steam system operating at 10 bar, with make-up water from 20 to 60 °C. The heat remaining is calculated by using the following formula:

$$\text{Heat Remaining in Condensate } [\%] = \frac{h_C - h_{MU}}{h_{SS} - h_{MU}} \cdot 100\,\% \qquad (2.59)$$

where:
h_c = Condensate enthalpy, kJ/kg
h_{MU} = Make-up water enthalpy, kJ/kg
h_{ss} = Saturated steam enthalpy, kJ/kg

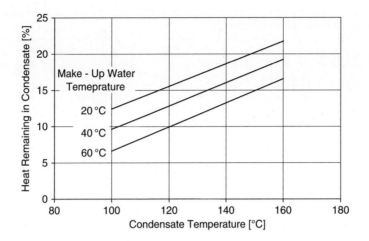

Figure 2.30   Heat Remaining in Condensate versus Condensate Temperature

2.10.1 Flash Steam Recovery

In many process applications, condensate enters a steam trap as saturated liquid with high pressure. The enthalpy of high-pressure saturated liquid is greater than the enthalpy of saturated liquid at reduced pressure; therefore, the low-pressure fluid exiting the trap cannot exist as liquid condensate alone. As a result, some of the liquid condensate 'flashes' to steam. The remainder of the liquid condensate exists as saturated liquid condensate at lower pressure. The low-pressure flash steam exists as saturated vapor at low pressure. This steam is available for use in low-pressure steam systems. The amount of flash steam per kilogram of condensate can be estimated using Figure 2.31.

A common and effective method used for flash steam recovery is represented in Figure 2.32. The system is identical to the blow-down flash steam recovery system described in section 2.6.3.3.

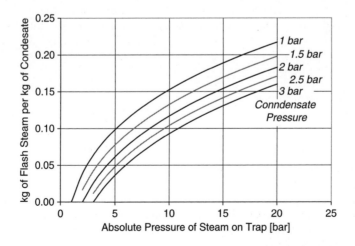

Figure 2.31   Quantity of Flash Steam Generated from Condensate Pressure Reduction

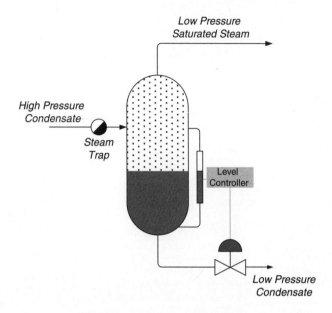

Figure 2.32 Flash Steam Recovery System

A flash steam recovery system not only reduces economic loss from the system but also reduces the steam flow in condensate return systems. The condensate flow out of the flash tank will have much less flash steam. If the discharge is pumped and the pressure that is maintained is greater than the flash vessel pressure, no flash steam will enter in the condensate discharge piping.

2.10.2 Performance Improvement Opportunities

In general, measures which can be applied to steam distribution and condensate return systems are technically simple and relatively low-cost, and should provide for an attractive return on investment.

The possible and recommended measures for energy savings in the steam distribution system are as follows:

- improvement of insulation;
- return of more condensate;
- improvement of steam trapping;
- repair of steam leaks;
- modification of piping;
- installation of condensate heat recovery;
- modification of operating pressure;
- recovery of flash steam from condensate.

The basic and most important rule when condensate is in question is to provide maximum condensate return into the boiler. This certainly implies the utilization of flash steam. The heat contained in condensate is significant (Fig. 2.30) which is a sufficiently convincing reason that it should be returned into the boiler, if economically justified.

2.10.3 Software 8 STEAM SYSTEM INSULATION

This software enables the calculation of losses in the steam distribution system and the quantification of the effects of insulation improvement. There are three possible cases for analysis:

- single pipe of a given diameter, thickness and type of insulation;
- part of a steam distribution system;
- single tank with given dimensions (vertical cylinder, horizontal cylinder and box shaped).

The software flow chart is presented in Figure 2.33.

For any temperature of fluid in the pipe and ambient air, it is possible to obtain a diagram of the relationship between the outside pipe diameter and heat loss for a meter in length of bare pipe and of five given insulation thicknesses. Such a diagram is presented in Figure 2.34.

Part III Toolbox 10 provides the computation elements used for the preparation of this software.

2.10.4 Example 3: Steam and Condensate System Insulation Improvements

The part of the steam and condensate installation with all the relevant data is shown in Figure 2.35. The current heat losses have to be calculated and the performance improvement measures have to be proposed. The insulation is made of mineral wool. The temperature of the condensate is 75 °C. This part of installation is used for 7 000 hours per year. The boiler uses heavy fuel oil and its efficiency is 0.75.

By using Software IV-8 *Steam System Insulation*, the losses of current steam distribution and condensate return systems can be calculated. This can be done by individual calculation pipe by pipe or by

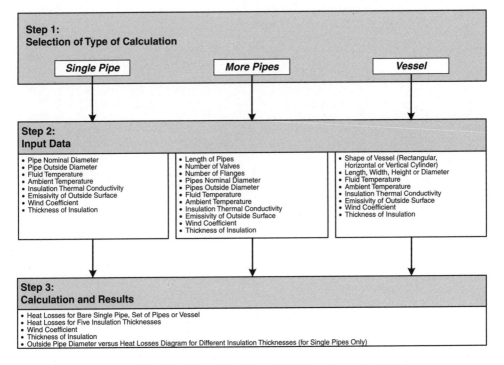

Figure 2.33 Flow Chart of the Software IV-8: Steam System Insulation

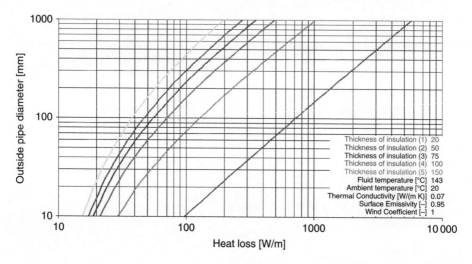

Figure 2.34   Outside Pipe Diameter versus Heat Loss for Bare Pipe and Five Different Insulation Thicknesses

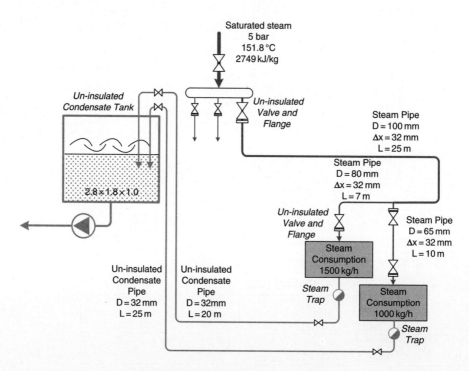

Figure 2.35   Scheme of Steam/Condensate Pipeline

Table 2.24 Heat losses for the current situation

No.	Description	D [mm] Diameter	L[m] Length	Δx [mm] Insulation Thickness	Q [W]
1	Steam	100	25	32	1 983
2	Steam	80	7	32	458
3	Steam	65	10	32	583
4	Valve & Flange	100	2V+2F	0	2 484
5	Valve & Flange	80	1V+1F	0	994
6	Valve & Flange	65	1V+1F	0	867
7	Condensate	32	20+25	0	4 471
8	Condensate	32	2V+2V	0	298
9	Condensate Tank	2.8 × 1.8 × 1.0 m		0	9 676
				TOTAL HEAT LOSSES:	**21 815 (21.8 kW)**

aggregate calculation of all pipes, valves and flanges. The complete calculation for the analyzed part of the steam and condensate systems is given in Table 2.24. The total heat loss of this part of the steam distribution system is 21.8 kW.

(a) Heat losses per year:

$$Q_{Year} = Q\,[kW] \cdot N\,[h/year] = \quad (2.60)$$
$$21.8 \cdot 7\,000 = 152\,600\,[kWh/year]$$

(b) The heavy fuel oil loss per year is:

$$\Delta M_F = \frac{\text{Heat Loss}}{\text{Boiler Efficiency} \cdot \text{GCV}} = \frac{\frac{152\,600 \cdot 3\,600}{1000}}{0.75 \cdot 42.9} = 17\,074\,[kg/year] \quad (2.61)$$

(c) If the fuel oil price per liter is 0.2076 $US, the money loss is:

$$\text{Money Loss} = \frac{0.2076\,[US\$/l]}{0.95\,[kg/l]} \cdot 17\,074\,[kg/year] = 3\,731\,[\$US/year] \quad (2.62)$$

(d) It is proposed to insulate the un-insulated part of the pipeline using mineral wool with an aluminum alloy casing (thickness of insulation 32 mm, ambient temperature 25 °C, insulation thermal conductivity 0.05 W/(m K), emissivity of outlined surface 0.9 and wind coefficient 1). The calculation in the case of the insulated pipeline is given in Table 2.25.

(a) The heat loss after lagging insulation of un-insulated part of the distribution system is as follows:

$$Q^*_{Year} = Q^*\,[kW] \cdot N\,[h/year] = 5.2 \cdot 7000 = 36\,400\,[kWh/year] \quad (2.63)$$

Table 2.25 Heat Losses after Improved Insulation

No.	Description	D [mm]	L [m]	Δx [mm]	Q* [W]
1	Steam	100	25	32	1984
2	Steam	80	7	32	458
3	Steam	65	10	32	583
4	Valve & Flange	100	2V+2F	32	38.2
5	Valve & Flange	80	1V+1F	32	34.1
6	Valve & Flange	65	1V+1F	32	87
7	Condensate	32	20+25	32	46.0
8	Condensate	32	2V+2V	32	60
9	Condensate Tank	2.8 × 1.8 × 1.0 m		32	1310
				TOTAL HEAT LOSSES:	**5 235 (5.2 kW)**

(b) The heavy fuel oil loss per year after lagging insulation is:

$$\Delta M_F^* = \frac{\text{Heat Loss}}{\text{Boiler Efficiency} \cdot \text{GCV}} = \frac{\frac{36\,400 \cdot 3\,600}{1\,000}}{0.75 \cdot 42.9} = 4\,073\,[\text{kg/year}] \qquad (2.64)$$

(c) If the fuel oil price per liter is 0.2076 US$, the money loss is:

$$\text{Money Loss}^* = \frac{0.2076\,[\$US/l]}{0.95\,[\text{kg/l}]} \cdot 4,073\,[\text{kg/year}] = 890\,[\$US/\text{year}] \qquad (2.65)$$

(d) Annual energy cost saving is: ECS = 3 731 − 890 = 2 841 $US/year

(e) If the cost of insulation (32 mm) is 11 $US per meter of length and 30 $US per square meter (for tank insulation), the cost of insulation will be:

$$\text{Cost of Insulation} = 55.5 \cdot 11 + 26.88 \cdot 30 = 1\,414.5\,\$US \qquad (2.66)$$

(f) Simple payback period is:

$$\text{Simple Pay Back} = \frac{1\,414.5}{2\,841} = 0.50\,\text{Year} \qquad (2.67)$$

2.10.5 Performance Monitoring of Steam Distribution and Condensate Return System

Monitoring of these two systems is considerably simpler although no less important than monitoring boilers. We have shown above that their individual efficiencies have an impact on the performance of the entire steam system. We have identified the main factors affecting the performance of steam distribution and condensate return systems. There is no need for any special additional measurements but only regular visual surveillance and simple control of steam traps which are an integral part of such systems. Simply, all leakages must be eliminated, defective steam traps either repaired or replaced and complete condensate returned to the feed water tank. Even polluted condensate should be recuperated for preheating the make-up water.

2.11 Environmental Impacts

2.11.1 Ambient Air Pollution

The most important polluting substances from the combustion process (depending on the fuel composition) are:

(1) sulfur dioxide (SO_2) and other sulfur compounds;
(2) oxides of nitrogen (NOx) and other nitrogen compounds;
(3) carbon monoxide *(CO)*;
(4) volatile organic compounds (*VOC*), in particular hydrocarbons (except methane);
(5) heavy metals and their compounds;
(6) dust, asbestos (suspended particulates (*PM*) and fibers), glass and mineral fibers;
(7) chlorine and its compounds;
(8) fluorine and its compounds.

The most environmentally desirable fuel is natural gas and red dye distillate fuel oil. Natural gas-fired boilers emit significantly lower levels of particular matters, nitrogen oxides (NOx), and sulfur oxides (SOx), than boilers fired with other fuels. An example of the differences in potential emissions is shown in Table 2.26. Although the use of red dye distillate fuel (0.05 % sulfur by weight) results in higher emissions than natural gas, it is significantly cleaner than other grades of oil or solid fuels. For example, the sulfur contents allowed in red dye distillate, light fuel oil, and heavy fuel oil are 0.05 %, 0.3 % and up to 2.2 % sulfur by weight, respectively.

2.11.2 Water Pollution

Boiler blow-down water is released periodically from a boiler to remove impurities and sediment. It is categorized as industrial wastewater. The purpose of releasing blow-down is to limit the buildup of

Table 2.26 Potential Emissions in Tons per Year (2.93 MW Capacity Boiler)*. Reproduced from: Boiler Environmental Certification Workbook, Massachusetts Department of Environmental Protection

Fuel	NOx	SOx	VOC	PM	CO
Natural Gas 12 Months	1.5	0.03	1.3	0.44	3.5
Red Dye Distillate*** 12 Months	6.6	2.4	1.3	0.88	3.5
Natural Gas-9 Months; Red Dye Distillate-3 Mo.	2.8	0.6	1.3	0.55	3.5
#6 Residual Fuel 12 Months	13.1	47.4	0.34**	2.14**	1.5**
Major Source Threshold	50	100	50	100	100

* Potential emissions mean the maximum capacity of a boiler to emit a pollutant while operating continuously, i.e., 8 760 hours per year.
** These potential emissions are calculated using EPA AP-42 emission factors. The other potential emissions are calculated using regulatory emission limits.
*** Red dye distillate fuel oil means a distillate fuel oil with very low sulfur.

contaminants in the boiler water that may degrade boiler performance and increase maintenance costs. Contaminants may include one or more of the following:

(1) dissolved/suspended minerals;
(2) heavy metals (iron, copper);
(3) corrosion inhibitors;
(4) oil or algaecides.

Although boiler blow-down is generally non-hazardous, if it is contaminated with oil, it must be stored and disposed of as hazardous industrial wastewater.

Table 2.27 Carbon Emission Factors

Fuel	kg C/kWh	kg CO_2/kWh
Grid electricity – Delivered[1] – Primary[2]	0.117 0.0453	0.43 0.1661
Natural gas	0.0518	0.19
Coal	0.0817	0.30
Coke	0.101[3]	0.37[3]
Petroleum coke	0.0927	0.34
Gas/diesel oil	0.068	0.25
Heavy fuel oil	0.0709	0.26
Petrol	0.0655	0.24
LPG	0.0573[4]	0.21[4]
Jet kerosene	0.0655	0.24
Ethane	0.0545	0.20
Naphtha	0.0709	0.26
Refinery gas	0.0545	0.20

1. The carbon emission factor for delivered electricity should be used when taking consumption as read from the meter.
2. The carbon emission factor for primary electricity should be used in calculations for Climate Change Agreements, where all energy use is reported in terms of primary energy.
3. Climate Change Agreement participants should use 0.117 kg C/kWh (0.43 kg CO_2/kWh) for Coke.
4. Climate Change Agreement participants should use 0.0627 kg C/kWh (0.23 kg CO_2/kWh) for LPG.

2.11.3 Carbon Emission Factors

Table 2.27[3] provides sample emission figures that can be used to assess reductions in the emissions of Carbon (C) and Carbon Dioxide (CO_2)[4] and potential performance improvement measures (PIM). Official figures are not available. The emission figure for, say heating fuels is calculated as kgC/kWh delivered energy in the form of heat.

2.11.4 Example 4: Practical Calculation of Carbon Emission Reduction

Example 3, described earlier, will be used to illustrate the practical application of the figures from Table 2.27. If all the measures proposed to improve performance in that example were to be applied, the expected fuel consumption would equal to:

$$\Delta M_F = 593.9 - 542.07 = 51.83 \, [\text{nm}^3/\text{h}] \tag{2.68}$$

As the fuel's GCV $= 36.32 \, [\text{MJ/nm}^3]$ (Table 2.15), it follows that expressed in energy units, the following reduction of fuel consumption is achieved:

$$\Delta M_F = 51.83 \cdot 36.32 = 1882.7 \, [\text{MJ/h}] \quad \text{or} \quad 522.9 \, [\text{kWh/h}] \tag{2.69}$$

The annual operation hours of the factory are 7200 h/a and fuel consumption before and after applying the proposed measures are medium values. Therefore, the estimate of overall annual reduction of the physical consumption of fuel is as follows:

$$\Delta M_{F,\text{annual}} = 51.83 \cdot 7200 = 373\,176 \, [\text{nm}^3/\text{a}] \tag{2.70}$$

or in energy units :

$$\Delta M_{F,\text{annual}} = 522.9 \cdot 7200 = 3\,764\,931 \, [\text{kWh/a}] \tag{2.71}$$

If this quantity of primary energy is not used, the emission of carbon is reduced by:

$$\Delta C = 0.0518 \cdot 3\,764\,931 = 195\,023 \, [\text{kgC/a}] \tag{2.72}$$

or carbon dioxide by:

$$\Delta CO_2 = 0.19 \cdot 3\,764\,931 = 715\,337 \, [\text{kgCO}_2/\text{a}] \tag{2.73}$$

As emissions are proportional to fuel consumption, this means that both observed emissions are reduced by around 3 %.

2.12 Bibliography

Boiler Environmental Certification Workbook, Massachusetts Department of Environmental Protection, www.mass.gov/dep/service/online/boilwbk.pdf.

Council Directive 84/360/EEC of 28 June 1984 on the combating of air pollution from industrial plants, www.europa.eu.int.

Council Directive 96/61/EC of 24 September 1996 concerning integrated pollution prevention and control, www.eper-prtr.kvvm.hu/docs/96L0061.pdf.

---

[3] The factors given in the table are taken from Guidelines for the measurement and reporting of emissions in the UK Emission Trading Scheme (http://www.actionenergy.org.uk).

[4] Conversion between C and CO_2 can be calculated using the relative atomic weights of the carbon and oxygen. The atomic weight of carbon is 12 and that of CO_2 is 44. To convert from C to CO_2 multiply by 44/12 and to convert from CO_2 to C multiply by 12/44.

Dockrill, P., Friedrich, F., Boilers and Heaters (2001) Improving Energy Efficiency, Natural Resources Canada, http://oee.nrcan.gc.ca

Dukelow, S.G. (1991) The Control of Boilers (Second Edition), Instrument Society of America.

Eastop, T.D., Croft, D.R.(1990) Energy Efficiency (for Engineers and Technologists), Longman Scientific & Technological.

Eastop, T.D., McConkey, A. (1993) Applied Thermodynamics (for Engineering Technologist - S.I. Units), Fifth Edition, ELBS with Longman.

Elonka, S.M., Kohan, A.L. (1999) Standard Boiler Operators' Questions & Answers, TATA McGraw-Hill Publishing Company, Ltd., New Delhi.

Elonka, S.M., Higgins, A. (1982) Standard Boiler Room Questions & Answers, TATA McGraw-Hill Publishing Company, Ltd., New Delhi.

EN 12953-11 (2003) Shell boilers, European Committee for Standardization.

Energy Consumption Guide (ECG092) (2004) Steam distribution costs, Action Energy, www.actionenergy. org.uk.

Energy Efficiency in Steam Systems, Energy Conservation and Efficiency Program (ECEP) (1990) RCG/Haler, Bailly, Inc., Washington.

Energy Technology Handbook (1977) editor-in-chief D. M. Considine, McGraw-Hill.

Fuel Efficiency Booklet No. 1 (1995) Energy Audit for Industry, Energy Efficiency Office.

Fuel Efficiency Booklet No. 2 (1993) Steam, Energy Efficiency Office.

Fuel Efficiency Booklet No. 8 (1993) The Economic Thickness of Insulation for Hot Pipes, Energy Efficiency Office.

Fuel Efficiency Booklet No. 14 (1984) Economic Use of Oil-Fired Boiler Plant, Energy Efficiency Office.

Fuel Efficiency Booklet No. 15 (1984) Economic Use of Gas-Fired Boiler Plant, Energy Efficiency Office.

Fuel Efficiency Booklet No. 18 (1983) Boiler Blowdown, Energy Efficiency Office.

Fuel Efficiency Booklet No. 19 (1993) Process Plant Insulation and Efficiency, Energy Efficiency Office.

Good Practice Guide No. 30 (1993) Energy efficient operation of industrial boiler plant, Energy Efficiency Office.

Good Practice Guide No. 369 (2004) Energy efficient operation of boilers, Action Energy, www.actionenergy.org.uk.

Harrell, G., Steam System Survey Guide, www.ntis.gov/support/ordernowabout.htm.

Improving Steam System Performance (a sourcebook for industry) (2002) DOE/GO-102002-1557, June.

Kaupp, A.(1997) Performance Testing of Industrial Combustion Systems for Efficiency Improvements, Seminar held at King Mongkut's Institute of Technology, Bangkok.

Nadaški, M., Gvozdenac, D.(2003) The Frequency Regulation Control System installed at 9.3 MW Hot-Water Boiler (Case Study), PSU-UNS International Conference 2003: Energy and the Environment, Prince Songkla University, Hat Yai, Thailand, 11–12 December.

Perry's Chemical Engineers' Handbook (1984) (sixth edition).

Petrović, J, Gvozdenac, D, Perunoviæ, P (1990) Monitoring of the Operating Thermal Performances in Water Heating Boiler - Case Study, in The Euro-Arab Workshop on Energy Conservation in Industry, Vol. 2, edited by D. Gvozdenac, V. Ishchenko, UNDP/UNIDO Project RAB/89/022.

The Boiler House, Spirax Sarco, http://www.spiraxsarco.com.

The Energy Saver (the complete guide to energy efficiency) (1994) Gee Publishing Ltd, London.

Witte, L. C., Schmidt, P. S., Brown, D. R. (1988) Industrial Energy Management and Utilization, Hemisphere Publishing Corporation.

3

Industrial Electric Power System

3.1 Introduction

Today, almost 30 % to 40 % of all conventional energy resources in the world are transformed into electric energy (hydropower plants are not included here). An electric power system that supplies electricity to final consumers may be divided into four parts: generation, transmission, distribution and consumption. Figure 3.1 describes the losses associated with each part of a thermal-electric power system. Losses are expressed as percentage of input energy in an observed section of the energy supply chain. The 'load' represents the portion of input (fuel) energy available in the observed part of the supply chain. Typically for a traditional thermal power plant only slightly more than 30 % of input energy is delivered to final consumers.

Therefore, improvements at performance level on the demand side have a significant impact on the entire power supply industry. The considerations here will focus only on the demand side of the power supply, i.e., the performance of electricity use in an industrial plant. We will cover the performance issues from the industrial power system perspective, i.e., the issues that are common to all industrial electricity end-users. User or application specific issues will be covered in subsequent chapters.

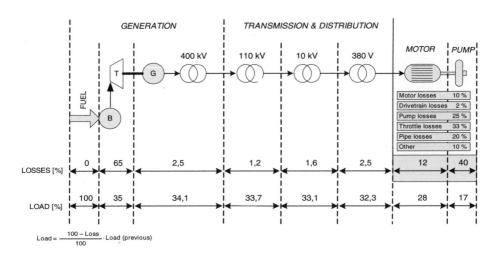

Figure 3.1 Electric Power System

Applied Industrial Energy and Environmental Management   Zoran K. Morvay and Dušan D. Gvozdenac
© 2008 John Wiley & Sons, Ltd

The approach adopted in this chapter assumes that the reader has very little or no background in electrical engineering, so everybody can get at least get a basic idea of how to improve the performance of electricity use in an industrial environment.

3.2 Description of Industrial Electric Power Systems

In most cases, an industrial electric power system starts from one or more medium voltage transformers located within a company's perimeter. Some companies may be connected to the grid on high voltage (110 kV) or low voltage (380 v), and some may have their own power generation or co-generation unit on site (see Part II Chapter 6).

In any case, an industrial distribution network branches out to form the busbars of a transformer sub-station (Fig. 3.2). The feeders (usually electric cables) will take electricity flows to the main departments and units of a company, where local electrical installations will branch out further to reach every end-user.

Other equipment at a transformer substation usually includes circuit breakers, capacitors, protection devices, fuses and metering instrumentation. In addition to the main commercial meters, generally there are meters at every feeder supplying a department or a larger consumer. It is rarely the case that existing metering arrangements are sufficient to cover the needs of performance measurement systems as required by the designation of energy cost centers. Therefore, further sub-metering is necessary for particular users or processes throughout a factory.

3.3 Basic Terms

In order to analyze the performance of an industrial electric power system, we have to be familiar with the basic concepts of power and energy, then load diagram, load factor, peak demand, concepts of active and reactive energy and power factor, and finally, the definition and importance of quality of power supply.

3.3.1 Active and Reactive Power and Power Factor

In an electrical power system, two components of power are present. The resistive component is real or active power which is found, for example, in electric motors converted from electrical into mechanical power and into heat in the form of motor losses. The second component, reactive power, is caused by the inductive elements (like windings) of the machine.

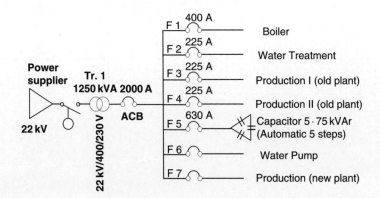

Figure 3.2   Simplified Single Line Diagram of an Industrial Electric Power System

The total power requirement of a load is made up of these two components, namely the active part and the reactive part. The resistive portion of the load cannot be added directly to the reactive component since it is essentially 90 degrees out of phase with the other. The pure active power is expressed in watts, while the reactive power is expressed as reactive volt amperes. To compute the total volt ampere load, it is necessary to analyze the power triangle indicated in Box 3.1.

Box 3.1: Active and reactive Power Triangle and Related Terms

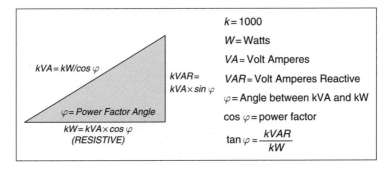

The power of a balanced 3-phase load can be expressed as:

$$\text{Power} = \sqrt{3} \cdot V_L \cdot I_L \cdot \cos \varphi$$
$$(\text{Watts} = \text{Volt Amperes} \times \text{Power Factor})$$

(3.1)

where V_L and I_L are line voltage and current respectively, while for 1-phase load, the active power P equals:

$$\text{Power} = V_L \cdot I_L \cdot \cos \varphi$$

(3.2)

The reactive power Q is expressed as:

$$Q = V_L \cdot I_L \cdot \sin \varphi$$

(3.3)

Active and reactive powers are connected in the power triangle with so called apparent power S, which may be expressed as:

$$S^2 = P^2 + Q^2$$

(3.4)

The power factor is defined as the ratio between active (P) and apparent power (S), and is also the cosine of the angle φ.

$$PF = \frac{P}{S} = \cos \varphi$$

(3.5)

The power factor *(cos φ)* can vary between 0 and 1, or between 0 and 100 %. When *cos φ* = 1 (PF = 100 %), that is an ideal case where there is no reactive power, and hence no losses on that account. Most of the equipment utilized by modern industry causes a poor power factor. One of the worst offenders is the lightly loaded induction motor. Examples of the equipment that cause poor PF and the approximate power factors follow:

- *100 % power factor or near to it*: heating systems used in ovens or dryers present a PF of close to unity.
- *80 % power factor or better:* electric motors in air conditioners (correctly sized), pumps, fans or blowers.

- *60 % to 80 % power factor:* induction furnaces, standard stamping machines and weaving machines.
- *60 % power factor and below:* single stroke presses, automated machine tools, finish grinders, welders.
- *50 % power factor is typical for lighting:* the PF for most incandescent lamps is unity. Fluorescent lamps usually have low power factor; fluorescent lamps are sometimes supplied with compensation devices to correct low power factor. Mercury vapor lamps have low PF; 40 %–60 % is typical if without compensation devices.

The PF of a transformer varies considerably as to the function of load applied to a device, as well as in the design of the transformer. An unloaded transformer will be highly inductive and therefore exhibit low PF.

The greater the angle φ becomes, the lower the power factor. The $cos\ \varphi$ or the power factor is very important in the transmission and distribution of electric energy. This is because when the power factor is 'bad or low' (values around 0.7 or less) total losses in transmission and distribution will increase. Therefore, the power factor is one element on the basis of which a charge is paid if excessive quantities of reactive power are required by a facility.

3.3.2 Load Diagram

Electric power will be consumed by an end-user as and when needed. This means that the load on the system will keep on changing with time. The curve showing the load demand of a device against time, on a daily basis, is known as a daily load diagram or load profile. If the variation of the load is considered over a period of one week or a month, then the plot of the load demand against time is known as weekly or monthly load diagram. If such a diagram is plotted for the load requirements of all electricity users in the factory, it will be useful in determining the characteristics of power requirements and in understanding power supply economics. Actually, the shape of a load diagram is the source of major billing items on electricity bills, and the basis for various tariff system charges, as explained later.

A sample load diagram is shown in Figure 3.3. The load diagram could have varying profiles depending on working arrangements. If a company operates three shifts, the load diagram would be fairly flat, while

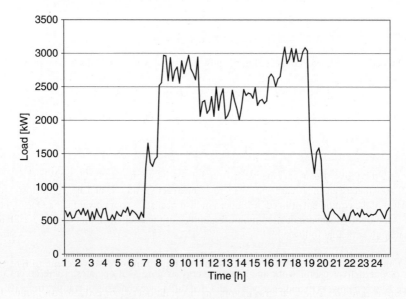

Figure 3.3 Typical Daily Load Diagram

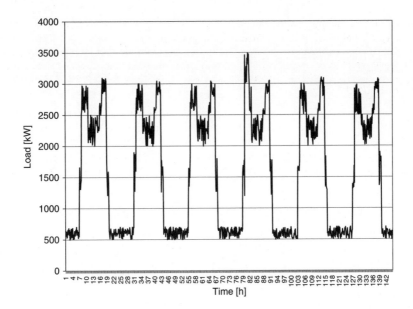

Figure 3.4 Sample Weekly Load Diagram (five days a WEEK, two shifts a day operation)

in case of one shift operation, it would have peaks and valleys. Figure 3.4 represents the weekly load diagram of a company working two shifts, six days a week. The load diagram can be plotted based on the measured demand for every 15 minutes or at half-hour intervals.

3.3.3 Peak Demand

It can be seen in Figure 3.3 that there are several peaks over observed intervals, but there is always only one maximum peak on the load profile, which corresponds to the maximum demand on the supply system. This maximum demand is also called peak demand. For every industrial consumer, peak demands are measured and consumers are invoiced for so called 'demand charges', which are based on the highest average power recorded over a 15 (or sometime 30) minute interval during one month.

3.3.4 Power and Energy

The terms power and load are basically synonymous, differing only in the point of view, so that 'load' mostly relates to demand, and 'power' to supply side considerations. The unit in both cases is watt [W] which may be preceded with prefixes 'k=kilo' or 'M=mega' denoting 1000 or 1 000 000 watt, respectively. An electric motor of say 10 kW capacity represents the load to supply side of 10 kW and it will require securing the connecting power of 10 kW. This requirement translated into sizing and installing needs adequate supply cable and securing capacity for the supply transformer. Whether the motor will work at full capacity all the time, part time or no time at all, is not important because the installation must be sized to the full capacity of connected device.

The connecting power of a device represents its capacity to do work. Only when it actually operates, does the device deliver useful work and it uses energy as long as it operates. There comes the main difference between power (P) and energy (E), and that difference concerns operating 'time' (t):

$$E = P\,[\text{Watt}] \cdot t\,[\text{hours}] \qquad (3.6)$$

The unit for electric energy is watt-hour, which usually comes with the prefixes 'kilo' or 'mega', as already indicated, and which is then written as kWh or MWh.

3.3.5 Load Factor

The load factor is the ratio of average load to the maximum load on the supply system. The average load on the system can be determined based on the total energy consumed over the observed period of time. The ratio of energy consumed to the total time is the average load on the system. The load factor can also be defined as the ratio of energy consumed during a given period to the energy which would have been consumed if maximum demand had been maintained throughout that period.

$$\text{Load Factor} = \frac{\text{Average Load}}{\text{Maximum Load}} = \frac{\text{Energy Consumed during a Period}}{\text{Maximum Demand} \times \text{Time under Consideration}} \quad (3.7)$$

If the consumer's maximum demand equals its average load, its load factor is 100 % and it uses the total installed capacity all of the time, which is almost never the case. However, if the average load of the consumer is less than the maximum demand, the load factor is less than 100 % and the total cost of electricity will be higher. The lower the load factor of the consumer, the higher the total cost of electricity will be.

A sample load profile of a plant working three shifts is represented in Figure 3.5. The respective load factor is calculated as follows:

$$\text{LF} = \frac{\text{Average Load}}{\text{Maximum Demand}} \times 100\,\% = \frac{63.8}{109.7} \times 100\,\% = 58.2\,\% \quad (3.8)$$

It is an example of a poor load factor.

3.3.6 Quality of Power Supply

Power supply quality is an issue that is becoming increasingly important to electricity consumers at all levels of usage because of the increasingly sensitive equipment used in an industrial environment. Occurrences on the supply network that were once considered 'normal' by electricity companies and users

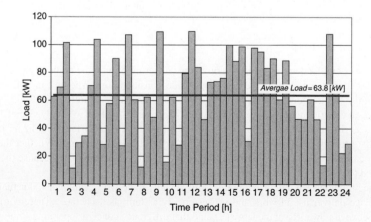

Figure 3.5   Load Profile without Demand Control

alike are now considered to be a problem for the users that have more sensitive equipment. Therefore, standards for the quality of power supply have been introduced nationally and internationally. [1]

These standards define number of terms for quality and permitted ranges of variations, such as:

    (i) under-voltage and over-voltage;
   (ii) voltage dips and swell;
  (iii) short interruptions;
  (iv) transients;
   (v) harmonics;
  (vi) flickers;
 (vii) voltage imbalances;
(viii) frequency deviations.

Sometimes, the sources of problems are within companies' own power systems due to equipment such as power electronic devices, variable speed drives, electronic ballasts, arcing devices, load switching, etc., but sometimes an outside distribution network supplies electricity of poor quality. The general effects of a poor quality of power supply are summarized in Table 3.1.

For sensitive equipment in particular, the main areas of concern are:

- reduced equipment operating life;
- instantaneous equipment malfunction;
- equipment malfunction leading to data corruption;
- reduced process quality;
- process stoppage;
- equipment damage;
- economic damage;
- safety issues.

In the context of electricity market liberalization (see Introductory Chapter), and competing electricity suppliers, power quality issues are standard items defined within supply contracts. Therefore, the quality

Table 3.1 Power Quality, Quantities and Effects

Quantity	Effect
Voltage dips	Machine/process downtime, clean up costs, product quality and repair costs all contribute to make these types of problem costly to the end-user.
Transients	Component failure, hardware reboot required, software problems, product quality.
Harmonics	Transformer and neutral conductor heating leading to reduced lifespan. Audio Hum, Video 'flutter', software glitches, power supply failure.
Flicker	Visual irritation.
Unbalanced voltage	Motor overheating, excessive losses, shorter insulation life time.

---

[1] For an excellent tutorial and a list of standards visit http://www.powerstandards.com/tutor.htm.

of power supply has to be checked regularly in the same way as the monthly electricity bill and discussed with the supplier if there are deviations from the contracted standards.

3.4 Tariff System

Electricity suppliers derive their income from customers through electricity bills. The different methods of charging customers are known as tariffs. The tariff should fulfill the following objectives and requirements:

(1) The cost of capital investment in generation, transmission and distribution equipment must be recovered.
(2) The cost of operation, supplies, maintenance and losses must be recovered.
(3) The cost of metering, billing, collection and miscellaneous services must be recovered.
(4) A satisfactory net return on capital investment must be ensured.
(5) It should provide an incentive for using power during the off-peak hours.
(6) It should have a provision for higher demand charges for high loads demanded at system peak.
(7) It should have a provision of penalties for low power factors.

The deregulation of electricity markets increases the importance of the relationship between suppliers and consumers. On the supply side, generators, network operators and suppliers have to make investments and pricing decisions taking into account competitors and the new opportunities available to customers. On the demand side, customers are increasingly aware of opportunities for savings arising from the new flexibility in supply. Indeed, freedom in choosing suppliers stimulates consumers to look for the deal which best fits their specific electricity needs.

The demand sector usually distinguishes between eligible and franchised customers. In some markets, each non-domestic customer is defined as eligible and has direct access to the electricity wholesale market (see Introductory Chapter). The franchised (or regulated) customers are entitled to enter into supply contracts with the distributor in their territory only.

In each case, electrical charges are based on the following main elements:

- *PEAK DEMAND (peak load) [kW]* is the maximum average loading through 15 (or 30) minutes in the period of high daily tariff.
- *ACTIVE ENERGY DEMAND [kWh]* is the measured energy consumed during a certain period (usually one month).
- *REACTIVE ENERGY DEMAND [kVA_r · h]* is the measured reactive energy consumed during a certain period. A part of the consumption is free of charge (most commonly representing one-third to one-half of the active energy used);
- *TIME OF USE* refers to the varying tariff charges for energy and peak demand depending on the time of day or season;
- *NETWORK RELATED CHARGES* are transmission or distribution system related charges, which are expressed explicitly in the bill in liberalized electricity markets.

The specific price of electricity is an economic parameter that can be calculated and used for the objective evaluation of the effects of performance improvement measures taking into account all the applied charges for the electricity as billed:

$$C_{SEE} = \frac{\text{Total Cost of Electricity}}{\text{Total Active Energy Demand}} \left[\frac{\$US}{kWh} \right] \tag{3.9}$$

Plant energy managers must understand the basis on which their electricity bills are formulated. Understanding electricity rate structures is essential because monetary savings will accrue by lowering the specific charges of electricity bills. Based on such an understanding, appropriate strategies can be developed for reducing both electricity consumption and charges on the utility bill. This is particularly important for load management considerations.

Nowadays, tariff structures are increasingly complex, with various suppliers offering a number of specific 'products' according to the characteristics, type and size of electricity users. Therefore, most suppliers offer a customer support service in selecting the product or tariff that best suits particular end-user needs. Alternatively, third party independent consultants may be engaged in order to assist in determining the optimum electricity supply tariff, a supplier and a product.

Buying electricity at the lowest available rate is definitely the first step in managing the performance of an industrial electric power system.

3.5 Main Components of Industrial Electric Power Systems

Although an industrial power system involves numerous components, from an operational aspect of performance management at an existing plant, we are actually interested just three of them:

  (i) transformers;
 (ii) cables with connecting points;
(iii) motor drives.

All electrical devices have rated powers or capacities and they incur some losses, whether they are operated at full capacity or just switched on and idling. It is important to emphasize that electrical devices are quite efficient (Table 3.2) and very reliable, and that their performance does not change too much over time. It is the way they are operated that determines the performance of electricity use.

3.5.1 Transformers

For instance, it is a common case that a factory has two (or more) transformers installed as a reserve or as spare capacity for future enlargement. Only one transformer has sufficient capacity to supply existing demand, but often both operate at partial load. For a transformer of 1000 kVA average losses (depending

Table 3.2 Electric Losses of Typical Electric System Components

Component	Energy Loss [%]
Transformers	0.40–1.90
Cables	1.00–4.00
Motors:	
0.8–8 kW	14.0–35.0
8–150 kW	6.0–12.0
150–1100 kW	4.0–7.0
1100 kW	2.3–4.5

on the load) can amount to 10 kW, so if the transformer is connected year around (8700 h), total losses incurred amount to 87 000 kWh/a. These losses can be avoided just by switching off one transformer and operating with only one instead of two units.

3.5.2 Cables

The rated capacity for a given cable is the amount of current it can carry under some specified condition (ambient temperature) without overheating and seriously affecting (reducing) insulation life-time. Cable losses are essentially ohmic losses which must dissipate as heat. Generally, cables are sized on the basis of electric current and voltage drop considerations. Cable losses can always be reduced by choosing a larger cable with a greater cross-sectional area, since cable resistance varies inversely with the cross-sectional area. Losses can be reduced but at the expense of the cost of investment in cables.

3.5.3 Electric Motors

In a typical industry, motors are used for a variety of applications in order to provide (among other things), compressed air, chilled water or just water, ventilation, cooling and transportation. An average industry may contain literally hundreds of motors and their collective energy use can account for as much as three-quarters of industrial electricity costs, with the pumps and fans alone drawing on around two-thirds of this amount.

Electric motors can be classified as single-phase or 3-phase. Single-phase motors are mostly less than 1 kW of power. They are commonly the induction type for typical applications such as fans, small pumps, compressors, commercial appliances, business equipment and farm machinery. 3-phase motors are available for powers over 1 kW and larger and are usually of the induction (asynchronous) type. Electric motors always drive some kind of mechanical equipment and operate as part of a process:

(a) pumps and fans for air-conditioning plant;
(b) compressors for refrigeration plant;
(c) pumps for purification plant;
(d) hoisting equipment, lifts and other industrial uses;
(e) air compressors;
(f) conveyer belts, etc.

Electric motor losses consist of two components: variable dependent on load and other that is constant. Their share is around 70 % for variable and 30 % for the constant component relevant to total losses at the full load. Figure 3.6 gives an overview of losses that occur in stator and in rotor of asynchronous (induction) electric motors.

High-efficiency motors feature an improvement in efficiency ranging from roughly 0.5 % to 1.5 % at cost increase of about 15 % to 25 %. The efficiency improvement typically comes from an increased use of laminated steel, increased core length, use of more copper in stator windings and rotor bars and more rigid quality control measures. The typical efficiencies of a standard motor and of a high efficiency motor are shown in Figure 3.7.

Under certain operating conditions, an electric motor that drives a pump or fan will consume electricity every year worth 10 times its capital cost. For this reason, extra costs in a more efficient motor system can often be recovered quickly and given the long life span of most motors, system efficiency improvements can rack up huge savings over the 10 to 15 years that they may be in service.

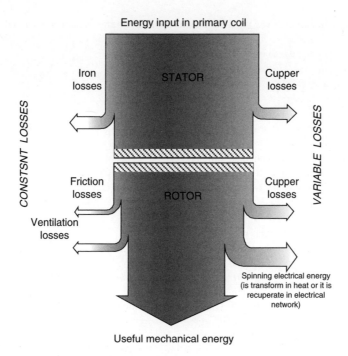

Figure 3.6   Losses in Asynchronous (Induction) Electric Motor

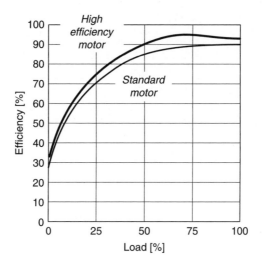

Figure 3.7   Efficiency of Standard and High Efficiency Motors

The efficiency of a motor driven process depends upon several factors which may include:

- motor efficiency;
- motor speed controls;
- proper sizing;
- power supply quality;
- distribution losses;
- transmission;
- maintenance;
- end-use (pump, fan, etc.) mechanical efficiency.

Over-sizing of electric motors is a common problem that varies from industry to industry, and from application to application. Generally, experience shows that average loads are only 65 % of the motor's nominal power or designed values. In many cases, users have no influence on the motor's power as the motor has been delivered together with the equipment. Equipment suppliers usually take into consideration the worst possible scenarios for equipment operation and accordingly they oversize the motors themselves.

As a consequence, a heavily oversized motor will have lower efficiency and power factor (Fig. 3.8) than at nominal load. We can see that the efficiency is approximately constant and close to the maximum up to approximately 75 % of the full load and it falls by around 5 % at a load of 50 %. At loads smaller than 50 %, efficiency declines dramatically.

The reduction of load also has a negative impact on the power factor. Figure 3.8 shows that the power factor declines even faster than efficiency. Therefore, over-sizing of electric motors has the following consequences:

- increase of investment costs for the motor itself;
- increase of investment costs for accompanying equipment (switches, cables, etc.);
- increase of investment costs for capacitors for power factor correction;
- increase of electric energy costs due to lower efficiency.

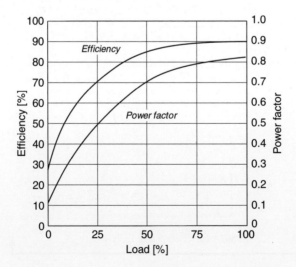

Figure 3.8   Efficiency and Power Factor to Motor Load Relation

Idling of motors occurs frequently where motors run for no apparent purpose or need. Compressors, conveyer belts, mills and other working machines and production lines are very often left in motion even when they do not perform any useful production work. An example has been given for the load profile of a hammer mill (Fig. 3.4, Part I Chapter 3).

3.6 Performance Assessment of Industrial Electric Power Systems

As already emphasized, the performance assessment of industrial power systems should not focus only on the efficiency of individual devices because in such a case most opportunities for performance improvement will remain undetected. Instead, we suggest a systematic approach which takes into consideration not only the performance of individual equipment and machines, but also their performance within a system where these machines deliver specific work or functions. In that case, the performance assessment of industrial electric power systems will have two distinctive aspects:

(1) supply side;
(2) demand side.

On the supply side, we should be concerned both with the commercial and technical aspects of electricity supply. Commercial considerations should dictate a review of which tariffs are contracted, what the overall costs paid for electricity are and how we choose a supplier if a choice is available. Then, one should proceed with analyzing monthly bills and establishing the amounts paid for active and reactive energy and for demand charge. Such analyses and considerations may result in identifying the opportunities to reduce the costs for electricity even before taking a look at the technical aspects of electricity use.

From the technical point of view, it is advisable to check the quality of electricity supply regularly (monthly) in order to avoid potential damage to the equipment and the energy losses that might occur should the quality of supply become compromised. There is a variety of instruments called *power quality analyzers* that help to locate, predict, prevent and troubleshoot problems in industrial power distribution systems. These are easy-to-use handheld tools that measure actual and peak voltages and currents, frequency, voltage dips and swells, transients, interruptions, power and energy, peak demand, harmonics, inter-harmonics, flicker and unbalance, thus alerting energy managers to potential problems with quality of power supply.

On the demand side, we start from the fact that a company does not need electricity as such, but rather the work or services that electrical devices deliver. Therefore, we have to consider the performance of all parts and of the entire system that delivers useful work to end-users according to their actual demand (Fig. 3.9).

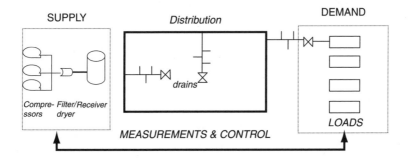

Figure 3.9 Scope of Consideration at Systematic Approach to Performance Assessment of Compressed Air System

For every such system, the basic balance equation must hold

$$\text{Supply} = \text{Demand} + \text{Losses} \tag{3.10}$$

whether it is a water supply, or a compressed air or a ventilation system, or any other that delivers required work or service. System performance can be expressed as the ratio of useful work delivered and electric energy used to deliver this work.

$$\text{Efficiency}(\eta) = \frac{\text{Useful Work Delivered}}{\text{Electric Energy Used}} \tag{3.11}$$

The options for performance improvement should be investigated according to the following priorities:

   (i) establish real demand;
  (ii) minimize losses;
 (iii) optimize supply.

First and foremost, the quality and quantity of real demand should be established. Even the most efficient supply of unnecessary or unproductive demand would be a waste. For instance, water quantities that are pumped should correspond only to the actual process needs, or the quality of compressed air should match the process requirements, excessive air purity should be avoided, or a ventilation system should provide only the required number of air exchanges per hour during the required time intervals and should not work around the clock irrespective of process needs, etc. This brings us on to the simple fact that the largest savings of electrical energy in an industrial environment will come from the basic principle:

'WHEN YOU DON'T NEED IT, TURN IT OFF'.

This is so simple, easy and cost-free, but in spite of this is all too often neglected and time and again we can find machines idling, lights left on, fans running, and pumps pumping, etc., without any apparent need. The *change in people's behavior*, as suggested in Part I Chapter 1, is especially important for improving that aspect of performance of electricity use. It starts with strict adherence to operational procedures.

Figure 3.10 reminds us of the main stages and events in a machine's operation. For every piece of equipment a similar table exists or should exist, prescribing the operational routines which lead to maximum productivity at minimum electricity use. Performance is especially 'vulnerable' during product change or retooling when a number of devices are left running idle. Similar incidents are experienced during breaks and inter-shift times. Such incidents could be avoided by the careful planning of operations, where care is taken not only about the aspects of a single machine's operation (Fig. 3.10), but also about the whole production chain so that machines idling while waiting for input material is eliminated.

Even when all operational procedures are in place, it still takes willing, skilled and motivated operators to apply them. Therefore, we come across again the *importance of the human factor* in achieving best operational practice.

A useful aid for analyzing the operational performance of electricity use may be a 'load duration' diagram (curve). The load duration diagram can be constructed from a load diagram so that the cumulative durations of any particular load over the observed period are plotted in sequence together (Fig. 3.11).

The useful features of a load duration diagram are as follows:

(1) It indicates not only the peak load but also the duration of peak loads over the observed time interval, which is important for consideration of demand control strategies.
(2) It provides an insight into variable and fixed demand, which provides a basis for determining operational performance. Ideally, variable demand is caused only by production variability, while fixed

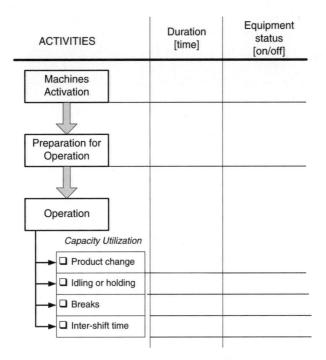

Figure 3.10 Planning Production Operations

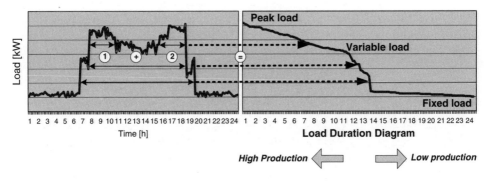

Figure 3.11 Construction a Load Duration Diagram

demand reflects the unavoidable minimum consumption that occurs irrespective of production output. In reality, neither would be the case. There is always a great deal of potential to reduce both the fixed and variable portions of demand. There are the common cases of so called 'false' fixed demand, when certain loads are operated as fixed although in reality such operation is not required. Again, a good example, as already mentioned, is a molding machine (Fig. 3.12) where hydraulic pump and heaters will usually operate constantly after switching on, while in fact their operation can vary with the production.

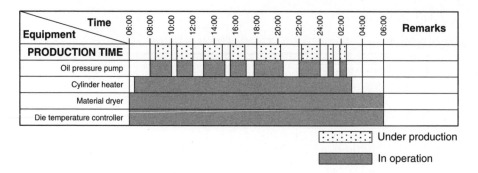

Figure 3.12   'Variable' Operational Procedure for a Molding Machine

When best operational practices are achieved and through that real demand determined, then other technical measures can be explored as a means for further performance improvement. Next, although simple, a very important step is to identify and eliminate losses all over the system. These are, for instance water leaks, compressed air leaks, etc. After these two steps are completed successfully, real demand and unavoidable losses will be established. What remains then is to optimize the supply side of the system, so that required supply is delivered with minimum cost.

3.7   Performance Improvement Opportunities

Previous considerations have emphasized the importance of a systematic approach and the human factors in achieving performance improvement. Therefore, under the 'umbrella' of the Energy and Environmental Management System (EEMS), both aspects of performance improvement should be addressed simultaneously (Fig. 3.13).

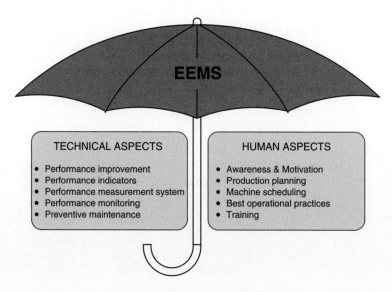

Figure 3.13   'Umbrella' of EEMS

3.7.1 Load Management

An electric utility incurs costs in providing electrical services to its customers. Consequently, customers are billed not only for the total energy consumed in a given period, but also for their usage of the utility-owned equipment required to serve them. The first charge is an energy charge and the second charge is a demand-related charge based on measured peak load. It is in the interests of both utility and customer to minimize maximum demand. This is achieved by demand side or load management. There are two areas of load management:

- supply industry demand-side management;
- demand control by consumers at their own location.

Supply industry load management assumes an agreement between an electricity supply company and a customer to reduce load at a particular time by a specified amount. It is usually performed by a customer whose peak demand is bigger than 5 MW and this restriction is rarely longer than 1.5 hours per day. This is the subject of negotiation with a supplier about a particular power supply agreement where such an option is available.

3.7.2 Demand Control

Demand control at the consumer's location is carried out by the consumer on its own loads. Demand control is nothing more than a technique for leveling out the load diagram, i.e., 'shaving' the peaks, and 'filling' the valleys (Fig. 3.14). The main advantage of good local load management is the significant reduction of electricity charges and the opportunity to use more electric energy for the same or lower total cost without the need to invest in expansion of the power supply system.

If the demand limit line is constructed so that it is greater than or equal to average power consumption over the billing period, it is possible to 'shave the peaks' and 'fill in the valleys' in such a way that the demand limit line is never exceeded and energy requirements are still satisfied (Fig. 3.14). In practice, the demand limit will be greater than the average power requirement since it is rarely possible to operate

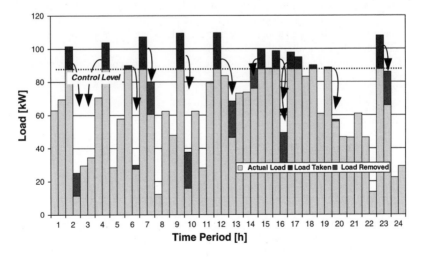

Figure 3.14   Load Profile with Load Factor Control (LF = 80 %)

with absolutely constant power consumption. The first step in assessing the merits of demand control schemes is to analyze the utility rate structure and past history of power demand at the location.

The concept of the load factor is an especially useful tool in this type of analysis. Maximum demand and total kilowatt-hours are easily obtained from past billing statements, permitting determination of past performance in terms of load factor.

The first choice and the simplest method for reducing peak loads is to schedule production activities so that the big electrical power users do not operate at the peak times at all, or at least that a number of them do not operate at the same time. *Machine scheduling* is the practice of turning equipment on or off depending on time of day, day of week, day type, or other variable or process need. Improving equipment schedules is achieved through better *production planning*, one of the most common and most effective opportunities for avoiding machine idling and reducing peak demands. In the absence of a production plan and an optimized schedule, it is not unusual for electrical equipment to run for 24 hours when only a 12-hour operation is required.

The second method relies upon automatic controls which shut off non-essential loads for predetermined time intervals during peak periods by means of some load management device such as:

- simple switch on-switch off devices: time clocks, photoelectric relays, night setback thermostats, etc.;
- single load control device;
- demand limiters;
- computerized load management systems that perform demand control based on feedback data from the remote metering of total load and electric power and the switching statuses of individual loads.

Any practical demand control scheme requires the identification and ranking of loads suitable for shedding. Production line machinery is generally excluded although electrical heating devices used in production may be amenable to some degree of demand control. Air-conditioning, fans and facility-heating equipment are loads generally considered suitable for shedding and restoring in a demand control scheme. In cases where required disconnection times are long, the *load rotation* can be applied, that is, reconnecting disconnected loads after a predetermined elapsed time while disconnecting others. The required data for designing a demand control based on a load rotation are specified in Table 3.3. Particular applications may require specific data not mentioned in Table 3.3. but Table still illustrates the main parameters for demand control.

When designing a demand control scheme, one should be aware that motor life is reduced if a motor is subjected to repeated starting and stopping. Manufacturers typically specify the maximum permissible number of starts per hour for large motors in order to avoid degradation of insulation's economic lifetime and this parameter if therefore referred to in Table 3.3.

The following examples demonstrate the structure of electricity cost and illustrate the effects of potential demand control measures.

Table 3.3   Required data for designing a demand control based on a load rotation

Type of load	Power [kW]	Maximum time off	Minimum time on	Max. number of switching	Priority [1-10]
1.					
2.					
3.					
. . .					

3.7.2.1 Example 1: Demand Control by Stand-by Diesel Generators

In our case study of Tuna Canned Factory, the following tariff system is applied:

Energy charge: 2.7557 cents $US)/kWh
Demand charge: 7.4232 $US/kW (On Peak Period)
1.5333 $US/kW (Partial Peak Period)
0 $US/kW (Off Peak Period)

The peak periods depend on the time of day as follows:

On peak   6:30 p.m.–9:30 p.m.
Partial peak   8:00 a.m.–6:30 p.m.
Off peak   9:30 p.m.–8:00 a.m.

Billing demand is defined as the maximum 15-minute integrated demand over the monthly billing period measured to the nearest whole kW discarding the fraction of 0.5 kW.

If in any monthly billing period the customer's maximum 15-minute kVAr demand exceeds 63 % of its maximum 15-minute integrated demand, the power factor charge of 0.3906 $US is made on each kVAr in excess.

Elements of monthly bills for a sample year are shown in Table 3.4. It is obvious that maximum demand during on-peak period is very small in comparison to partial and off-peak periods. This is the

Table 3.4 Energy Consumption and Cost

Month	Energy	Peak			Total Electricity Cost	Unit Price
		On	Partial	Off		
	kWh	kW	kW	kW	$US	cents $US/kWh
Jan	765 180	36	2017	1654	34 549	4.52
Feb	756 840	695	1391	1666	38 145	5.04
Mar	866 460	30	2051	1819	38 685	4.46
Apr	841 080	66	2124	1890	34 067	4.05
May	788 220	30	2010	1890	31 720	4.02
Jun	705 900	53	1889	1581	28 739	4.07
Jul	750 420	78	1880	1554	30 483	4.06
Aug	761 460	30	1908	1524	31 651	4.16
Sep	787 560	60	1992	1428	32 980	4.19
Oct	808 620	30	2009	1781	33 571	4.15
Nov	922 380	30	2160	2018	38 061	4.13
Dec	913 560	36	2160	1926	42 153	4.61
TOTAL	9 667 680				414 806	
AVERAGE	805 640	97.8	1966	1728		4.29

consequence of the decision to use four diesel generators (total power of 2200 kW$_e$), already existing at the location, during the on-peak period. This measure has dramatically reduced the peak load, except in February when some generators were undergoing general overhaul. Without generators, the average monthly peak would be 1917 ± 195.9 kW$_e$.

The use of diesel generators incurs certain fuel and maintenance costs. The calculation of costs for the operation of the generator during the on-peak period is presented in Table 3.4. When electricity costs (Table 3.5) are taken into account and the costs of running diesel generators, the average annual electricity cost was 4.62 cents $US/kWh, which is significantly lower than the cost which would have been incurred should the total required energy had been taken from the local distribution company (5.40 cents $US/kWh). The operating and maintenance cost of diesel generator is 0.01 $US/kWh. The average price

Table 3.5 Comparison of Electricity Cost

Month	Real Peak	All Electricity is bought from Local Distribution	Unit Price of Electricity (Local Distribution Only)	DIESEL GENERATORS			Current Price of Electricity (Local Distribution and Generators)
				Electricity Generated by Generators	Unit Price of Diesel Fuel	Diesel Cost	
	kW	$US/m	cents $US/kWh	kWh	cents $US/l	$US/m	cents $US/kWh
Jan-99	1828	48 739	5.64	98 866	16.85	5742	4.66
Feb-99	2152	50 197	5.92	90 503	15.03	4962	5.09
Mar-99	2114	55 209	5.62	115 388	17.29	6791	4.63
Apr-99	1803	47 766	5.10	94 733	19.61	6712	4.36
May-99	1485	43 237	4.98	80 216	19.61	6097	4.35
Jun-99	1769	42 293	5.29	93 997	19.09	5890	4.33
Jul-99	1854	44 467	5.25	96 461	20.49	6867	4.41
Aug-99	1868	46 221	5.36	101 656	21.72	7272	4.51
Sep-99	1920	47 664	5.36	101 779	23.02	7786	4.58
Oct-99	1952	48 804	5.33	106 318	21.85	8489	4.60
Nov-99	2117	54 611	5.26	115 565	23.91	9618	4.59
Dec-99	2142	58 842	5.71	116 410	26.80	12 262	5.28
TOTAL		588 050		1 211 892		88 489	
AVERAGE	1917		5.40		20.44		4.62
St Dev	195.9				3.28		

of diesel fuel was 20.44 cents $US/l. The annual diesel cost and operating and maintenance cost was 88 489 $US and generated electricity was 1 211 892 kWh.

Finally, electricity costs reduction is as follows:

$$\text{Cost Reduction} = 588\,050 - (414\,806 + 88\,489) = 84\,755 \text{ US\$/y} \qquad (3.12)$$

or 14.4 % compared to the electricity cost which would have been paid to a local distribution Company if total electricity had been supplied by it. Of course, such a performance improvement measure is very sensitive to the relationship between electricity and fuel prices, hence whenever any of these change, the economics of the measure has to be checked.

3.7.2.2 Example 2: Relation between Load Factor, Demand Charges and Total Electricity Costs

Demand charges and total energy charges are determined by the following rate schedule:

Fuel adjustment charge   0.02872 $US/kWh
Energy charge   0.01152 $US/kWh
Demand charge   3.22 $US/kW

where demand charge is based on the highest demand recorded in a billing month.

Past records indicate that a total monthly charge of 10 659 $US occurred when kilowatt hour consumption for the month was 220 968 kWh and the maximum demand over the billing period was 511.5 kW. If we assume that the same monthly energy is consumed but at various load factors, we can determine total charges as the function of load factor.

We can start by determining the average demand for the recorded month. The average demand D_{avg} is as follows:

$$D_{avg} = \frac{\sum_{i=1}^{n} E_i}{n \times T_d} = \frac{\text{Total energy consumption}}{\text{Total hours billing period}} = \frac{220\,968 \text{ kWh}}{30 \times 24} = 306.9 \text{ kW} \qquad (3.13)$$

The load factor for the recorded month is the ratio of D_{avg} to the maximum demand. Thus

$$LF = \frac{D_{avg}}{\max [D_i]} = \frac{306.9}{511.5} = 0.6 \qquad (3.14)$$

Now, we can assume that the total energy consumption remains the same while the load factor varies. If total energy consumption remains constant, D_{avg} must remain constant in accordance with Eq. (3.13). The maximum demand is determined from the definition of load factor and all of the information required to determine total monthly charges as the function of load factor is available (Table 3.6).

The tabulated results are displayed in Figure 3.15. It is easy to see that as the load factor approaches 1.0, total cost approaches 10 000 $US per month.

The potential benefits of demand control schemes can then be weighed against the cost of implementation of various schemes. If the load factor is characteristically low, a small improvement in load factor significantly reduces costs, as seen from the figure. Obviously, the calculations depend heavily on existing and future rate structures. It is also important to recognize that a utility may offer more than one rate structure to a large industrial user. The technique described in this example is helpful in ascertaining the most favorable rate structure.

Table 3.6   Load Factor and Maximum Demand

Load Factor	Energy Consumption [kWh]	Average Demand [kW]	Energy Charge [$US]	Maximum Demand [kW]	Demand Charge [$US]	Total Charge [$US]
1.0	220 968	306.9	8891	306.9	988	10 000
0.9	220 968	306.9	8891	341.0	1098	10 110
0.8	220 968	306.9	8891	383.6	1235	10 247
0.7	220 968	306.9	8891	438.4	1412	10 424
0 6	220 968	306.9	8891	511.5	1647	10 659
0 5	220 968	306.9	8891	613.8	1976	10 988
0.4	220 968	306.9	8891	767.3	2471	11 483
0.3	220 968	306.9	8891	1 023.0	3294	12 306
0.2	220 968	306.9	8891	1 534.5	4941	13 953
0.1	220 968	306.9	8891	3 069.0	9882	18 894

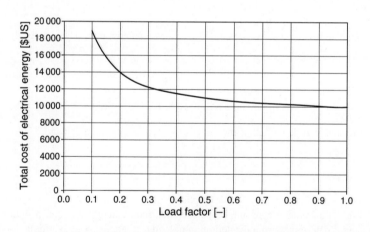

Figure 3.15   Total Cost versus Load Factor for an Assumed Rate Structure

3.7.3 Power Factor Correction

The first step in correcting for low PF is to determine the current situation with a *cos* φ at a given facility. Monthly data can be retrieved from the utility bills, and clamp-on power factor meters can be used for site-specific analyses. The power factor can be improved in two ways:

- Reduce the amount of reactive energy by eliminating unloaded motors and transformers.
- Apply external compensation capacitors to correct the low-PF condition.

PF correction capacitors perform the function of an energy-storage device. Instead of transferring reactive energy back and forth between the load and the power source, the reactive energy is stored in a capacitor at the load. Capacitors are rated in kVARs, and are available for single- and multiphase loads. Usually, more than one capacitor is required to yield the desired degree of PF correction. The capacitor rating required in a given application may be determined by using tables provided by PF capacitor manufacturers or by performing specific calculations (see example below). Installation options include:

- individual capacitors placed at each machine;
- a group or bank installation for an entire area of the plant;
- a combination of the two approaches.

If rectifier loads creating harmonic currents are the cause of a low-PF condition, the addition of PF correcting capacitors will not necessarily provide the desired improvement. When adding capacitors for PF correction in such case, be careful to avoid any unwanted voltage resonances that might be excited by harmonic currents.

An economic evaluation of the cost versus benefits, plus a review of any mandatory utility company limits that must be observed for PF correction, will determine how much power factor correction, if any, may be advisable at a given facility. Correction between 85 % and 95 % will satisfy most requirements. No economic advantage is likely to result from correcting PF to 100 %.

The installation of PF correction capacitors at a facility may be a complicated process (depending on the types of load present) that requires a knowledgeable consultant and licensed electrician. The local utility company should be contacted before any effort is made to improve the PF of a facility.

3.7.3.1 Example 3: Power Factor Correction from 0.6 to 0.9

1. Existing situation	
Active power P	550 kW
Exiting power factor ($\cos \varphi_1$)	0.6
Resulting apparent power S_1: $S_1 = \dfrac{P}{\cos \varphi_1} = \dfrac{550}{0.6} = 917\,\text{kVA}$	917 kVA
Apparent current I_1 $I_1 = \dfrac{1}{\sqrt{3}} \cdot \dfrac{S_1}{U} = \dfrac{1}{\sqrt{3}} \cdot \dfrac{917 \cdot 1000}{400} = 1324\,\text{A}$	1324 A
2. Improved situation	
Target power factor ($\cos \varphi_2$)	0.9
Required capacitor rating Q_C: $Q_C = P \cdot (\tan \varphi_1 - \tan \varphi_2) = 550 \cdot [\tan(53.1) - \tan(25.8)]$ $= 467\,\text{kVA}$	467 kVA

Apparent power S_2: $S_2 = \dfrac{P}{\cos \varphi_2} = \dfrac{550}{0.9} = 611\,kVA$	611
Apparent current I_2: $I_2 = \dfrac{1}{\sqrt{3}} \cdot \dfrac{S_2}{U} = \dfrac{1}{\sqrt{3}} \cdot \dfrac{611 \cdot 1000}{400} = 882\,A$	882 A
$\dfrac{S_1 - S_2}{S_1} \cdot 100[\%] = \dfrac{917 - 611}{917} \cdot 100 = 33.4\,\%$	33.4 %
$\dfrac{I_2^2 - I_1^2}{I_1^2} \cdot 100[\%] = \dfrac{1324^2 - 882^2}{1324^2} \cdot 100 = 55.6$	55.6 %
As the result of PF correction from cos $\varphi_1 = 0.6$ to cos $\varphi_2 = 0.9$, active power transmission capacity isincreased by 33.4 %, and transmission losses are reduced by 55.6 %.	

The significant effect of improving the power factor, beside cost savings, is the reduction of currents flowing through corresponding feeder(s), which in turn leads to the following benefits:

- increased supply capacity of the feeder(s);
- reduced voltage drop;
- reduced power losses in the distribution system.

The power losses are proportional to the square of the current, hence Eq. (3.15) can be applied to calculate the loss reduction resulting from power factor improvement.

$$\% \text{ reduction of power losses} = \left\{ 1 - \left[\frac{\text{Current} \cos(\varphi)}{\text{New} \cos(\varphi)} \right]^2 \right\} \times 100[\%] \tag{3.15}$$

3.7.4 Electric Motor Drives

Electric motors are used as sources of power for driving a number of mechanical devices, which in turn provide useful work. That is why we usually refer to the set of an electric motor and the attached mechanical device as an electric motor drive (EMD). EMDs have to be analyzed as a system, which includes the motor, its controls, and the connection between the motor and the equipment it drives, and the useful work of the attached mechanical device.

3.7.4.1 Variable Speed Drives (VSD)

The key to achieving peak efficiency of electric motor drives is that a system is running only when necessary and at the capacity needed to meet current demand, operating in a stable and precise control range that matches system demand. The applied controls should optimize the amounts of stopping and restarting and avoid idle running. We have already considered the example of an injection molding machine (Part I Chapter 3, and section 3.6 in this chapter), where it was pointed out that a piece of equipment is not necessarily a single machine. The molding machine is also an EMD which uses electricity to pump hydraulic oil for the press and water for cooling the oil. Molding machines are notoriously inefficient users of electricity with a saving potential up to 50 %. The reasons for such inefficiency vary from the overcapacity of a machine, softer raw material than anticipated, fewer cycles, idling, etc. When any combination of these factors exists, there is an opportunity to consider installing a variable speed drive (VSD), based on energy audit findings (Part I Chapter 4).

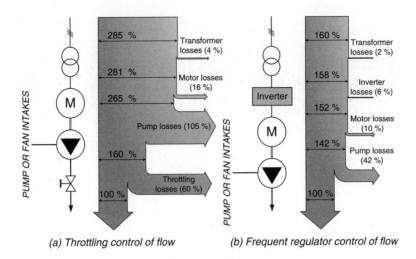

(a) Throttling control of flow   (b) Frequent regulator control of flow

Figure 3.16   Overview of Losses in Classical Pump System and in System with VSD

To truly minimize the energy use of an EMD system, we need to consider how its components operate as a system rather than looking at them individually. For example, focusing solely on motor efficiency may yield a 1 % to 5 % reduction in operating cost, but improvements that take into account the entire EMD system may provide savings of 50 % or more as shown in the molding machine example.

Until recently, VSDs were used only in industrial processes which required the variable speed of working machines. The technological development of these regulators, the dramatic fall of their prices and the growing costs of electrical energy have given room for a much more widespread application of VSDs in all EMD applications. At the moment, VSD application represents the greatest potential for reducing electric energy consumption by replacing the traditional regulation of pressure and fluids flows by throttling, by adjusting vanes and blades of pumps and fans and by by-passing lines, overflow tanks, etc.

Figure 3.16 shows the typical losses and their possible values in a system with traditional flow regulation and in a system with a VSD regulator. This example indicates the large savings potential. However, each individual case should be carefully researched and real savings identified based on actual process requirements for the useful work that EMD delivers. If these requirements vary considerably in frequency and amounts of deliveries of, let us assume, a fluid flow rate, then we have a case for considering VSD control. If the flow rate needs to be constant (100 % open valve in Figure 3.16a), the consumption of electric energy with or without VSD would be approximately the same. Figure 3.16 illustrates the performance improvement effects of VSD application for pump and fan operation.

3.7.4.2 Correct Operational Procedure

Another important issue for EMD operation is *capacity utilization*. Capacity utilization can be defined as a ratio between the actual production output and maximum production capacity. Even when all components are at peak performance, if machines are running at partial capacity, energy effectiveness will decrease and energy cost per unit of product will increase.

One solution may again be the application of VSD so that motor power is adjusted to work requirements, but before such a decision can be made, the very cause of low capacity utilization must be examined. If that cause can be eliminated and the machine can be operated at full capacity, which would definitively be the preferred solution then VSD would not be required. It can be illustrated by the real case of a hammer mill driven by an electric motor of more than 200 kW, which displayed a particularly inefficient

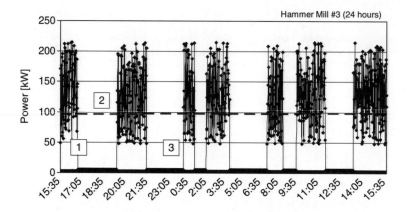

Figure 3.17   Actual Daily Load Profile of a Hammer Mill

operation. Figure 3.17 shows the daily load profile as measured during an energy audit. It reveals several points of concern:

(1) frequent idling (i.e. no load operation);
(2) low average capacity use – mostly at half of the nominal output level;
(3) frequent start/stop.

The main reason for these problems is that the operational procedures for the mill operation are not followed and strictly applied. As a result, the mill uses, on average, 40 % more electricity per unit of production than is required by the actual design! After a detailed analyses of the whole system operation, it is discovered that

(1) The mill feeder does not work correctly owing to partial loading. Due to the lack of proper production planning a storage tank containing the mill feed is often becoming empty, and this causes frequent starts and stops.

This example emphasizes that we should investigate every case of EDM performance carefully before proposing a VSD application. This investigation must address the whole system associated with EMD – in this case production planning, feed preparation, feed stock, feeder, electric motor, mill and controls. In the hammer mill case, VSD can be applied, but in fact it would be the wrong solution that would just accommodate bad operational practice.

Instead, the energy audit has proven that the mill can operate at full capacity when standard operational procedures are adhered to. This solution requires changes in production planning and fixing the mill feeder. In the end, the mill indeed operates at full capacity and there are no direct energy savings, even the opposite – the mill uses more electricity because it works at the full load all the time. But the real performance improvement comes from the dramatically reduced cost of electricity in the unit of production.

The understanding of the EMD function and its role within a larger production context is essential for performance analyses and improvement. Standardized operational procedures (SOPs) for EMD must be followed anyway and performance monitoring should confirm that this is the case. If the SOPs do not exist, they need to be established. That is one of the aims and results of energy and environmental management system (EEMS) – to establish a *best practice operation* that is embodied in standardized operational procedures and to assure that the best practice operation is not an exception but has become a routine.

3.7.4.3 Example 4: Comparison of Fan Operations with Constant and Variable Speed Control

This example will demonstrate the effect that can be accomplished with the use of VSD.

The asynchronous motor with the following characteristics:

Speed:	1500 1/min
Nominal power:	11.0 kW
Efficiency at nominal load:	92 %

drives a centrifugal fan with the following characteristics (requested work point):

Flow:	18 000 m³/h
Head:	1500 Pa

The operational characteristics of the fan which are guaranteed by the manufacturer are presented in Figure 3.18 and can be very precisely expressed in the analytical form by the following relations:

$$\frac{H}{H_r} = -0.44799 \cdot \left(\frac{V}{V_r}\right)^4 + 0.94980 \cdot \left(\frac{V}{V_r}\right)^3 - 0.98149 \cdot \left(\frac{V}{V_r}\right)^2 + 0.35881$$
$$\cdot \left(\frac{V}{V_r}\right) + 1.12184 \tag{3.16}$$

$$\frac{\eta}{\eta_r} = -1.50212 \cdot \left(\frac{V}{V_r}\right)^4 + 4.64734 \cdot \left(\frac{V}{V_r}\right)^3 - 5.74118 \cdot \left(\frac{V}{V_r}\right)^2 + 3.62077$$
$$\cdot \left(\frac{V}{V_r}\right) - 0.02653 \tag{3.17}$$

$$\frac{P}{P_r} = 0.24038 \cdot \left(\frac{V}{V_r}\right)^4 - 1.07178 \cdot \left(\frac{V}{V_r}\right)^3 + 1.17213 \cdot \left(\frac{V}{V_r}\right)^2 + 0.29880$$
$$\cdot \left(\frac{V}{V_r}\right) + 0.36348 \tag{3.18}$$

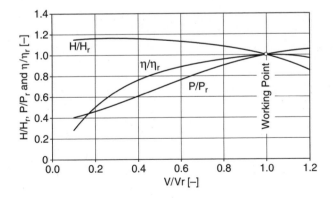

Figure 3.18 Operational Curves of the Fan

Where:
  V = Fan flow, [m³/h]
  H = Head, [Pa]
  P = Power, [kW]
  η = Efficiency, [-].

Index **r** with individual values designates nominal values.
Fan efficiency at the nominal load is:

$$\eta_{ven,r} = \frac{V_r[m^3/s] \cdot H_r\,[Pa]}{1000 \cdot P_r[kW] \cdot \eta_{em,r}} = \frac{\frac{18\,000}{3600} \cdot 1500}{1000 \cdot 11 \cdot 0.92} = 0.74 \tag{3.19}$$

The hydraulic characteristics of the air distribution ducts are represented in Figure 3.19.

As it has already been mentioned, the critical factor in assessing the effectiveness of VSD application is knowledge of the actual load profile for EMD. In this case, it is given in Figure 3.20.

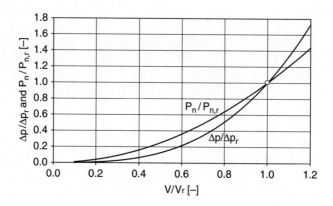

Figure 3.19   Hydraulic Characteristic of Air Ducts

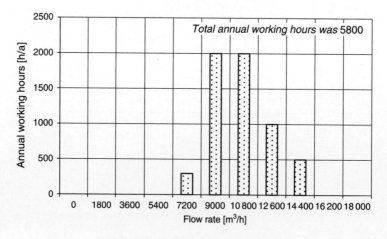

Figure 3.20   Fan Required Load Profile

In the case of constant speed, the flow regulation is carried out by throttling (with a damper). It means that the pressure drop in the duct system is increased intentionally in order to reduce the flow. This also increases the requested fan discharge head as represented in Figure 3.21.

Electric energy consumption for this EMD is calculated and given in Table 3.7. The overall annual consumption of electric energy in the case of constant speed and flow regulation by throttling is equal to **48 162 kWh/a.**

In the case of *variable speed control*, the required motor power can be calculated as follows:

$$P_{VSD}[kW] = \frac{V[m^3/s] \cdot \Delta p_{System}[Pa]}{1000 \cdot \eta_{Ven} \cdot \eta_{EM}} \tag{3.20}$$

The result of the calculation in this case is given in Table 3.8. The overall consumption of electrical energy is 15 940 kWh/year. We can conclude that by applying VSD control instead of throttling, in the given operational regime of relevant EMD, electrical energy savings of 67 % are achievable.

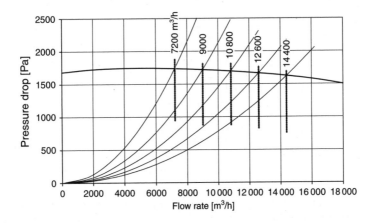

Figure 3.21 Changes of Hydraulic Characteristics with Regulation by Throttling

Table 3.7 Consumption of Electric Energy for Constant Number of Revolutions

Hours of Operations [h]	Flow [m³/h]	Discharge Head [Pa]	Power [kW]	Consumption of Electric Energy [kWh]
300	7200	1736	6.69	2007
2000	9000	1720	7.56	15 120
2000	10 800	1696	8.42	16 840
1000	12 600	1665	9.22	9220
500	14 400	1625	9.95	4975
5800	**TOTAL**			**48 162**

Table 3.8 Consumption of Electric Energy for Variable Number of Revolutions

Hours of Operations [h]	Flow [m³/h]	Discharge Head [Pa]	Power [kW]	Consumption of Electric Energy [kWh]
0	0	0	0.00	0
0	1800	15	0.04	0
0	3600	60	0.18	0
0	5400	135	0.45	0
300	7200	240	0.93	278
2000	9000	375	1.65	3296
2000	10 800	540	2.68	5358
1000	12 600	735	4.07	4071
500	14 400	960	5.88	2938
0	16 200	1215	8.16	0
0	18 000	1500	11.04	0
5800	TOTAL			15940

The required power for the electric motor in both of the presented cases is shown in Figure 3.22. Investment costs are estimated as follows:

(a) Purchase of VSD and pressure sensors   2750 $US
(b) Wiring (10 %)   275 $US
(c) Assembly (10 %)   275 $US

               TOTAL: **3300 $US**

The price of electricity is 0.0625 $US/kWh. The corresponding costs of electrical energy for the cases with constant and variable speed are:

- Constant   3010 $US/a
- Variable   996 $US/a

Reduction of energy costs   **2014 $US/a**

The simple payback period of this investment is:

$$\text{Payback} = \frac{3300}{2014} = 1.64 \text{ years} \tag{3.21}$$

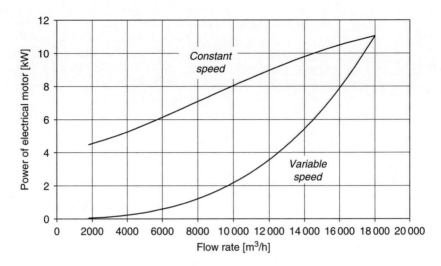

Figure 3.22   Electric Motor Power in Case of Constant and Variable Speed

Table 3.9   Approximate Payback Periods for Various Motor Powers

Motor power	1500 hrs/yr	3000 hrs/yr	4500 hrs/yr	8500 hrs/yr
< 10 kW	7.6 years	3.8 years	2.5 years	1.3 years
>10 kW < 100 kW	3.8 years	1.9 years	1.3 years	0.7 years
> 100 kW	2.3 years	1.1 years	0.8 years	0.4 years

Generally, payback periods for VSD control depend on the motor's power and on operation time per year. Table 3.9 provides the approximate payback periods for various motor powers and annual operating hours.

High efficiency motors are also an interesting alternative that can be combined with VSD control if the motor is to be replaced anyway. In that case, the additional cost for a high efficiency motor is paid back in less than two years by 3000 hours of operation per year.

3.8 Maintenance Considerations

Electrical devices are usually very reliable with minimal performance changes over time. But one of the most frustrating things that can happen are unanticipated outages due to equipment or electric system failures. Such outages disrupt production, require extensive repairs, **work-in-progress might** need to be scrapped and they give rise to all sorts of associated costs for unscheduled downtime. Therefore, proper preventive maintenance will positively affect literally every energy and environmental performance indicator related to electricity use in industry.

On the industrial power supply side, the main concern is to prevent the gradual erosion of power supply quality. *On the equipment side*, the main concerns are not only electrical motors, but also mechanical rotary equipment driven by motors. The performance of this equipment (pumps, fans, compressors, etc.) directly affects overall system efficiency. For example, the erosion of the pump impeller causes pump efficiency to drop sharply which in turn reduces the overall system (motor + pump + piping + controls) performance. The key motor's maintenance issues are summarized in Box 3.2.

Box 3.2: Key Maintenance Issues for Electric Motors

1. **Lubrication** – *if lubrication is not adequate, motors may use 4 %–8 % more energy;*
2. **Loose electrical connections** – *result in increased temperature and therefore, increased ohmic resistance. If resistance is increased by 0.1 ohm at the connection points of the motor of a hammer mill from our example in Part 1 Chapter 3, which operates 6000 h/y, and at a full load takes approximately 300 amperes (A), the following losses may occur:*

$$Power\ loss = I^2 \times R = 300^2 \times 0.1 = 9000\ W = 9\ kW$$

$$Money\ loss = 9\ kW \times 6000\ h/y \times 0.05\ \$/kWh = 2700\ \$/y$$

3. **Mechanical alignment** – *if mechanical alignment is not adequate, then:*
   - *Energy consumption may increase up to 6 %;*
   - *Bearing life may decrease up to eight times;*
   - *Break down maintenance may go up to 50 %;*
4. **Belt drives** *and conveyers regular tension control may reduce slip and give savings of 5 %–10 %.*
5. *If a motor is* **clean and cooling** *not obstructed, so that overheating is avoided, the savings could be 2 %–4 % compared to the case of a motor covered by dust and scale.*

There is no doubt that the cornerstones for successful industrial energy and environmental management are well conceived and regularly conducted preventive maintenance practices.

3.9 Performance Monitoring

In the context of this chapter we want to emphasize the aspect of EEM and performance monitoring that leads to the avoidance of outages and drives the planning of preventive maintenance based on operational data monitoring and interpretation. This is the approach of using EEM for equipment and system diagnostics, thus anticipating the need for preventive maintenance in a timely manner. We can devise, as one of EEMS functions, the daily monitoring of information such as temperatures, flows, pressures and actuator positions so as to gain the data necessary to determine whether equipment is operating incorrectly or inefficiently. This enables the avoidance of outages and enables the planning of preventive maintenance based on power quality monitoring, operational data recording and interpretation instead of on the basis of prefixed schedules.

3.9.1 Power Quality Indicators

Power quality issues have been described, as well as the potential consequences of a corrupt power supply. Therefore, a regular schedule for power quality monitoring has to be adopted (Table 3.10).

Table 3.10   Power Quality Monitoring Schedule

Performance indicators	Frequency
Harmonics and THD (V,A,W),	Monthly or weekly
Inter-harmonics	Monthly or weekly
Voltage Dips and Swells	Monthly or weekly
Flickers	Monthly or weekly
Unbalances:	
Voltages	Monthly or weekly
Currents	Monthly or weekly

3.9.2 Electric Motor Drives – Maintenance Needs Indicators

Generally, electric motor drives are usually very reliable and predictable devices where performance changes a little during their lifetime. However, over time physical and mechanical demands will adversely affect overall performance. Steady temperature or pressure increases or decreases over time do indicate a need for maintenance. Equipment suppliers usually provide instructions on the important indicators that need to be monitored depending on the size and type of motor and the mechanical equipment it drives.

For instance, monitoring the *air-end discharge temperature* of rotary screw compressors can be a critical element in reducing downtime by allowing for timely preventive maintenance.

For the performance of centrifugal compressors effective inter-cooling is critical. Compressors that are water-cooled are very sensitive to the performance of a heat exchanger which erodes over time due to fouling and scaling. This performance erosion may be detected by continuous monitoring of cooling *water temperature*.

With larger heavy duty motors operating in dirty environments, *casing temperatures* should also be monitored alongside the overheating of *cable connections*. Non-contact infrared thermometers can be used for these purposes. Generally, any steady temperature increase over time of an electric device operating at nominal loading conditions should be interpreted as a warning sign.

In all these cases, the performance indicators can be expressed in a so called 'per unit' form:

$$PI = \frac{\text{Measured Value}}{\text{Target Value}} \tag{3.22}$$

It is called 'per unit' because performance indicators assume that the values that will vary around a 'unit', i.e., around 1. When PI is equal to 1, this is an instance of perfect performance because measured value is exactly the same as target. However, this is seldom the case, so usually we establish significance limits (see Part I Chapter 3) to distinguish between normal performance variations and cases of substandard performance. It is the task of performance monitoring to detect gradual but steady deviations of indicators from defined target values (Fig. 3.23). Random fluctuations around targets and within significance limits are of no concern.

In most cases, data recoding frequency of once per shift will suffice.

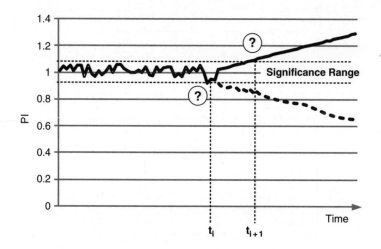

Figure 3.23   Random Fluctuations and the Case of Deviation

3.9.3 Electric Motor Drives – Operational Practice Indicators

The hammer mill example has illustrated the consequences of machines running at partial capacity, where energy effectiveness decreases and energy cost per unit of product increases. The analysis of mill operation has revealed specific instances of undesirable operational practice:

(1) frequent idling (i.e. no-load operation);
(2) low average capacity use;
(3) frequent start/stop.

We can design performance indictors that will help in the monitoring of EMD operational practice from that perspective. Generic indicators are impossible to define because of the various characteristics and requirements of the production systems that EMDs are part of. For instance, a molding machine operation is characterized by cycles, while, a hammer mill operates continuously, but with more or less frequent changes of product types. However, some sample performance indicators can be proposed:

$$PI_1 = \frac{\text{Number of Machine Cycles}}{\text{Shift Time}} \tag{3.23}$$

$$PI_2 = \frac{\text{Machine Run Time}}{\text{Shift Time}} \tag{3.24}$$

$$PI_3 = \frac{\text{Machine Stop Time}}{\text{Shift Time}} \tag{3.25}$$

These three indicators ($PI_1 - PI_3$), when compared with targets, will provide information on excessive (or not) idling and the start/stop frequency of machines.

$$PI_4 = \frac{\text{Total Energy Consumed during the Shift [kWh]}}{\text{Machine Nominal Capacity (P)} \times \text{Shift Time (h)}} \tag{3.26}$$

The PI_4 indicator provides information on capacity utilization: the closer to 'one' the better capacity utilization provided it is not water pumping for an overflow or a similarly wasteful operation.

Table 3.11 Summary of PIs for Electricity Performance Monitoring

Performance indicator	Recording frequency
$PI_1 = \dfrac{\text{Number of Machine Cycles}}{\text{Shift Time}}$	once/shift
$PI_2 = \dfrac{\text{Machine Run Time}}{\text{Shift Time}}$	once/shift
$PI_3 = \dfrac{\text{Machine Stop Time}}{\text{Shift Time}}$	once/shift
$PI_4 = \dfrac{\text{Total Energy Consumed during the Shift [kWh]}}{\text{Machine Nominal Capacity (P)} \times \text{Shift Time (h)}}$	once/shift
$PI_5 = \dfrac{\text{Actual Amount of Material Use}}{\text{Planned Amount of Material Use}}$	once/shift
$PI_6 = \dfrac{\text{Amount of Final Product}}{\text{Amount of Material Use}}$	once/shift
$PI_7 = \dfrac{\text{Amount of Shift Products Rejected}}{\text{Total Shift Products Made}}$	once/shift

The following indicators ($PI_5 - PI_7$) provide information on adherence to production schedules, on productivity of material use and on quality of products. For performance instances declared as substandard, causing factors should be investigated and identified. Typically, they are either machine related or human related, as described earlier.

$$PI_5 = \frac{\text{Actual Amount of Material Use}}{\text{Planned Amount of Material Use}} \tag{3.27}$$

$$PI_6 = \frac{\text{Amount of Final Product}}{\text{Amount of Material Use}} \tag{3.28}$$

$$PI_7 = \frac{\text{Amount of Shift Products Rejected}}{\text{Total Shift Products Made}} \tag{3.29}$$

All PIs are given in Table 3.11 together with their respective recording frequencies.

3.10 Environmental Impacts

The use of electricity as a clean final energy does not produce any direct environmental impact. If electricity is generated at the site by a thermal power plant or cogeneration plant, then the environmental impacts thereof are considered in the Part II Chapter 2 and Part II Chapter 6.

But mostly, electrical energy is produced outside the factory by a power system with a mix of thermal, hydro and sometimes nuclear power plants. Thermal power plants produce significant air emissions, but electricity used at a particular location does not come from an individual 'green' or polluting plant but from the power system as a whole. Therefore, on a system basis equivalent emission factors are usually established for production of 1 MWh of electric energy (see Table 5 in the Introductory Chapter). These factors can be used to quantify indirect emission reduction effects due to electricity use performance improvements in a company.

For instance, the implementation of VSD, as shown in Example 4, resulted in electricity savings of roughly 32 MWh per year. By applying an emission factor of 420.12 kgCO$_2$/MWh, as given in Table 5 of the Introductory Chapter, the equivalent annual emissions reductions are as follows:

$$\Delta\text{Emission CO}_2 = 420.12\,[\text{kgCO}_2/\text{MWh}] \cdot 32\,[\text{MWh}] = 13443.8\,[\text{kgCO}_2] \qquad (3.30)$$

3.10.1 Polychlorinated Biphenyls (PCBs)

The greatest environmental concerns associated with electrical equipment are polychlorinated biphenyls (PCBs). Most transformers and capacitors use dielectric fluid based on polychlorinated biphenyls (PCBs). PCB oils were initially proposed as dielectric fluids for use in electrical equipment such as transformers, capacitors, circuit-breakers, voltage regulators, heat transfer systems, hydraulic systems, etc. because of their excellent dielectric properties and also because of their very low flammability. PCB oil can absorb rapid changes in electric fields with very little heating up, i.e. with very little loss of energy. Also, PCBs have a low flash point and no fire point, meaning that they are stable in changing temperatures. They only burn if placed in contact with an open flame. However, the disadvantages of PCB fluids are now seen as considerable, since they are

- non-biodegradable;
- persistent in the environment;
- able to accumulate in fatty tissues in the body; and
- suspected of being carcinogen.

The effects of PCBs on humans can be serious:

- leading to failure of kidneys and other human organs;
- producing headaches, sickness, etc. if inhaled; and
- causing chlor-acne if absorbed through the skin.

Most environmental regulations govern the following aspects of materials that contain polychlorinated biphenyls (PCBs):

- procedures for equipment use and servicing;
- registry of PCB containing equipment;
- regular inspections;
- reporting of spills and fire-related incidents;
- labeling;
- cleanup requirements for suspected contaminations;
- proper disposal of PCBs
- controlled storage;
- record keeping and reporting.

According to applicable regulations, the management and disposal of fluids containing PCBs may be categorized into three groups according to PCB concentrations where various requirements may apply:

- greater than 500 parts per million (ppm) PCBs (PCB-containing);
- between 50 and 500 ppm (PCB-contaminated);
- less than 50 ppm (not generally regulated, although some requirements may apply).

3.11 Bibliography

Eastop, T.D., Croft, D.R. (1990) Energy Efficiency (for Engineers and Technologists), Longman Scientific & Technological.

Economic Use of Electricity in Industry, Best Practice Programme (1993) www.actionenergy.org.uk.

IEEE Bronze Book: IEEE STD 739: IEEE Recommended Practices for Energy Management in Industrial and Commercial Facilities.

Laughton, M.A. (2002) Electrical Engineers Reference Book, Newnes.

Mehrdad, M. (2005) Savings Through Power Quality, Electenergy Technologies Inc. Columbia, Missouri.

Morvay, Z., Gvozdenac, D., Košir, M. (1999) Systematic Approach to Energy Conservation in the Food Processing Industry, Manual for Training Programme 'Energy Conservation in Industry' (training course performed in Thailand), supported by Thai ENCON Fund, EU Thermie, ENEP program.

Pantaleo, A., Pellerano, A. (2005) Load Management Strategies for Eligible Customers in Electricity Markets: An Application to a Wood Casing Firm, Riv. di Ing. Agr., 4, pp. 71–78.

PCB Transformers and Capacitors: from Management to Reclassification; UNEP Chemicals (2002) May.

Saving energy with motor systems, Best Practice Programme (2002), www.energy-efflciency.gov.uk.

Variable Flow Control, General Information Report 41 (1996) Best Practice Programme Publications, BRECSU and ETSU, (1996).

Warned, D.F. (2000) Newnes Electrical Engineer's Handbook.

Witte, L.C., Schmidt, P.S., Brown, D.R. (1988) Industrial Energy Management and Utilization, Hemisphere Publishing Corporation.

4

Compressed Air System

4.1 System Description

Compressed air is a safe and reliable power source that is widely used throughout industry. In fact, approximately 70 % of all companies use compressed air for some aspect of their operations. Compressed air is generated on-site, giving users greater control over usage and air quality.

On average, compressed air represents approximately 10 % of industrial electricity consumption. The versatility, flexibility and safety of compressed air as an energy transmitting medium ensure its use as an important service. The total cost of compressed air over a ten-year period can be estimated as: 75 % energy; 15 % capital and 10 % maintenance. The large participation of energy cost in the cost of compressed air indicates that an energy efficient system can be a highly cost effective measure even if it increases the cost of capital slightly.

The basic components of the compressed air system are presented in Figure 4.1. The presence of many filters, oil separators and air dryers indicates that the main problems in relation to air compressing are associated with removing particles, oil and moisture. Air distribution systems are more or less always designed in a way that follows the position of end-users. However, the slope of the pipeline and its size have to meet standards.

The represented system uses water to cool the air, but air is also used frequently as a cooling medium.

The cost of generating compressed air is controlled by the efficiency of compressors[1] and is influenced by the following factors:

- operational procedures;
- compressor configuration;
- individual compressor control system;
- number of compressors used to meet the demand;
- overall control system;
- location of compressor room;
- temperature of the inlet air;
- quality of the cooling systems;
- quality of maintenance.

---

[1] Compressors can be tested on site according to the International Standards Organization standard ISO 1217 or equivalent national standards.

Applied Industrial Energy and Environmental Management   Zoran K. Morvay and Dušan D. Gvozdenac
© 2008 John Wiley & Sons, Ltd

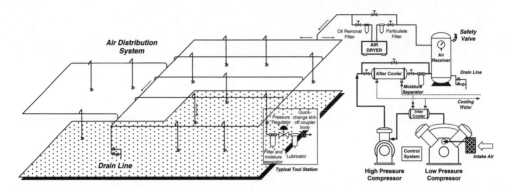

Figure 4.1 Basic Components of the Compressed Air System

Table 4.1 Advantages and Disadvantages of Different Types of Compressors

Item	Reciprocating	Rotary vane	Rotary screw	Centrifugal
Noise level	Noisy	Least noisy	Low if enclosed	Low if enclosed
Size	Least compact	Compact	Compact	Compact
Oil carry over	Moderate	Low-medium	Low	Low
Vibration	High	Very low	Very low	Medium
Maintenance	Many wearing parts	Few wearing parts	Very few wearing parts	Low
Capacity	Low-High	Low-Medium	Low-High	Low-High
Pressure	Medium-Very High	Low-Medium	Medium-High	Medium-High
Efficiency at full load	High	Medium-High	High	High
Efficiency at partial load	High	Poor below 60 %	Poor below 60 %	Poor below 70 %

4.1.1 Compressors

Each compressor type has its advantages and disadvantages. Generally speaking, the main advantages and disadvantages for compressors are presented in Table 4.1.

Energy efficiencies for specific compressors are given in Table 4.2.

Reciprocating compressors can be partially loaded in several ways. Those compressors with two or three step inlet suction valves operate more efficiently at partial loads than other types. Those with variable inlet throttle valves, which modulate the capacity over a narrow pressure range, are not efficient at loads below 70 % of full output.

Modern rotary compressors can be switched either manually or automatically to automatic 'stop-start' control (that is, the compressor will stop after a period of no load running and automatically restart when there is a demand for air). Off-load running time helps to protect the motor drive from too many start-ups. Rotary machines can also be fitted with modulating control. However, the modulation option should only be used if the load is over 70 % of full output.

Table 4.2 Energy Efficiency of Compressors. Reproduced from: Improving Compressed Air System Performance (a sourcebook for industry), U.S. Department of Energy, Energy Efficiency and Renewable Energy, Industrial Technologies Program

COMPRESSOR TYPE	Range of free air flow rate l/s	**UNLOAD** Power as % of full load	**FULL LOAD** Power required at 7 barg (pressure gauge)
Reciprocating Compressor		10–25 %	Single Stage
– Single stage – Double stage	< 50 50–3000		0.38–0.43 kW per l/s Two Stage
			0.30–0.35 kW per l/s
Rotary Vane Compressor	50–1500	30–40 %	0.40–0.45 kW per l/s
Rotary Screw Compressor	< 600	25–60 %	0.35–0.40 kW per l/s
Centrifugal Compressor	600–9000	20–30 %	0.30–0.35 kW per l/s

A centrifugal compressor can provide excellent control and reduce power consumption at partial load, provided that the machine operates within its design envelope. This is between 60 %–100 % of full output depending on the aerodynamics of the entire through passages.

4.1.2 Air Pressure

Although the air requirements can vary between factories, the most common operating compressed air pressure is about 6–7 barg. As pressure is increased, the flow rate is reduced (although this depends on the pressure and the required air flow). The formula below relates the *volume* of air before and after compression to pressure:

$$\text{Compressed Air Flow Rate} = \text{Free Air Flow Rate} \cdot \frac{\text{Initial Pressure}}{\text{Final Pressure}} \tag{4.1}$$

Free air volume is taken at standard conditions: 20 °C; 50 % relative humidity; and absolute pressure of ambient air as 1.013 bar. Initial and final pressures in Eq. 4.1 are absolute.

4.1.3 Air Receiver

The main function of an air receiver is to act as a buffer between the factory and the compressor by storing a large volume of compressed air. It also balances the pressure changes in the air distribution system. An air receiver has to be installed if the total volume of the compressed air mains is not large enough to keep the *Load/Unload* interval longer than 45 seconds. This interval can be checked by a simple experiment during the operation of the system.

4.1.4 Water Removal

Proper water removal from compressed air prevents:

- damage to air tools;
- corrosion of equipment.

Water should be removed by an after-cooler and separator, or a refrigerated dryer (Fig. 4.1). Compressed air system performance is typically enhanced by the use of dryers, but since they require added capital and operating costs (including energy), drying should only be performed to the degree that it is needed for proper functioning of the equipment and for end use.

The main types of dryers are:

- *Refrigerant dryers.* These can achieve a pressure dew point[2] of 3 °C and are adequate for many standard applications. An energy efficient version is available which incorporates a control system that matches the cooling/drying load with the air demand.
- *Desiccant dryers.* These can produce virtually moisture-free compressed air with pressure dew points down to −70 °C. They are designed in such a way that while one desiccant column is in use to dry the air the other is regenerated (i.e. dried out for re-use). Desiccant dryers fall into two categories – heated and heatless. Desiccant dryers use water absorbing chemical (e.g. silica gel), which can absorb water out of the air stream. The standard models can add 15 %–20 % to overall compressed air running costs. However, more and more energy efficient desiccant dryers are available.
- *Membrane dryers.* These achieve dryer air than refrigerant dryers but, if much lower pressure dew points (−40 °C to −100 °C) are required, membrane dryers become less efficient than the desiccant dryers described above. Membrane dryers should not be used for the supply of breathing air as they can cause oxygen depletion unless specifically designed to avoid this.

To minimize energy use, the type of water removal equipment depends on how the compressed air will be used:

- For high quality dry air (e.g. pneumatic controls), a desiccant air dryer should be used with a fine filter.
- If only a portion of the factory's air has to be of a high quality, then only a small air dryer for this use has to be installed.

In order to remove significantly more water from the compressed air than can be achieved by after-cooling, a dryer is necessary. Whichever dryer is used this implies the cost of compressed air and because of that the real quality of air has to be analyzed very carefully.

Let us discuss the problem of water removal with a simple example. Standard ambience air (20 °C; 50 %; 1.013 bar) contains 0.00726 kilogram of water per kilogram of *dry* air[3]. All of the water is still a vapor. If a compressor processes 1000 liters per second of such humid free air, the inlet and outlet compressed air will contain 31.095 kilograms of water per hour. If the air is compressed at 8 bar absolute, the water will remain partially as vapor and the rest of water will be condensed. The rise in air temperature during compression prevents condensation, but when the air passes through the after-cooler a large amount of vapor condenses. The typical temperature of the air following the after-cooler is approximately 35 °C, but the air pressure is much higher than ambient one. The saturated absolute humidity of air at 35 °C, 8 bar and 100 % is 0.00440 kg/kg and the density is 9.0192 kg/m^3. It means that 0.00726 − 0.00440 = 0.00286 kg/kg of vapor will be condensed. For the same free air flow rate, this means that 12.25 kg of water can be removed from compressed air following the after-cooler.

---

[2] 'Pressure dew point' means the dew point temperature at the pressure of compressed air.
[3] Part III Toolbox 13 Software 6: Thermodynamic Properties of Moist Air can be used for this or similar calculations.

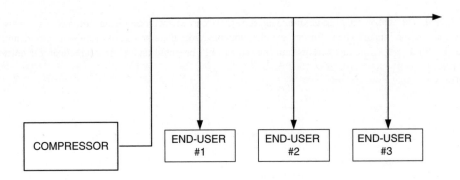

Figure 4.2   Single Pipe Supply System

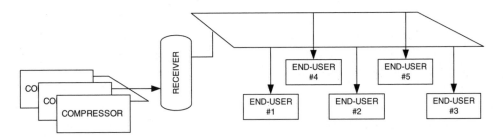

Figure 4.3   Ring Main Supply System

4.1.5 Pipe Work

A single pipe supply system is typical for small compressed air systems (Fig. 4.2). A ring main system is usually used to allow the sections of the pipe work to be isolated without stopping air supply to other users (Fig. 4.3).

Small pipe sizing can cause high air velocities, which result in a large pressure drop due to friction inside the pipes. Most pneumatic equipment requires a minimum of 6 bar for maximum operating efficiency. The distribution of air should be done with a minimum pressure drop, say about 0.50 bar including all fittings and connections. This should result in a compressed air speed no more than 6–10 m/s.

4.1.6 Ventilation of Compressor Room

The temperature of the compressor room has to be less than 40 °C. The ventilation system of a compressed air room has to be checked by a simple experiment which assumes the measurement of the air temperature in the compressed air room and the current conditions outside and inside the room (ambient temperature, load of compressors, doors on/off, etc.).

4.2 Performance Analysis

For an existing compressed air system, the most convincing and the most complete performance indicator is the quantity of energy needed for the production of a compressed air unit. The unit of compressed air is usually expressed as free air volume flow rate at standard conditions: 20 °C, 50 % humidity,

1.013 bar. The reason for this is simple because it is easiest to measure the volume of air flow at the compressor intake. Then, further computation of variable parameters of ambient air is simple Part III Toolbox 13 Software 6. This indicator should be computed on a daily basis for larger units as follows:

$$PI_1 = \frac{\text{Free Air Volume Flow Rate}}{\text{Energy Consumed}} \; [m^3_{FA}/kWh] \tag{4.2}$$

Another important performance indicator is the unit price of compressed air. A dominant share in the cost of compressed air production is taken by the price of electricity. However, maintenance costs should not be neglected. Thus, this indicator can be computed on a monthly basis, and is defined as follows:

$$PI_2 = \frac{\text{Free Air Flow Rate}}{\text{Cost of Electricty} + \text{Maintenance Cost}} \; [m^3_{FA}/\$US] \tag{4.3}$$

The factors that cause changes of these indicators over time are numerous. Some of them are:

- increased uncontrolled leakage;
- changed price of electricity (PI_2);
- increased maintenance costs (operating materials, spare parts and the like);
- significant changes in the volume of production.

The monitoring of these indicators is part of energy management and implies undertaking measures to keep them within allowed limits. The other factors that can influence performance changes and that need to be checked are listed below:

- whether it is possible to reduce pressure drops for all elements of the system;
- whether operating pressure is optimum and whether it is possible to reduce this pressure;
- whether the measurement of compressed air consumption is adequate;
- what the time variation of air consumption is and whether system control is adequate;
- whether the quality of air is satisfactory;
- whether recuperation of waste heat is possible;
- whether maintenance is carried out according to recommendations.

In addition to the above-mentioned performance indicators, it is also important to monitor the compressor's efficiency. However, this figure will not change significantly for the corresponding load if the preventive maintenance recommended by the manufacturer is carried out properly.

4.2.1 Compressor's Performance Graph

A compressor performance graph shows free air delivery [l/s] versus power of compressor [kW] (Fig. 4.4). The maximum rated free air delivery, or FAD, is the compressor design capacity [l/s] and can be found in the technical manuals. This is typically quoted under standard conditions (i.e. 1.013 bar absolute, 20 °C and relative humidity of about 50 %). The values of power input and maximum free air delivery are plotted and that point is denoted as **1**. No air flow rate condition is denoted by point **2** and real working conditions by point **3**.

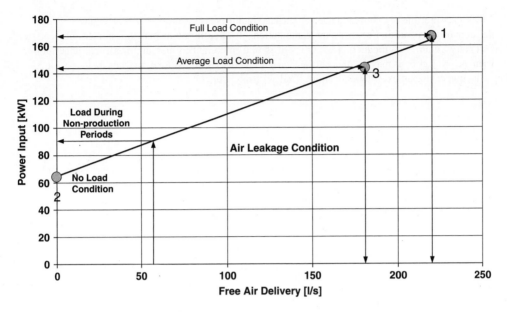

Figure 4.4   Power Input versus Free Load Delivery Graph

This graph could be very useful for the estimation of real capacity if the average power is known.

4.2.2 Measurements

Measurements at a compressed air system undertaken in industrial conditions are related primarily to compressors. An air distribution system is highly diversified and compressed air flow rates differ considerably over time. This is the reason why measurement for certain users or groups thereof is very strenuous and sometimes pointless (constant turning on and off). Where an air distribution network is concerned, the most important thing is to perform regular inspections of installations and establish whether there are any possible leakages. However, the measurement of pressures and pressure drops per certain branches is a very good indicator showing also leakages and the current flows in branches. One of the most important factors influencing energy consumption in the production of compressed air is the discharge pressure of air which is in turn dictated by end-users and by the hydraulic characteristics of the distribution network.

Compressors are the main energy consumers in a compressed air system. Modern compressors enable the measurement of the current consumption of electrical energy, discharge pressure, efficiency, suction parameters, time of operation, etc. Where these single measurements are concerned, the analysis also includes confirmation that the data is within the limits prescribed by the manufacturer. This involves the temperature of oil in the compressor, cooling and the like.

The measurements which should be performed in the distribution network are as follows:

- Start-up pressure for each branch which carries more than 10 % of the total compressed air flow. Flow assessment per branche is possible in accordance with the pipeline diameter. The diameter is established on the basis of recommended pipeline flows (Part III Toolbox 7).
- Pressures at the remotest parts of the branch.

Values measured in the above manner define maximum pressure drops per branch and are the best indicator of the condition of this part of the system. It is desirable to make these measurements for customary volumes of production activities. It is also important to measure these pressures not only when the production is maximum but also when it does not exist. Pressure drop when there is no ongoing production can be an excellent indicator of leakages and poor control of production machines.

4.2.2.1 Power

Power is measured in order to assess, based on the number of the system's operating hours, the annual electricity consumption of the compressed air system.

Typically, compressors cycle between *load* conditions (the outlet pressure gauge rises) and *unload* conditions (the outlet pressure gauge falls). For both load and unload conditions, a clip-on ammeter can be used to measure the current of each of the three phases, with an average value calculated for both load and unload conditions (I_{ave}). The power of a three phase electric motor for both load and unload conditions can be calculated using the formula below.

$$P[kW] = \sqrt{3} \times I_{ave} \cdot PF \cdot \frac{U}{1000} \qquad (4.6)$$

where:
I_{ave} = average phase current, [A]
U   = supply voltage in volts (usually 380 [V])
PF = power factor (typically 0.85 at full load, 0.7 at half load and 0.6 at no load for reciprocating compressors)

If the compressor is not cycling, the following procedure can be adopted to achieve load and unload conditions:

Full load conditions: Open the discharge valve from the air receiver and measure the power consumed by the compressor as it builds up pressure.
No-load conditions: Shut the valve between the air compressor and the air receiver and measure the power.

It is important to ensure that these tests will not affect production.

4.2.2.2 Pressure

Pressure is measured in order to provide feedback for control adjustments and to determine pressure drops throughout the system. A calibrated pressure gauge is required. The required pressure levels must take into account system losses from dryers, separators, filters and piping.

The following pressure measurements should be taken:

- inlet to compressor (monitor inlet filter);
- differential across air/lubricant separator for a lubricated rotary compressor;
- inter-stage on multistage machines;
- pressure differentials, including, after-coolers, treatment equipment (dryers, filters, etc.) at various points in the distribution system.

4.2.2.3 Flow Rates

Flow meters are necessary to measure total flow and to determine air consumption. Flow should be measured

- during various shifts;
- after implementation of energy saving measures;
- for leaks during non-production periods.

Flow meters should be of the mass flow design to compensate for pressure and temperature variations and, if practical, should be suitable for measuring the output of each individual compressor in the system. The mass flow is based upon standard reference conditions, which should be checked for the specific instrument used.

4.2.2.4 Temperature

Temperature measurements help to indicate whether equipment is operating properly. Generally, equipment that runs hotter than specified parameters requires servicing. The following temperature measurements should be taken:

- after-cooler and intercoolers in and out temperatures;
- inlet and outlet temperatures of water for air cooling (if water cooling is applied);
- inlet air temperature.

4.2.2.5 Compressed Air Leakage Determination

The number of uncontrolled cracks and holes in the system is numerous. Small leakages are very difficult to notice but when multiplied by their large numbers a significant volume of air is lost from the system without any control. Figure 4.5 shows the dependence of the free flow air leak and energy wasted

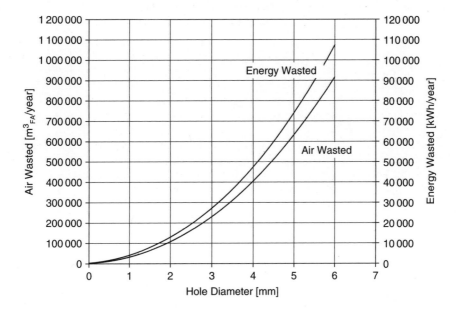

Figure 4.5 Air and Energy Wasted Versus Equivalent Hole Diameter of Air Leaks

The curves are based on the following assumptions: Discharge pressure = 7 bar; 1 kWh = 8.5 m^3_{FA} air delivered; 8000 hours operation per year

versus the equivalent hole diameter. This example is sufficiently convincing to indicate the importance of eliminating unwanted leakage from installations.

The best way to quantify the leakage in a compressed air system is by measurement. There are two methods that are sufficiently exact and not too complicated and which can be suggested as a regular procedure in industry. Common to both tests is that they can be performed only during a non production period. These are as follows:

(a) LOAD/UNLOAD TEST

The procedure for this test is as follows:

- Shut down all machines and equipment using compressed air.
- Select one of compressors for testing and close the discharge valves of the others.
- Start the selected compressor and operate it to full line pressure. If this compressor cannot reach full line pressure alone, then the leakage exceeds the full capacity of that compressor. An additional or bigger compressor has to be started.
- Over a number of cycles, make a note of the average load time (T_{LOAD}) and average un-load time (t_{UNLOAD}). Air leaks will cause the pressure to fall and the compressor(s) will switch in load mode again.
- Total leakage can be calculated by using the following formula:

$$V_{Leak}[l/s] = V_{CC}\,[l/s] \cdot \frac{T_{LOAD}}{T_{LOAD} + t_{UNLOAD}} \qquad (4.7)$$

In this formula, V_{CC} is compressor air capacity. If the compressor operates in an on/off control mode, V_{CC} represents its nominal capacity. But, if it operates in any of modulating control modes, dependence on average power versus free air delivery must be known (Fig. 4.4). The average power during the test can be calculated or measured.

(b) PUMP-UP TEST

This test assumes the installation of the pressure gauge downstream of the receiver and the calculation of the volume of the system which is tested. The test procedure is as follows:

- Calculate the tested system volume (air mains and receivers). For this calculation the blue prints of the system or its physical inspection could be used.
- Pump up the system to operating pressure (p_1) and then close the isolating valve.
- Record the time (τ) necessary for air pressure to drop from p_1 to p_2.
- Leakage can be calculated as follows:

$$V_{Leak} = \frac{V_{MAIN}}{\tau} \cdot \frac{p_1 - p_2}{B} \qquad (4.8)$$

where:
V_{MAIN} = Volume of the air mains and receivers, [m^3]
p_1 = Pump-up pressure, [bar]
p_2 = Final pressure after time (τ) of testing expired, [bar]
τ = Time for pressure drop from p_1 to p_2, [s]
B = Barometric pressure, [bar]

By knowing the leakages, which are calculated by applying any of the presented tests, it is possible to calculate their cost. Of course, what is most interesting is to calculate the cost of leakages per year. For this calculation it is necessary to know the total annual electrical energy consumed by the compressed air system and the number of working hours. These values can be used for the calculation of the average power of the compressed air room. By establishing the relation between power input and free air delivery for all compressed air rooms tested, it is possible to determine the average free air delivery flow rate. If we denote this value as $V_{C/A}$, the percentage of leakages, the loss is as follows:

$$L_{C/A} = \frac{V_{Leak}}{V_{C/A}} \cdot 100 \, [\%] \tag{4.9}$$

4.3 Performance Improvement Opportunities

The most frequent reason for increased costs of compressed air is uncontrolled leakage of air and its inadequate usage. All of this is an integral part of the activities related to maintenance. Another important group of activities in a factory which provides performance improvement opportunities is related to improvement of the operational practice of the compressed air systems.

4.3.1 Operation and Maintenance

Maintenance is essential in keeping compressed air costs to a minimum. This includes:

- *Better Housekeeping*:

    - Leakages from hoses can be identified by listening to 'hissing' noises during non-production time.
    - The cost of repairing leaks is small compared to the energy wasted.
    - Leaking pipe joints require replacement of a joint.
    - Repair or replace faulty connections to end-users.
    - Ensure water drains on receiver and filters work correctly or install automatic drain traps.

- *Compressor Maintenance*:

    - Lubricating oil reduces friction and prevents wear on seals and other rubbing parts – units should be maintained regularly and only the oil recommended by the manufacturer should be used.
    - Filters remove dirt, water and oil from the air supply – these should be cleaned regularly or replaced as filters can get dirty very quickly and create excessive pressure drops.

- *Improved Air Usage*:

    - Avoid air blasts for cleaning and drying products as blowers are just as suitable.
    - Do not use compressed air for burner combustion air – use a fan instead.

4.3.2 Control Systems

4.3.2.1 Compressors

Controls are the best way to manage the operation of the system in a proper way. A modern control system can save 5 %–20 % of the total generation costs for a relatively small capital outlay. As operation conditions sometimes vary greatly (much more than assumed during the design process), it is important to check whether the existing control system operates properly. The compressor output should be controlled by a controller to match the air demand. This can be done by:

(i) Measuring pressure in the receiver, so that the compressor operates at full load until the pressure in the air receiver reaches a pre-set limit, then the compressor stops or 'unloads' – this is the most common method.

(ii) Throttling the air flow at the inlet of the compressor by using a valve to restrict air flow often results in the compressor power in the unload condition being about 20 %–60 % of full load power, depending on the type of compressor (Table 4.2).

The energy cost of a multi-compressor system can be high if all compressors load and unload together. Sequencing controls are available from most compressor manufacturers. They provide automatic control and achieve the optimum operation of the compressed air system by ensuring that:

(i) The correct number of air compressors operate (i.e., one compressor is fully loaded before starting another).

(ii) The correct types of air compressors operate (i.e., the more efficient compressor operates better than less efficient compressors) for a particular load.

The latest technology entails using a frequency regulator to control the load by adjusting the speed of the compressor to the instantaneous requirements of the process for compressing air. From an energy point of view, this control system is the best. Controllers can also be programmed to ensure that compressors with a high energy use during no-load conditions operate as fully loaded as possible.

A real case example will illustrate the opportunities provided by a modern automatic control system. In this case study, seven compressors with various capacities operate in a network and they are used to supply the air to a production process. In the first two shifts (6:00–22:00), the production process is carried out under full capacity and compressed air needs are expressed through the power of compressors varying from 690 to 1125 kW (Fig. 4.6). Compressors Nos 1 to 3 are larger than the other four relevant to their capacities and they are used to cover the base consumption of the compressed air. This is the reason why they have been set to higher discharge pressures enabling them to operate constantly. Their overall power is 690 kW.

The total power of compressor Nos 4 to 7 is 435 kW. Compressor No. 7 is set to operate in the range from 5.9 to 6.3 bar and this is also the required pressure range (Fig. 4.6). The represented operational

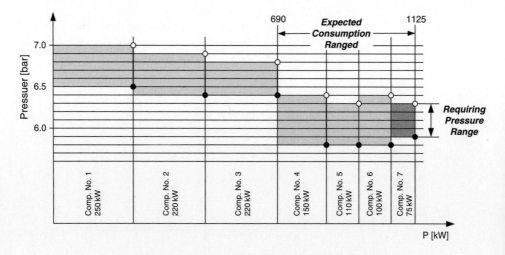

Figure 4.6   Daily Air Compressors Operational Schedule

scheme is used for daily production activities (from 6:00–22:00). Compressors Nos 6 and 7 have time logic switch off control activated after 10 minutes of compressor motor unload operation. In this way, it can be expected that in the case of reduced consumption of compressed air, only compressor No. 5 will be in unload operational mode and compressors Nos 6 and 7 will be in such an operational mode only occasionally. Compressor No. 4 is set to a somewhat higher discharge pressure (6.4 bar) and therefore does not have time logic switch off control.

Figures 4.7 to 4.9 show the change of power of compressors Nos 4, 6 and 7. It can be seen that compressors Nos 6 and 7 have operated only under load conditions (except compressor No. 7 which has been turned off once). Only compressor No. 4 has had higher load and unload periods.

In this way, the unload operational mode includes only to one compressor with medium power. Before such an operational strategy was deployed, several compressors were at the same time in the unload mode

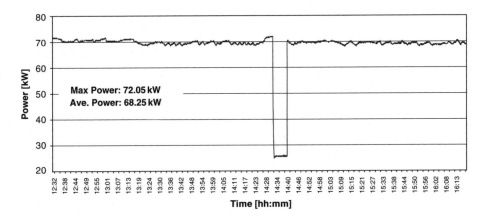

Figure 4.7   Average Power of Compressor No. 7

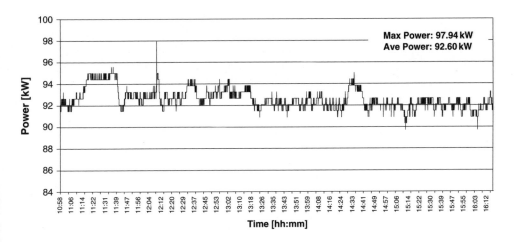

Figure 4.8   Average Power of Compressor No. 6

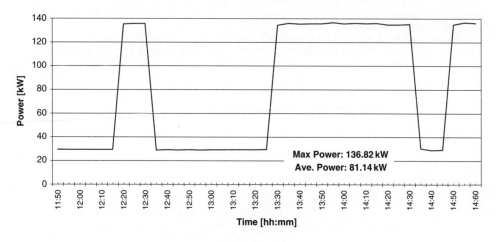

Figure 4.9   Average Power of Compressor No. 4

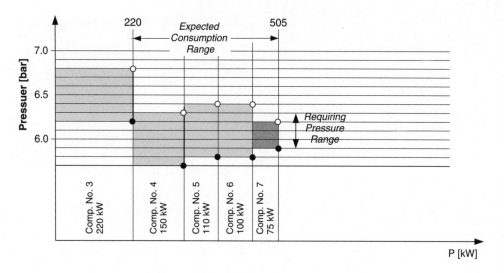

Figure 4.10   Night Air Compressors Operational Schedule

operation which, knowing that the compressor power in unload operational mode is 30 % of nominal power, represented a significant energy loss.

The compressed air consumption during the night shift (22:00–6:00) is low and the expected power of the compressors is within a range from 220 to 505 kW. Lower consumption allows pressure reduction of about 0.1 bar as the losses in the main pipelines are also reduced. Pressure control is set in the range from 5.9 to 6.2 bar (Fig. 4.10). At night, compressors Nos 1 and 2 are simply turned off.

4.3.2.2 Control of Compressed Air Quality Standard

The current international standard for compressed air quality is specified by ISO 8573.1: 2001. Within this standard, a number of quality classes is shown, specifying the recommended maximum amount of

Table 4.3   Specification of Contamination Levels. Reproduced from: High Quality Compressed Air from Generation to Application (A Guide to ISO 8573.1:2001 Air Quality Classes)

Class	Solid Particle [Maximum number of particles per m³]			Water Dew Point Temperature at 7 barg [°C]	Oil (including vapor) [mg/m³]
	0.1–0.5 micron	0.5–1.0 micron	1.0–5.0 micron		
1	100	1	0	−70	0.01
2	100 000	1,000	10	−40	0.1
3	–	10 000	500	−20	1
4	–	–	1000	3	5
5	–	–	20 000	7	–
6	–	–		10	–

allowable dirt, water and oil per cubic meter of compressed air. The specification of contamination levels proposed by this standard is presented in Table 4.3.

Compressed air contains contaminants such as: (a) water vapor, (b) condensate, (c) particulate matter which can be air-borne and/or pipe-scale, (d) oil in vapor or liquid state inhaled by the compressor from the atmosphere or added during compression and (e) microbes. The treatment of air depends on the users needs. However, it is recommended that one follows the real needs of end-users rather than install more sophisticated and more expensive equipment for air purification. A generally recommended standard for different manufacturing applications is given in Table 4.4.

4.3.2.3 Cooling Water Quality Control

If a compressor's cooling system use water, it is very important to keep water properties within the range prescribed by the manufacturer. Fouling problems with heat transfer surfaces could significantly decrease the efficiency of the compressor and increase the energy consumption for the same compressed air flow rate. Gradually, it will reduce the flow rate. The main problem with water quality's influence on the process of compressor cooling is that fouling acting gradually over a long period of time will slowly change the performance of the compressor. A similar problem occurs with air cooled machines if the cooling system becomes dirty.

If there are no manufacturer's recommendations on cooling water quality, the data in Table 4.5 can be used.

4.3.3 Performance Improvement Measures

4.3.3.1 Compressed Air Leakage Reduction

So far, we have mentioned leakage problems several times and have indicated the need for their reduction or elimination. This should be a continuous process and an integral part of regular maintenance.

4.3.3.2 Control System

Modern technology offers today a variety of opportunities to improve the control system of existing compressed air systems. Very often, even the existing control system is not used in an optimum way.

Table 4.4   Recommended Standard for Different Applications

Application	Typical quality class		
	Oil	**Dirt**	**Water**
Air agitation	3	5	3
Air bearings	2	2	3
Air gauging	2	3	3
Air motors	4	4–1	5
Brick and glass machines	4	4	5
Cleaning of machines parts	4	4	4
Construction	4	5	5
Conveying, granular products	3	4	3
Conveying, powder products	2	3	2
Fluidics, power circuits	4	4	4
Fluidics, sensors	2	2–1	2
Foundry machines	4	4	5
Food and beverages	2	3	1
Hand operating air tools	4	5–4	5–4
Machine tools	4	3	5
Mining	4	5	5
Micro-electronic manufacture	1	1	1
Packing and textile machines	4	3	3
Photographic film processing	1	1	1
Pneumatic cylinders	3	3	5
Pneumatic tools	4	4	4
Process control instruments	2	2	3
Paint spraying	3	3	3
Sand blasting	–	3	3
Welding machines	4	4	5
General workshop air	4	4	5

Table 4.5 Recommended Quality of Cooling Water. Reproduced from:Reproduced from: Improving Compressed Air System Performance (a sourcebook for industry), U.S. Department of Energy, Energy Efficiency and Renewable Energy, Industrial Technologies Program

Item	units	A	B
ph at 25°C	–	6.5–8.0	6.5–8.0
Conductivity at 25°C	μS/cm	< 800	< 200
Total hardness as $CaCO_3$	ppm	< 200	< 50
M-Alkalinity as $CaCO_3$	ppm	< 100	< 50
Chlorine ion, Cl^-	ppm	<200	< 50
Sulfuric acid ion, SO_4^{2-}	ppm	<200	< 50
Total ion, Fe	ppm	<1.0	< 0.3
Silica, SiO_2	ppm	< 50	< 30
Sulfur ion, S^{2-}	ppm	0	0
Ammonium ion, NH_4^+	ppm	0	0

Note:
– Column B should be applied to the make-up water for the circulating water system.
– Inspect the quality of circulating water regularly and confirm satisfying values in column A.
– Sea water cannot be applied to the compressor unit because it is corrosive.
– Solid particles such as mud, dust should be removed from the supplied water.

Taking into consideration the fact that modern control systems are not expensive in relation to the energy costs of the compressed air system, the potential for upgrading control system should be investigated when there are any changes in the compressed air system.

4.3.3.3 Discharge Pressure

Each machine which uses compressed air has a given pressure and a necessary connection. The pressure should be within specified limits, for example, from 3–5 barg or the like. Unfortunately, designers often start from the highest pressure and add up accompanying losses thus unjustifiably increasing initial pressure in the compressor room in order to reduce risks. The practice has shown that the discharge pressure of compressed air is increased unnecessarily at the compressor's end and then just before the end-users this pressure is reduced. Figure 4.11 shows the percentage of reduced power of the compressor versus reduced discharge pressure. Let us consider the meaning of this diagram with the following example.

In Figure 4.11, locate the reduced discharge pressure on the abscissa, draw a vertical line to the curve which represents the initial discharge pressure and draw (from the point of intersection) a horizontal line to read as the approximate decrease in kW from the ordinate. The percentage of power reduction is calculated regarding initial discharge pressure. For the above conditions, the savings realized by lowering the discharge pressure from 8 to 6.5 bar is 13 %.

4.3.3.4 Waste Heat Recovery

There is a potential to use up 50 % to 90 % of the heat which is drained off by cooling the air during compression. The most energy is recoverable with screw compressors and the least with turbo compressors.

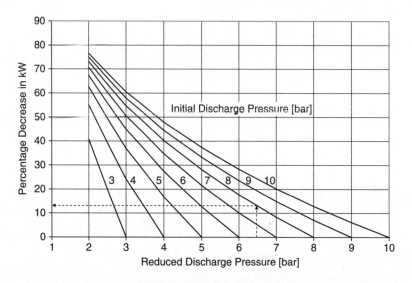

Figure 4.11   Percentage Decreases in kW with Discharge Pressure Reduction
(Single Stage Compressors)

However, realistically waste heat recovery can be utilized economically up to approximately 30 % of the total energy used for air compression. This energy can be used for some low temperature purposes within the factory (heating up boiler make up water, sanitary water and the like).

4.4 Performance Monitoring

The permanent monitoring of the two main performance indicators, as defined earlier, is essential for the proper surveillance of the system and adequate action in the case of variation from given target values. This implies the measurement of electrical energy and air flows. Modern compressors measure these values as well as others which are used by an operator, although the measured indicators are not often processed further. The task of management is to set up procedures for data gathering and processing and to establish corrective actions or trigger preventive maintenance when performance indicators are not within the prescribed limits.

Modern electronic monitoring systems log the running condition and, if anything goes wrong, signal any abnormalities and shut the machine down before any damage occurs. Electronic monitoring units can be fitted onto existing compressors as well.

4.4.1 Values to be Monitored

The data which have to be known for the proper monitoring of the compressed air system assume the following:

- energy consumption [kWh/TI] (TI is time interval: h, day, week, . . .);
- free air flow rate, [m^3/TI] (calculated for standard air at 20 °C, 1.013 bar and 50 %);
- average ambient air parameters;
- discharge pressure of compressed air [bar];
- pressure at point of use to establish the pressure drop across the system [bar];
- parameters of water if the system is water cooled;

- hours-run meters;
- air quality.

In addition to the above parameters which are indispensable for estimating the energy efficiency of the compressed air system operation, it is also necessary to ensure the monitoring of other values which fall within the domain of maintenance (oil level and pressure, oil temperature, filter pressure drops, etc.). These measurements provide an opportunity to check the causes of faults and identify the maintenance needs.

The values listed above refer primarily to compressors but it is also necessary to monitor other equipment and devices such as filters, dryers, etc. In particular, end-users should be monitored carefully for leaks.

4.5 Example: Detailed Energy Audit of Compressed Air System

This example demonstrates the process of analysis and measurement performed during an energy audit in one factory which runs 132 machines with compressed air.

(a) Basic Facts about the Factory and C/A System

Total electricity cost	2.18	m$US
Total electrical energy consumed	39.91	GWh
Electrical energy unit price	0.0546	$US/kWh
Cost of electricity for compressed air	0.37	m$US
Total electrical energy consumed by compressed air system	6.49	GWh
Total share of compressed air system	16.26	%
Average power of C/A system	773	kW
Installed power of C/A	1092	kW
Load factor of the C/A systems	0.71	

(b) Facility Description

There are two separated rooms (in the first one there are five units and in the second two units of compressors). All of them are connected together by a piping system which functions as a buffer (air tank) in order to reduce the frequency of load and unload of the compressor.

The general scheme of the compressed air system is presented in Figure 4.12 and the system specifications are shown in Tables 4.6 and 4.7.

(c) Scope of Work

Three distinctive activities are performed:

- inspection, testing and fault finding (before the factory's shut down);
- air leak test (before and during shut down);
- development and recommendations of C/A systems performances improvement.

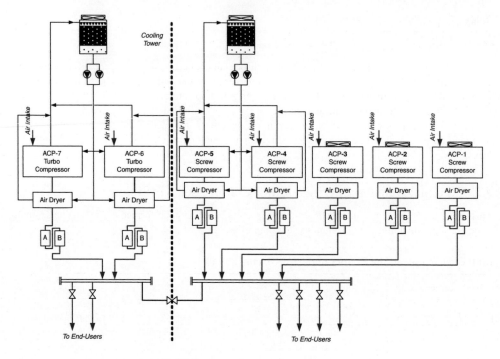

Figure 4.12   Scheme of Compressed Air System.

Table 4.6   Specifications for Air Compressors

Compressor	Type	Capacity* m³/min	Motor Power kW	Cooling System	Air Dryer kW
ACP–1 (Stand by)	Screw	17.8	132	Air	5.4
ACP–2 (Stand by)	Screw	17.8	132	Air	5.4
ACP–3	Screw	17.8	132	Air	5.4
ACP–4	Screw	19.5	132	Water	4.6
ACP–5	Screw	19.5	132	Water	6.6
ACP–6	Turbo	21.6	150	Water	6.6
ACP–7	Turbo	21.6	150	Water	6.6
TOTAL:		**135.6**	**960**		**40.6**

* Free air volume flow rate at standard conditions: 20 °C, 50 % humidity, 1.013 bar

Table 4.7   Specifications for Cooling Tower and Cooling Water Pumps

	Water Flow Rate l/min	Power kW	Type
CT–5 (100 RT)	1300	1.5 (fan)	Counter Flow
CT–7 (100 RT)	1300	1.5 (fan)	Counter Flow
CDP–5–1	1300	22	Centrifugal
CDP–5–2 (Stand by)	1300	22	Centrifugal
CDP–8–1	1300	22	Centrifugal
CDP–8–2 (Stand by)	1300	22	Centrifugal
TOTAL:		**91**	

4.5.1 System Inspection, Testing and Fault Finding

In order to establish the full scope of work for the planned maintenance and the parts replacement plan, an inspection and testing program was carried out prior to the plant shut down. Inspection and testing refer to visual and audible inspection with the added testing of components, such as pressure and temperature testing, leakage testing, rpm checks, etc., in order to verify operational performance and identify faults where and if they exist.

In this stage of work, the following checks were planned:

- specification of main end-users;
- set pressure;
- specified flow rate;
- specified quality of air.

The results of the inspection follow below:

(a) *Compressors and Air Dryers*:

- Servicing of these machines is performed on regular basis every month by an authorized company (lubrication, condition of rotors, presence of vibration, excessive discharge pressure and temperature, tension in belts and adequateness of cooling).
- The daily recording of relevant periodical measurements is managed properly.
- The instruments are in good condition, but their calibration has not been performed in a requested period. The management of the technical department has to do that within 30 days.
- The entering air temperature is in the range recommended by the manufacturer.
- The leakage in the condensation drain is recognized and repairs suggested.

(b) *Distribution Network Inspection*:

- Pipe sizing and drainage are adequate. During the load period, the pressure drop in the longest pipeline is less than 0.5 bars or 0.25 mbar/m. This result is related satisfactorily to the design standards and the manufacturer's recommendations.

- To avoid the air wastage that frequently occurs with current manually operated drain valves, automatically operated drain valves are recommended. The slope of the pipeline is not sufficient. In general, the main pipeline should be installed with a fall of not less than 1 meter in 100 meters of ring pipe-work in the direction of air flow.
- For the sake of energy efficiency, automatic drain traps have to be fitted to the bottom of the drain line within the compressed air system. Reliable electronic condensation traps are available that ensure that water is drained away regularly. The ball float type of drain trap is most common because it provides a positive shut off, opening only if water is present and closing immediately the water has cleared.
- Leakage is the largest single waste of energy associated with compressed air usage. Leakage rates exceeding 20 % of site consumption are common. The sources of leakage are numerous, but the most frequent problems concern:

  – condensing drain valves left open;
  – shut-off valves left open;
  – leaking pipes and joints;
  – leaking hoses and couplings;
  – leaking pressure regulators;
  – air cooling lines left open permanently;
  – air-using equipment left in operation when not needed.

After performing an inspection, the action plan is prepared. Some of the proposed actions do not require further analyses as they are so obvious. But the following actions, assuming also a leak test, are recognized as very promising and they are performed immediately without waiting for the shut-down of the factory.

4.5.1.1 Activities before the Plant's Shut-down

1. All covered ducts on the floor for pipeline distribution are cleaned. They were heavily polluted with production waste materials and dust.
2. The joints and valves are tested with soapy water. The number of machines in production departments, which use compressed air is 132. They are connected to the main pipelines by two valves and plastic pipe. The soapy water test is performed at 40 points. The results are shown in Table 4.8. Leaking is recognized at 33 spots or on 82.5 % of the machines' junctions to the main distribution lines. The repair of all junctions is strongly recommended. A leakage intensity that is estimated as high is denoted with $+++$, medium with $++$, and low with $+$.

Table 4.8 Soapy Water Test

No.	Machine number	Leakage		No.	Machine number	Leakage	
		Yes	No			Yes	No
1	69-1	+++		21	67-2	+	
2	70-2		+	22	67-1	+++	
3	70-4	+		23	66-1	+	
4	70-6	+		24	66-2		+
5	71-7	++		25	66-4	++	

Table 4.8 (*continued*)

No.	Machine number	Leakage		No.	Machine number	Leakage	
		Yes	No			Yes	No
6	72-5		+	26	66-5	++	
7	71-4		+	27	66-6	+++	
8	72-2		+	28	66-8	++	
9	71-1		+	29	64-8	++	
10	73-2	++		30	64-7	+++	
11	74-3	+++		31	64-6	+++	
12	74-6	+++		32	64-5	+++	
13	76-7	+++		33	64-3	+++	
14	76-4		+	34	64-2	+++	
15	75-2	+		35	64-4	+++	
16	68-3	+++		36	65-1	++	
17	68-8	++		37	68-2	+++	
18	67-6	+++		38	69-5	++	
19	67-5	+		39	69-8	+++	
20	67-3	+++		40	73-6	++	
21	67-2	+			**TOTAL:**	**33**	**7**

Immediately after the test was performed, the system was repaired.

4.5.1.2 Air Leak Test (during Shut-down)

(a) Measurements, Technical Calculations, Analyses and Findings
The ower of all compressors, pumps and fans in the production period is represented in Figure 4.13. The Individual consumption of compressors is presented in Figures 4.14 to 4.17.

- Total installed power on the electric motor shafts   = 960 kW
- Total annual number of operating hours of factory: = 8400 h
- Power used only by compressors   = 664.8 kW
- Install free air delivery of all compressors   = 135.6 m³/min
- Real free air delivery (no temperature and atmospheric pressure correction):

$$\text{Free Air Delivery} = \frac{\text{Real Electrical Power of Compressors}}{\text{Install Shaft Motor Power/Electical Motor Efficiency}}. \quad (4.11)$$

$$\text{Install Free Air Delivery Flow Rate} = \frac{664.8}{960/0.93} \cdot 135.6 = 87.33 \, \text{m}^3/\text{min}$$

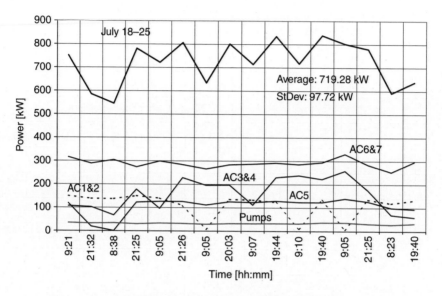

Figure 4.13   Power of Compressors versus Time

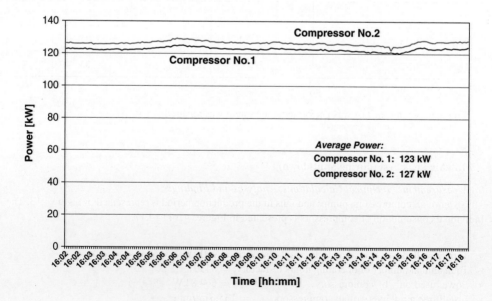

Figure 4.14   Power of Compressors No. 1 and No. 2

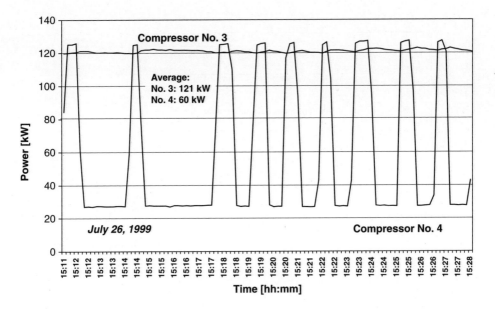

Figure 4.15 Power of Compressors No. 3 and No. 4

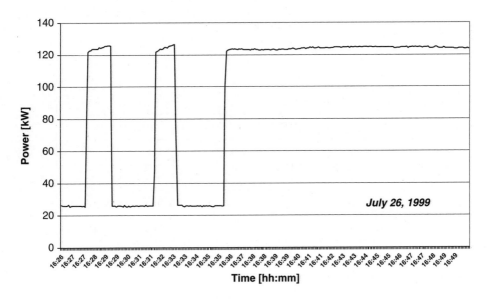

Figure 4.16 Power of Compressor No. 5

(b) Total Running Time and Loading Time in Analyzed Period (Table 4.9)
In this period, the master compressor was No. 4 with 52.5 % in load condition. For maintenance purposes it is advisable to change the master compressor (one of the screw compressors) every month.

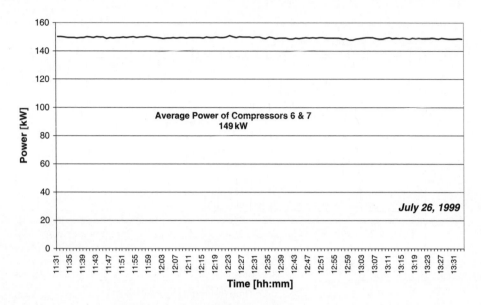

Figure 4.17   Average Power of Compressors No. 6 and No. 7

Table 4.9   Running and Loading Times for Analyzed Period

Air Compressor	Total operating hours per week [h]	% of 168 hours (7 days)	Load time [h]	% of load time	% of unload/ running
No. 1 (SCREW) – Air Cooled	1.33	0	1.32	0	0
No. 2 (SCREW) – Air Cooled	139	82	138.15	82	0
No. 3 (SCREW) – Air Cooled	111.67	66	105.51	62.8	5.5
No. 4 (SCREW) – Water Cooled	150.76	90	88.3	52.5	41.4
No. 5 (SCREW) – Water Cooled	150.27	90	145.49	86.6	3.2
No. 6 (TURBO) – Water Cooled	166	99	166	98.8	0
No. 7 (TURBO) – Water Cooled	167	99	166	98.8	0

4.5.1.3 Average Power of Compressor

$$\text{Average Power of Compressors} = \frac{\text{Energy Consumed by Compressor in Analysed Period [kWh]}}{\text{Total Hours of Operation [h]}} = 664.8\,[\text{kW}] \qquad (4.12)$$

The unload power of the compressors is approximately 22 % of rated power.

The measured average *electric power* of the motors is:

- Compressor No. 1 = 121.1 kW
- Compressor No. 2 = 125.1 kW

- Compressor No. 3 = 120.8 kW
- Compressor No. 4 = 125.8 kW
- Compressor No. 5 = 123.2 kW
- Compressor No. 6 = 149.1 kW
- Compressor No. 7 = 149.1 kW

4.5.1.4 Air Leak Test of Distribution Network (Non-Production Period)

- all valves directly before the machines are closed;
- compressor No. 3 is started;
- after reaching the set pressure (7 barg) the average flow rate of compressed air is (Table 4.6):

$$V_{Leak1} = V_{Nominal\ COMP.3} \cdot \frac{T_{LOAD}}{T_{LOAD} + t_{UNLOAD}} = 17.8 \cdot \frac{27\,[s]}{27\,[s] + 70\,[s]} = 4.95\,m^3/\,min \qquad (4.13)$$

- By measuring the velocities of free air entering the compressors (air intake) the average velocity of the air during the load period is 4.04 m/s. As the diameter of the inlet duct is 300 mm, this means that the free air flow rate is 0.285 m³/s or 17.13 m³/min. This indicates that this compressor works very closely to the nominal capacity.
- The air leak loss is:

$$Air\ Leak\ Loss = \frac{Mesured\ Leaks}{Average\ Annual\ Compressed\ Air\ Flow\ Rate}$$

$$= \frac{4.95}{87.33} = 0.057\,(5.7\,\%) \qquad (4.14)$$

- The energy cost for this compressed air production was 0.057 · 370 000 = 21 090 $US.
- The power and pressure profile during this test is presented in Figure 4.18. The full load, air leak load and unload power of compressor No. 3 versus free air flow rate is presented in Figure 4.19.

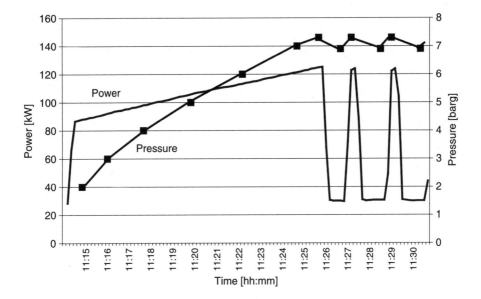

Figure 4.18   Power and Pressure Profile versus Time during Air Leak Test (Distribution Network)

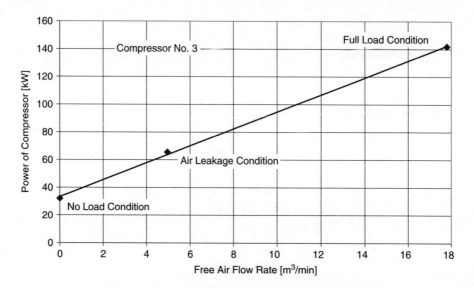

Figure 4.19   Power versus Free Flow Rate for Compressor No. 3

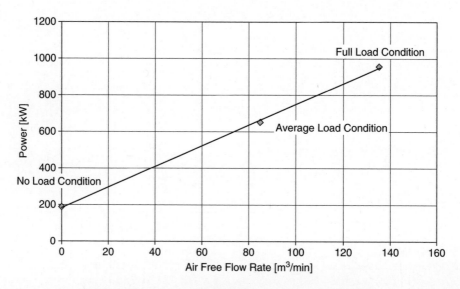

Figure 4.20   Power versus Free Air Flow Rate for all of Compressors

The power of all compressors versus free air flow rate for all compressed air rooms is presented in Figure 4.20.

4.5.1.5 Pressure Drop

Critical end-users are defined as physically the most distant bearing in mind the length of the pipeline. At these spots, the pressure is measured. The pressure drop is (during the factory operation period) only

0.05 bar. This means that the distribution network is well designed (the recommended pressure drop to end-users is less 0.2 bar).

For each air dryer two filters were installed in *series* (Fig. 4.7). By measuring the pressure drop for each of them it is found that average pressure drops are as follows:

Filter A	=	0.35 bar
Filter B	=	0.25 bar
TOTAL	=	0.60 bar

This pressure drop is considered to be too high and the filters need to be checked.

4.5.1.6 Ventilation of Compressor Room

In the case of this factory, one side of the compressor room is open to the surroundings and the temperature in the room is 30.6 °C. For three of the air cooled compressors, the connection with the surroundings is designed well.

4.5.1.7 Recommendations

A. Installation of Parallel Set of Filters (A and B)

By installing a *new parallel* set of filters, the single flow rate of air per set of filters will be reduced by 50 % and the adequate pressure drop will be reduced by 75 %. This means that the pressure drop will be only 0.15 bar $\left(0.6 \cdot \left(^1/_2\right)^2 = 0.15\right)$. Also, the pressure discharge of the compressors can be reduced by 0.45 bar. The energy consumption can be reduced as follows:

$$\text{Energy Saving} = \frac{\text{Discharge Pressure Reduction}}{\text{Nominal Discharge Pressure}} \cdot \text{Annual Energy Consumption} \qquad (4.15)$$

$$= \frac{0.45}{8} \cdot 6.49 = 0.36\,\text{GWh}$$

The current set of filters was too small for the requested flow rate.

B. Discharge Pressure Reduction

The majority of the machines need pressure lower than that at which the system was operating. By analyzing the end-users' recommended pressures at machine input, it was estimated that 18 % of machines needed 7 barg pressure and that the rest could operate with a pressure of 5 barg or less. The energy consumption reduction can be:

$$\text{Energy Saving} = \frac{\text{Discharge Pressure Reduction}}{\text{Nominal Discharge Pressure}} \cdot \text{Percentage of} \qquad (4.16)$$

$$\text{Machines} \cdot \text{Annual Energy Consumption}$$

$$= \frac{2}{8} \cdot 0.82 \cdot 6.49 = 1.33\,\text{GWh}$$

It was recommended that the factory separate air supply for the machines into two groups according to the pressure required: one group of the machines which needed a pressure of 7 barg and the other with 5 barg.

C. Eliminate Leakage of Machines

The percentage of compressed air loss by leakage has been already calculated (Eq. 4.16) and the energy loss is as follows:

$$\text{Energy loss} = \frac{4.95}{87.33} \cdot 6.49 = 0.37 \, \text{GWh} \tag{4.17}$$

D. Installation of solenoid valves for each machine in the case when the machine is not in operation and the manual valve is not closed (break downs, repairs, etc.)

Based on experience, it can be estimated that 3 % of the total annual operating time of machines can be switched off electronically but not disconnected from the compressed air distribution network. This value is tested in one of the departments with 33 machines over the period of one shift.

This leakage can be estimated as follows:

$$\text{Idle leakage} = 0.03 \cdot 82.38 = 2.47 \, \text{m}^3 / \text{min} \tag{4.18}$$

The energy loss is:

$$\text{Energy loss} = \frac{2.47}{87.33} \times 6.49 = 0.18 \, \text{GWh} \tag{4.19}$$

4.5.1.8 Expected Energy Savings

	Unit	A	B	C	D	TOTAL
Energy reduction	GWh	0.36	1.33	0.37	0.18	**2.24**
Energy cost reduction	$US/a	19 656	72 618	20 202	9 828	**122 304**

Note: Average factory's electric energy cost was 0.0546 $US/kWh.

The total annual energy cost reduction is 122 304 $US or 5.6 % of the total factory electrical energy costs.

4.6 Example: Comparison of Load/Unload and Pump-up Tests

This real case illustrates a comparison of two types of tests: (a) Load/Unload (b) Pump-up. They were performed in the same factory and on the same day and under the same conditions. In the factory there were 19 compressed air units. The total installed capacity was 151.1 nm³/min and the total motor electric power was 931 kW. The set air pressure in the compressed air system was 6 barg.

4.6.1 Load/Unload Test

The air free flow rate of compressors used for testing is 12.4 nm³/min with power input of 75 kW. The control system is on/off.

Cycle:	Time: **running load** T [min]	Time: **running unload** t [min]
1	3.6	2.0
2	3.2	2.5
3	3.5	2.05
4	3.1	2.1
TOTAL:	13.4	8.65

$$\text{Leakage} = Q \cdot \frac{T}{T+t} = 12.4 \cdot \frac{13.4}{13.4 + 8.65} = 7.54 \, \text{nm}^3/\text{min} \tag{4.20}$$

$$E = H \cdot \frac{T}{T+t} \cdot P_c = 8400 \cdot \frac{13.4}{13.4 + 8.65} \cdot 75 = 382\,857 \, \text{kWh} \tag{4.21}$$

where:
E   =   Energy losses [kWh]
H   =   Annual working time (h) (for this factory = 8400 h)
P_c =   Compressor rating (kW)

If the electricity price is 0.0435 $/kWh the money loss is 31 190 $ US per year.

4.6.2 Pump-up Test

$$\text{Leakage} = \frac{V}{\tau} \cdot \frac{p_1 - p_2}{B} = \frac{17.74}{4.6} \cdot \frac{6.1 - 4.6}{1.013} = 5.71 \, \text{m}^3/\text{min} \tag{4.22}$$

where:
V = volume of the air mains and receivers, [m³] (in this case, V = 17.74 m³)
p_1 = pump-up pressure, [barg]
p_2 = final pressure after time (τ [s]) of testing expired, [barg]
B = barometric pressure, [bar]

Two tests were performed on the same day and under the same conditions for the same compressed air system. However, two different results were obtained: (1) in the case of load/unload test, the calculated leakage based on performed measurements was 7.54 m³/h and (2) in the case of the pump-up test, it was 5.71 m³/h. The difference is significant (24 % in comparison with the highest value). The reason could be measurement errors or that some influencing factors have not been taken into consideration. For example, the volume of the air mains and receivers were estimated roughly based on the blue prints, existing industrial type instruments were used, etc. But, both results are very useful as they point out that there are significant leakages that must be eliminated in order to reduce the energy cost for compressed air.

There is no doubt that more accurate measurement can be done, but the question is whether it is necessary. Obtaining more precise data can costs much more. We have already mentioned several times that energy management starts with people and that the introduction of any test on a regular basis can motivate people to change their attitude towards energy and trigger the practice of continuous performance improvement.

4.7 Bibliography

Energy, Energy Efficiency and Renewable Energy, Industrial Technologies Program, www.eere.energy.gov/industry.

Good Practice Guide No. 126 (1996) Compressing Air Costs, Energy Efficiency Office, Department of the Environment, UK.

High Quality Compressed Air from Generation to Application (A Guide to ISO 8573.1:2001 Air Quality Classes), http://www.domnickhunter.com.

Improving Compressed Air System Performance (a sourcebook for industry), U.S. Department of

Energy Efficiency and Renewable Energy, Industrial Technologies Program, http://www.eere.energy.gov/industry.

5

Refrigeration System

5.1 Description of System

A *refrigeration system* is a combination of components, equipment and piping, connected in a sequential order to produce a refrigeration effect. A *refrigeration effect* is the process of extracting heat from a lower temperature heat source, a substance or cooling medium and transferring it to a higher temperature heat sink, commonly atmospheric air and surface water, in order to maintain the temperature of the heat source below that of the surroundings.

Refrigeration systems can be classified in following categories:

- Chilling and chilled storage. The common temperature range is from 0 to 10 °C.
- Freezing, deep freezing, cold storage and deep cold. The common temperature range is from 0 to minus 40 °C.

As there are numerous and versatile industrial processes requiring refrigeration, there are also different manners of connecting the components and equipment in a refrigeration system. Figure 5.1 depicts a refrigeration system used in the textile industry for supplying air washers. This particular example will be used to describe the principles of the operation of refrigeration systems.

Box 5.1: Description of a sample refrigeration system

> *The task of this refrigeration system is to produce chilled water with an initial temperature of 10 °C. The design return water temperature is 15 °C. The flow of this water in the evaporators of chillers No. 1, No. 3 and No. 4 is 363 m³/h, and in the evaporator of chiller No. 2 it is 393 m³/h. The total installed refrigeration capacity is 8088 kW. This refrigeration capacity can be achieved with given temperatures of chilled water and condensers' inlet/outlet water temperatures of 30/35 °C. The flow rates of chillers No. 1, No. 3 and No. 4 condensers are 460 m³/h and in the condenser of chiller No. 2, it is 474 m³/h. The refrigerant is R-11. The electrical power of the electric motors for running the turbo compressors of chillers No. 1, No. 3 and No. 4 is 373 kW and for the compressors of the chiller No. 2, it is 470 kW the (total installed power of these electrical motors is 1,589 kW). In the regular plant, chillers No. 1, No. 3 and No. 4 are anticipated to operate while chiller No. 2 is used as a stand-by.*
>
> *In addition to that, the power of the pumps for running water for the cooling tower (5 × 45 = 225 kW; one is stand-by), evaporators pumps (4 × 22 = 88 kW; one is stand-by) and water supply pumps (4 × 55 = 220 kW; one is stand-by and 2 × 22 = 44 kW; one is stand-by) is also important. The total installed power of the pumps is 577 kW. The installed power for driving the ventilators of the cooling towers is 4 × 15 = 60 kW and 1 × 22 = 22 kW. One of cooling towesr is a stand-by.*

Applied Industrial Energy and Environmental Management   Zoran K. Morvay and Dušan D. Gvozdenac
© 2008 John Wiley & Sons, Ltd

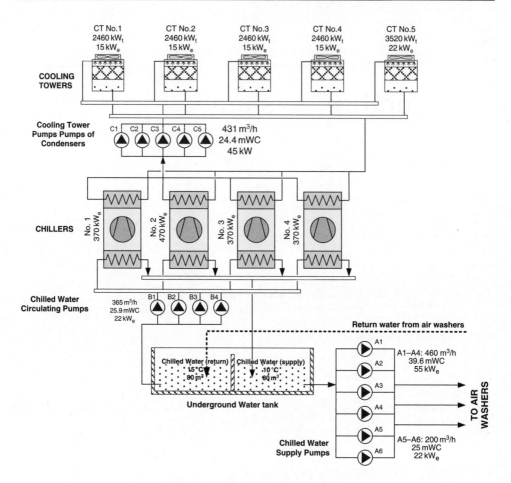

Figure 5.1 A Sample Refrigeration System

The illustrated refrigeration system is certainly a considerable consumer of electric power. We are interested mainly in the overall energy efficiency of the whole refrigerating system, which besides chillers also includes pumps, cooling towers, maintenance, manner of administration, outside temperatures and relative humidity, load variations, etc. All of these factors have a significant impact on the energy performance of the system. In the following we will define the performance indicators of refrigeration systems and consider the opportunities for improving performance by increasing energy efficiency.

5.1.1 Principles of Operation

Refrigeration systems that provide cooling for industrial processes including air-conditioning are classified mainly within the following categories:

(a) *Vapor compression systems.* In these systems, compressors compress the refrigerant to a higher pressure and temperature from evaporated vapor at low pressure and temperature. The compressed refrigerant is condensed into liquid form by releasing the latent heat of condensation into the condenser water or ambient air. Liquid refrigerant is then throttled to a low-pressure. The formed vapor-liquid mixture produces the refrigeration effect during evaporation. Vapor compression is often called *mechanical refrigeration* as the refrigerant vapor is compressed by using mechanical compressors. The compressors are driven mostly by an electric motor and rarely by an engine, gas turbine or steam turbine.

(b) *Absorption systems.* In an absorption system, the refrigeration effect is produced by means of thermal energy input. After liquid refrigerant produces a refrigeration effect during evaporation at low pressure, the vapor is absorbed by an aqueous absorbent. The solution is pumped and heated by a direct-fired gas or waste heat and the refrigerant is again vaporized, but at higher pressure, and then condensed into liquid form. The liquid refrigerant is throttled to a low pressure and is ready to produce the refrigeration effect again. The absorption chiller uses heat to compress the refrigerant vapors to a high-pressure, therefore this 'thermal compressor' has no moving parts, except the pump.

These two refrigeration systems are presented in Figure 5.2. Practically, the difference between them is in the way of refrigerant vapor compression. Traditionally, vapor compression refrigeration systems are used more often in industry than absorption refrigeration systems. The particular advantage of absorption refrigeration systems is the opportunity to utilize waste heat of a relatively low temperature which under industrial conditions is frequently available.

5.1.1.1 Vapor-Compression Refrigeration

Heat can only flow naturally from a hot to a colder body. In the refrigeration system the opposite must occur. This is achieved by using a substance called a refrigerant which absorbs heat and hence boils or evaporates at low pressure to form gas. This gas is then compressed to higher pressure, so that it transfers heat it has gained to the ambient air or water and turns back into liquid (condenses). In this way, the heat is absorbed, or removed, from a low temperature source and transferred to one at a higher temperature.

The operation of a simple refrigeration system is shown in Figure 5.3. The cycle is presented in the pressure (bar)-enthalpy (kJ/kg) diagram of the refrigerant used.

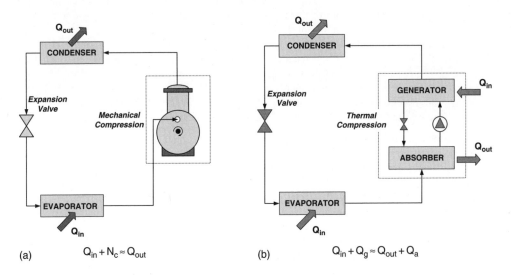

Figure 5.2 Schematics of Vapor-Compression (a) and Absorption Refrigeration Systems (b)

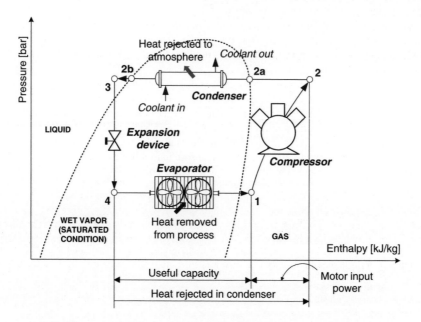

Figure 5.3   Single Stage Vapor Compressor Circuit and Pressure – Enthalpy Diagram

The refrigeration cycle can be broken down into the following stages:

1⟹2   The superheated vapor enters the compressor where its pressure is raised. An adequate increase of temperature will also occur there. Most of the energy put into the compression process is transferred to the refrigerant.

2⟹3   The high pressure superheated gas passes from the compressor into the condenser. The initial part of the cooling process (2⟹2a) de-superheats the gas before it is turned back into liquid (2a⟹2b). The cooling for this process is usually achieved by using air or water. Further reduction in temperature happens in the pipe-work and liquid receiver (2b⟹3), so that the refrigerant liquid is sub-cooled as it enters the expansion device.

3⟹4   The high pressure sub-cooled liquid passes through the expansion device, which both reduces its pressure and controls the flow into the evaporator.

4⟹1   Low-pressure liquid refrigerant in the evaporator absorbs heat from its surroundings, usually air, water or some other process liquid. During this process, it changes its state from liquid to gas, and at the evaporator exit it is slightly superheated.

It can be seen that the condenser must be capable of rejecting the combined heat inputs of the evaporator and the compressor; i.e. (4⟹1) + (1⟹2) has to be the same as (2⟹3). There is no heat loss or gain through (adiabatic process) the expansion device.

In such systems, the compressed gas is known as the primary refrigerant. In some cases, the cooling effects are indirect. The primary refrigerant is used to cool another medium, such as brine or water, which is then pumped to the point where it is required. This medium is called the secondary refrigerant (in the example shown in Figure 5.1, the primary refrigerant is R 11 and secondary one is water). The choice of refrigerants varies according to their use. Critical factors which govern the choice are: (a) cost; (b) potential hazards in use; (c) temperature ranges required; (d) size of plant.

The most common primary refrigerants in industry are ammonia and halogenated hydrocarbons (Freons). Typical secondary refrigerants are water, brine, ethylene glycol, and propylene glycol.

The types of compressors most commonly used in industrial refrigeration systems are:

(a) screw;
(b) centrifugal;
(c) reciprocating.

The screw and centrifugal compressors use rotating elements. The screw compressor is a positive-displacement machine and a centrifugal compressor operates based on centrifugal force. The reciprocating compressor consists of a piston moving back and forth in a cylinder with suction and discharge valves arranged in order to allow pumping to take place.

The type of compressor often depends on the size of the plant. The reciprocating type is used for smaller plants and the centrifugal type for very large units. There is, however, considerable overlap between the ranges covered by each type of compressor. In many situations, the appropriate choice of compressor can lead to energy savings.

Reciprocating compressors range from small single cylinder hermetic compressors (for freezers and refrigerated display cabinets) up to large 600 kW refrigeration capacity multi-cylinder models. The power consumption of modern reciprocating compressors is similar to other types of compressors under full load and is generally better at part-load.

Typical screw compressors range in size from 200 to 2000 kW (refrigeration capacity). Single screw compressors generally provide slightly lower efficiency than twin screw compressors although maintenance requirements are likely to be lower for the single screw type. The power consumption on full loads is comparable with reciprocating units but is likely to be higher under part-load conditions.

Centrifugal compressors normally range in size from 800 to 3000 kW (refrigeration capacity). Under part-load conditions, centrifugal compressors are generally less efficient than screw and reciprocating ones. However, their power consumption in full load is comparable to that of other compressors. The low maintenance requirements of centrifugal compressors make them particularly attractive in situations where continuous running is essential.

It is customary to talk about the refrigeration capacity of compressors. However, the compressor as such does not have any refrigeration capacity. It has only the power necessary for its drive. The compressor's refrigeration capacity is in fact the refrigeration capacity which can be obtained by this compressor if it is included in a certain pre-defined refrigeration cycle. Commonly, *standard refrigeration cycle* is defined with previously designated relevant parameters such as pressure of condensation and evaporation and temperature of vapor overheating after leaving the evaporator, as well as temperature of fluid refrigerant sub-cooling at the condenser's outlet. This cycle is shown in Figure 5.3. Table 5.1 presents the recommended cycle parameters. These are the so called standard parameters. They can differ from country to country but should always be known when talking about the refrigerant capacity of a compressor or

Table 5.1 Parameters of Standard Cycles

Regime	t_e [°C]	Δt_{sh} [°C]	t_c [°C]	Δt_{sc} [°C]
Standard Ammonia Refrigeration System	−15	−10	+30	+25
Standard Freon Refrigeration System	−15	+15	+30	+25
Freon Refrigeration System for Air Conditioning	+5	+15	+40	+30
Low Temperature Freon Refrigeration System	−35	+15	+30	+25
Low Temperature Ammonia Refrigeration System	−40	−30	+30	+25

when several compressors are compared. This is exceptionally important as the refrigeration capacity of a compressor changes dramatically with the variation of these parameters. Therefore, the comparison of several compressors is only possible if they operate under the same conditions which are defined by the temperature (pressure) of condensation and evaporation and by the suction temperature of the compressor and the sub-cooling temperature of the refrigerant entering the expansion device. The change of these parameters causes the change of the energy indicators of the compressor itself which should also be taken into consideration when computing the refrigeration capacity of the compressor for various cycle parameters.

The refrigeration capacity of a compressor is defined as the product of the mass flow rate of refrigerant and the difference between its enthalpy at the evaporator outlet and inlet. So, the refrigeration capacity is as follows (Fig. 5.3):

$$Q_e = m_r \cdot (h_1 - h_4) \qquad (5.1)$$

where:

Q_e = refrigeration capacity, [kW]
m_r = mass flow rate of refrigerant, [kg/s]
h_1 and h_4 = enthalpy of refrigerant at point 1 and 4, respectively, [kJ/kg]

5.1.1.2 Vapor-Absorption Refrigeration

The scheme of the simplest absorption device is shown in Figure 5.2b. The appropriate binary mixture (e.g. $H_2O - NH_3$) evaporates in the generator at a pressure (p) by heat input (Q_g). This produces more or less pure vapor of a more easily volatile component of the binary mixture. In the given example this is ammonia (NH_3). The vapor pressure should be high enough to become liquid in the condenser when heat is separated by means of cooling water. After condensation, liquid ammonia is throttled in the expansion valve to a lower pressure (p_o). This means that the boiling temperature of the liquid will be lower than the one in the condenser. Thus, the created conditions enable this liquid to evaporate at a lover temperature in the evaporator by receiving heat Q_{in} as a refrigeration effect. Therefore, conditions in the condenser and evaporator of the absorption refrigeration system coincide with those in the mechanical compression refrigeration system, but with a slight difference which concerns temperatures of condensation and evaporation of *pure* ammonia which are constant at stable pressures and, in case of the *mixture* $H_2O - NH_3$, they are more or less variable.

Cold vapor leaving the evaporator is taken into the absorber where it is absorbed by means of a cold mixture. This releases absorption heat Q_a. It is precisely this feature of binary mixtures which enables a warmer mixture (solution) to absorb colder vapor which is decisive for absorption device operation.

The flow of the solution is very simple. The weak solution obtained by evaporation leaves the generator. This solution is throttled in the second expansion valve to the pressure of the absorber and then is introduced into the absorber which is cooled by chilling water. In the absorber, vapor is introduced into the weak solution which it absorbs and obtains ammonia. The solution strengthened in the above manner is by means of a pump raised to higher pressure (p), where it will be evaporated again.

Commercially available absorption chillers in industry can run on:

- natural gas;
- steam;
- hot water; or
- hot exhaust gases.

Current absorber media/refrigerant for absorptions chillers are either:

- lithium bromide/water; or
- ammonia/water.

Lithium bromide/water is suitable for chilling down to 5 °C and ammonia/water plant is suitable for temperature less than zero.

Without getting too technical, it is worth pointing out that the number of 'effects' in the chiller reflects the number of times heat energy is used. A single-effect machine uses heat just once to produce a refrigeration effect. A double-effect machine contains heat exchangers in order to recover heat left over from the first stage of cooling so as to produce additional refrigerant vapor and more cooling. Double-effect is more efficient than single-effect.

The main differences between these two types of absorption chillers are as follows:

- *Single Effect*: Lower initial costs but less efficient and consuming more energy and thus becoming more expensive to operate.
- *Double Effect*: Higher initial costs but more energy efficient and thus require less energy to operate.

The components of the absorption chiller must be integrated much more closely than the components of the vapor-compression system. As a result, all absorption chillers are contained within a single compact package. For the same reason, absorption chillers have few variations. In all large absorption systems, cooling is distributed by chilled water. Similarly, all condensers are cooled by water, usually from the cooling tower.

In large absorption machines, the actual refrigerant is water is at very low pressure. An absorber, usually the salt of lithium bromide, is used to move water vapor through the system. Crystallization of the salt is a major operating problem that the design of an absorption chiller seeks to avoid.

As already mentioned in the case of vapor-compression refrigeration systems, here it is also important to know the conditions under which the information regarding refrigeration capacity applies. Changes of condensation and evaporation temperatures will change the refrigeration capacity significantly but there will also be change in the case of heat power supply to the generator.

5.2 Performance Definitions

In Part I Chapter 2, we considered some performance indicators which can be called general and applicable to all energy systems. Most frequently, it is specific energy consumption which is an exceptional indicator as it enables the comparison of achieved energy efficiency in previous periods within the factory itself or comparison with the same in other similar factories. These indicators have always been associated with the production process where the energy is consumed. Here, we are going to deal with technical performance indicators which are used for estimating the efficiency of energy conversion and which will not be linked with the productions process directly.

It is very important to understand the factors that affect performance and therefore operating costs. The savings opportunities discussed later are all concerned with the effective control of these factors.

One of the most important performance indicators is the ratio of cooling achieved to power used by the compressor and this is called the Coefficient of Performance (COP). It is always useful to first define theoretical COP as its difference in relation to the actual COP indicates how far we are from the best possible process. The coefficient of performance of an ideal thermodynamic refrigeration cycle (Carnot) is as follows:

$$\text{COP}_{\text{Carnot}} = \frac{Q_e}{P} = \frac{Q_e}{P_{\text{com}} - P_{\text{exp}}} = \frac{Q_e}{Q_c - Q_e} = \frac{T_e}{T_c - T_e} \tag{5.2}$$

where:

Q_e = refrigeration capacity is the amount of heat received by the refrigerant (working fluid) from the medium being cooled at the temperature of evaporation T_e, [kW]

Q_c = amount of heat given up by the refrigerant to water at the temperature of condensation T_c, [kW]

P_{com} = power spent in the compressor upon the isentropic compression of the refrigerant vapor, [kW]
P_{exp} = power received upon isentropic expansion of the refrigerant in the expansion machine, [kW]
P   = theoretical power spent in the cycle, [kW]

The thermodynamic cycle is presented in Figure 5.4.

Although the Carnot Cycle is the best in a thermodynamic sense, it is not used in practice. The reasons for that are of both an economic and technical nature. For example, isentropic expansion (3–4) is very difficult, is practically impossible to achieve technically without significant losses and the work obtained with this expansion is very small. Therefore, this isentropic expansion is in practice replaced by throttling. The situation is similar with isentropic compression. Losses in a real machine which operates in a two-phase domain are very large; therefore, compression in a real cycle is carried out in a superheated area. Thus, the simplest *real* cycle is, in fact, the one which is represented in Figure 5.3.

For the refrigeration cycle represented in Figure 5.3, the COP is as follows:

$$COP_{V-C} = \frac{Q_e\ [kW]}{P_c\ [kW]} = \frac{h_1 - h_4}{h_2 - h_1} \tag{5.3}$$

where h is the specific enthalpy of refrigerant at subscribed points, [kJ/kg].

A higher COP implies that the refrigeration system achieves more cooling for a given power input hence the system is more efficient. In practice, real cycles tend to have approximately 60 % or less efficiency releted to theoretical performance, and thus opportunities for increasing the efficiency of refrigeration systems are great. For most industrial applications COPs can be expected to range between

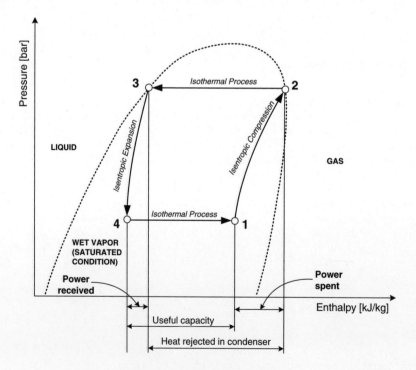

Figure 5.4   Carnot Refrigeration Cycle

about 2 (for plants at around $-40\,°C$ evaporating temperature) and 5 (for plants at around $0\,°C$ evaporating temperature). The COP can vary quite considerably as the ambient climate conditions and process requirements change.

The definition of COP given above is the one which is commonly used in industry to define the performance of a single refrigeration unit. However, it is not always the best parameter to use when evaluating the entire system. For that purpose we need to define the system boundary and then to define the SYSTEM COP as the ratio of *cooling achieved* ('useful' energy) and *total power input* to the system.

$$COP_{SYSTEM} = \frac{\text{Cooling Achieved}}{\text{Total Power Input}} \qquad (5.4)$$

The COP of each chiller for the refrigeration system represented in Figure 5.1 is as follows:

$$COP_{No.1} = COP_{No.3} = COP_{No.4} = \frac{\frac{363}{3.6} \cdot 4.186 \cdot (15 - 10)}{370} = 5.70$$

$$COP_{No.2} = \frac{\frac{393}{3.6} \cdot 4.186 \cdot (15 - 10)}{470} = 4.86 \qquad (5.5)$$

But, the system COP is (if chillers, pumps and cooling towers are in operation):

$$COP_{SYSTEM} = \frac{8088}{1589 + 577 + 82} = 3.60 \qquad (5.6)$$

Indeed, a large difference. This is the reason why when comparing COPs even for the same system, it is always necessary to know the conditions under which it has been computed and the boundaries of the system to which the COP refers. It is particularly important to point out that COPs computed in the above manner apply *only* for the temperature regimes indicated in Figure 5.1. If, for example, chilled water temperature is changed, the COP value will also be changed.

The main advantage of *absorption chillers* lies in the opportunity to use relatively low temperature heat energy for their operation. However, their efficiency is lower than that of compressor type refrigeration systems. The Coefficient of Performance (COP) of an absorption chiller is as follows:

$$COP_{abs} = \frac{\text{Refrigeration capacity}}{\text{Heat energy supplied to generator}} = \frac{Q_R\,[kW]}{Q_G\,[kW]} \qquad (5.7)$$

As the COP of an absorption chiller is approximately 0.75 (single stage) and approximately 1.1 (two stage) and the COP of classical vapor-compressor systems is approximately 3–3.5 or more, it is easy to calculate that the ratio between electric and heat energy prices has to be from 4 to 5. Practically, it means that electrical energy has to be four to five times more expensive than heat energy. This may be the case when some waste heat is available, which is often the case in industrial plants. The advantage of absorption chillers in comparison to vapor-compression chillers is that they operate more quietly.

Further, we are going to deal only with vapor-compression refrigeration systems. The reason for that lies in the fact that a large majority of industrial refrigeration plants use that type. However, absorption refrigeration systems will also be considered owing to their growing significance in utilizing what are often considerable quantities of waste heat in industry. In cases in which there is waste heat of a proper temperature level, it is always advisable to consider the use of the absorption refrigeration system as a potential measure for increasing the energy efficiency of the entire factory. Their utilization is possible in combination with existing vapor-compression refrigeration systems. Part II Chapter 6 provides greater technical detail related to absorption chillers in regard to their use in cogeneration plants.

5.2.1 Factors Affecting Performance

With same LEGO pieces we can construct several different figures, equally with the same elements of a refrigeration system, we can assemble several different installations. However, different connections will produce different plant performances. As we have already mentioned, we are not going to deal with the energy performance of certain elements in particular, but primarily with the performance of the entire plant. We can mention Figure 5.1 again as it represents one refrigeration system consisting of different elements. The boundaries of the plant are defined by the limits of the figure. The performance of the entire plant is the subject matter of our analysis with the aim of determining the conditions for its exploitation with the minimum consumption of energy.

Let us consider now the critical factors which impact upon the performance of a refrigeration plant.

5.2.1.1 Evaporating Temperature

A higher evaporating temperature will produce higher COP and lower running costs. It can be used as a guide that raising evaporating temperature by 1 °C reduces operating costs by 2 % to 4 %. Now, the question can be asked regarding the optimum evaporation temperature and what it is dependent on. Table 5.2 provides recommendations for the difference between evaporation temperature and that of the space which is cooled. It is obvious that the lesser temperature differences are the higher evaporation temperature will be, with an assumption that the given ambient temperature of the space which is cooled is constant. The direct consequence of increased evaporation temperature is the increase of the surface for heat exchange in the evaporator, which will in turn lead to the increase of initial investments. Although the increase of the evaporator's surface lowers operating costs, it increases investment costs for the machine. For exploitation, we still have to check and monitor the mentioned temperatures (ambient that is cooled and evaporation) and to maintain them within prescribed limits.

In the case of water or brine evaporators, the logarithmic temperature difference has to be around 5 °C. Lower values are desired but they will require increasing the flow rate of water or brine and increasing the pump power. The size of evaporator is constant.

For example, the relation of COP and leaving chilled water temperature for a chiller with a screw compressor (400 kW$_e$), air cooled condenser and water evaporator (cooler) is presented in Figure 5.5. The refrigerant is R 22. The condenser's entering air temperature is presented as a parameter.

In this case, the evaporation temperature is shown implicitly. It is defined by water flow through the evaporator and by the size of the evaporator's surface, as well as by its characteristics. This dependence has a single meaning for the given plant. In this particular real case, the water flow rate is constant. This

Table 5.2 Temperature Differences in Correctly Sized Evaporators for Refrigeration and Frozen Chambers. Reproduced from: Review of Energy Efficient Technologies in the Refrigeration Systems of the Agrofood Industry, Thermie Programme Action No I 185, http://edo.uniba.sk:8000

The heat exchanger heat transfer surface is correctly sized if:
Chamber around 0 °C: (chamber temperature) – (evaporation temperature) < **8 °C**
Frozen goods store chamber: (chamber temperature) – (evaporation temperature) < **7 °C**
Chamber freezers: (chamber temperature) – (evaporation temperature) < **6 °C**
Circulating air-flow through evaporator: (air input temperature to evaporator) – (air output temperature from evaporator) < **3 °C**

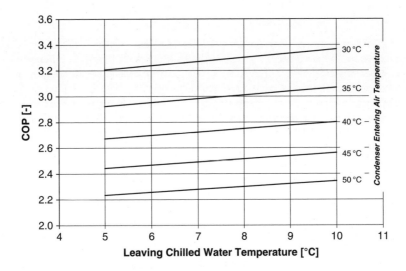

Figure 5.5   Effect of Leaving Chilled Water Temperature on System Performance

example clearly shows that it is necessary for the plant to be used only at the highest possible chilled water temperatures which still meet the process requirements.

5.2.1.2 Condenser Temperature

A refrigeration system operates most efficiently when the evaporation process is carried out at a temperature that is as *high* as possible (as we have already seen) and the condensation process is carried out at a temperature that is as *low* as possible.

For each degree centigrade by which condensation temperature is reduced, an energy saving of 2 %–4 % is made, according to the compressor's characteristics and required evaporation temperature.

Similarly, as in the case of evaporation temperatures, there are recommendations for condensation temperatures which are dependent on the type of condenser.

(a) *Evaporating Condenser*: $t_C = t_{WB,d} + 12\,°C$ where $t_{WB,d}$ is the design wet bulb temperature for the region where the condenser is to be installed.
(b) *Horizontal Multi-Tube Condenser Using Water*: $t_C = t_{W,in} + \Delta t + \Delta t_W$

   where:
   $t_{W,in}$ = inlet temperature of water in the condenser, [°C]
   Δt = temperature difference between output temperature of water and temperature of condensation, [°C]
   Δt_W = temperature difference of water at the condenser outlet and inlet, [°C]. Commonly it is around 5 and 8 °C.

(c) *Horizontal Multi-Tube Condenser with Cooling Tower*: $t_C = (t_{W,in} + 7\,°C) + \Delta t + \Delta t_W$ where the meaning of terms is the same as in the previous case.
(d) *Air-Cooled Condenser*: In these types of condenser, the condensation temperature can be set approximately to 15 °C above the design dry bulb air temperature for the region.

The design dry and wet bulb temperatures for the main cities in the world are presented in Figure 5.6.

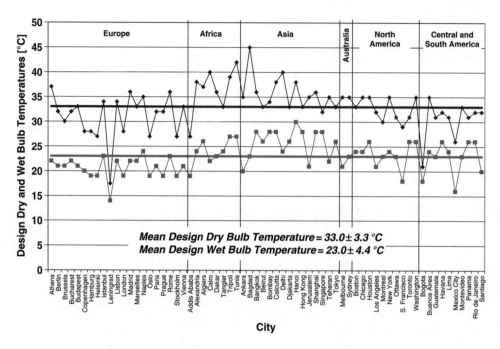

Figure 5.6   Design Dry and Wet Bulb Temperatures for some Cities in the World

Table 5.3   Condensing Temperatures of Different Types of Condensers (Example)

Type of Condenser	Condensing Temperature
Evaporative condenser	$t_C = 23 + 12 = 35\,°C$
Horizontal multi-tube condenser using well or surface water	$t_C = 20 + 5 + 5 = 30\,°C$
Horizontal multi-tube condenser with cooling tower	$t_C = (23 + 7) + 5 + 5 = 40\,°C$
Air-cooled condenser	$t_C = 33 + 15 = 48\,°C$

For example, taking mean design dry and wet bulb temperatures in the summer period of 33 and 23 °C respectively and well-water temperature of 20 °C, the following condensation temperatures will apply, according to the type of condenser which is used (Table 5.3):

This example makes it possible to conclude that there is a large difference between condensation temperatures depending on the type of condenser used. It is obvious from this example that it would be most favorable to use a horizontal multi-tube condenser with well or surface water. Unfortunately, this water is not always available or it is very expensive. Therefore, we have to choose a fluid which is available under the relevant conditions and in the given space. The choice of type of condenser is most often imposed by the surroundings in which the given installation will operate. Once the type of condenser is selected, it is important to select design temperatures carefully in order to obtain an optimal condenser in respect of investment and operating costs. During exploitation, it is necessary to monitor all of these parameters and maintain them within optimum limits.

Figure 5.7 illustrates the effect of a condenser's entering air temperature on the system performance for the previously mentioned chiller. Condensation temperature is here also determined implicitly by

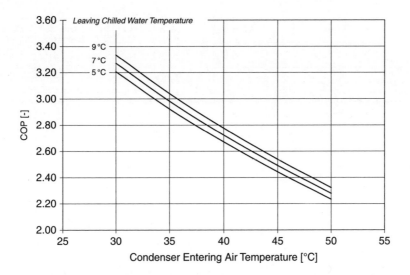

Figure 5.7 Effect of Condenser Entering Air Temperature on System Performance

input air temperature. The impact of leaving water temperature from the evaporator is also represented as parameter. The air flow rate is constant.

Air temperature is variable during the day and year which will cause the COP to vary. This will have an influence on the electrical energy consumption of the compressor. We cannot have an influence on this temperature, but we can certainly have an influence on proper maintenance in order to keep it within the desired limits.

5.2.1.3 Compressor Isentropic Efficiency

The efficiency of a compressor as the largest energy consumer of a vapor-compressor refrigeration system has certainly a decisive influence on the efficiency of the entire refrigeration plant.

The reversible adiabatic process (entropy remains constant) is called an isentropic process. Isentropic efficiency is the best measure of the compressor's efficiency. With most types of compressor, particularly centrifugal ones, efficiency falls significantly at part-load operation. The compressor's motor efficiency is also important. In general, high efficiencies can be maintained by avoiding part-load operation, by using the best compressors at any given time and by good maintenance. The isentropic efficiency of compression process is defined by:

$$\eta_{i,com} = \frac{\text{Isentropic Work}}{\text{Actual Work}} \tag{5.8}$$

Figure 5.8a illustrates the dependence of the COP on the isentropic efficiency of the compressor for one ammonia installation (condensing pressure 12 bar (30.9 °C), evaporating pressure 3 bar (−9.2 °C), superheating temperature difference 4.04 °C). Compressor discharge temperature versus isentropic efficiency is presented in Figure 5.8b.

5.2.1.4 Auxiliary Power

The energy consumption of auxiliary devices in the refrigeration plant can be estimated to around 25 %. In the example of the refrigeration plant presented in Figure 5.1, the participation of pumps and fans in the overall installed power is as much as 29 %.

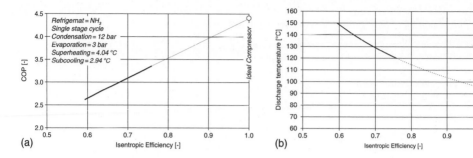

Figure 5.8   Effect of Compressor Isentropic Efficiency on System Performance

Reducing auxiliary power can dramatically increase the efficiency of the refrigeration plant. Good use of controls in order to avoid excessive operation of auxiliaries, avoidance of flow blockage and good maintenance of pumps and fans are especially good practice which can be explored.

5.2.1.5 Cooling Loads

There are two main categories of cooling loads: (a) process; and (b) auxiliary. Within process load, the following can be distinguished: (a1) sensible cooling of fluids and solids; (a2) latent cooling; and (a3) heat of reaction.

Auxiliary cooling loads are loads of the refrigeration system arising from auxiliary equipment such as fans, pumps, lighting, cold store doors, pipe and insulation, people, etc. Types of auxiliary load include: (b1) insulation; (b2) infiltration; (b3) fans in cold space; (b4) lighting in cold space; (b5) pumps of chilled fluids; (b6) defrosting in cold space, if any.

It is important to point out that many auxiliary loads are paid for twice. For example, lighting, pumps and electric defrosting systems consume electrical energy but at the same time release heat energy into the cold space from where the heat has to be removed by the refrigeration system.

Reducing load is a very efficient manner of increasing the energy efficiency of the refrigeration plant and thus, its consideration should be given top priority. Box 5.2 contains load factors and offers means for their reduction. The recommendations concern mainly cold stores and cabinets. For other types of refrigeration plant, a similar list of load influential factors should be prepared with the aim of reducing them.

5.2.1.6 Part-Load Operation

As a rule, the compressor and the entire refrigeration plant operate under full capacity for a very small number of hours. In these cases, the compressor's or plant's COP drop with the reduced load. The effect on COP due to part-load operation of compressor can be very significant. Figure 5.9 shows rapid reduction in the *COP* which occurs at part-load of centrifugal compressor. The outlet chilled water temperature (7.2 °C) and condenser inlet water temperature (32.2 °C) are constant. The compressor capacity control is regulated by inlet guide vanes.

The load influence on COP of screw compressors is shown in Figure 5.10. But this Figure shows how the temperature of evaporation and compression also influence the COP. All three compressors are with different capacities. Two of them are used as low pressure compressors in two stage cycles ($-45/-10$ and $-37/-10$ °C, evaporation and condensation temperatures, respectively) and one can be used as single stage ($-10/+38$ °C).

Box 5.2: LOAD (Relates Mainly to Cold Stores and Cabinets). Reproduced from: GPG347, Installation and Commissioning of Refrigeration Systems, www.actionenergy.org.uk.

Load Factors:	What to do?
Air change loads through doors or lids, or through doors/lids with damaged seals.	– Repair damaged doors, lids and seals; – Check and repair or replace damaged strip curtains, chiller strips and night blinds; – Advise on effective cold store door management; – Check that heaters/ventilation grilles are not sited so that warm air enters the cooled space.
Heat infiltration to cabinets and cold stores, e.g. through damaged insulation.	– Repair damaged cold store/cabinet insulation – quick repairs will also prevent further deterioration; – Advise user to load products into cabinets and cold stores as soon as possible
Occupancy, e.g. through people and machines.	– Advise management how to keep it to a minimum.
Ancillaries such as evaporator fans motors, defrost, in line coolers in beer cellars contribute twice to energy costs. They use electricity directly and also heat the cooled space, increasing the load.	– Make sure the system does not defrost longer than necessary.
Loading of products	– Advise on the best way of product loading in cold stores and cabinets so that air flow is not obstructed.
Suction line heat gains.	– Replace or repair damaged or missing suction line insulation.

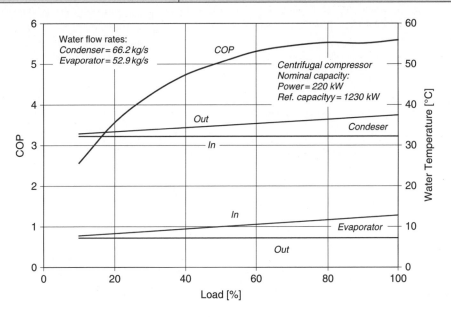

Figure 5.9   The COP versus Centrifugal Compressor's Load

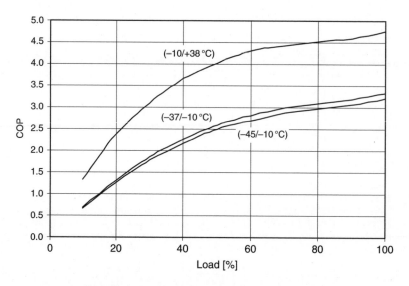

Figure 5.10   The COP versus Screw Compressor's Load

A variety of systems is used to provide compressor capacity control. Nearly all capacity control methods reduce the overall system COP to greater or lesser extent. The control methods are presented in Box 5.3.

Care should be taken when selecting the compressor for operating under part-load. Reciprocating compressors generally demonstrate better part-load performance than screw or centrifugal compressors and should be used whenever it is practical to provide modulation. Screw and centrifugal compressors are ideally suited to provide for base-load of refrigeration, preferably with systems having reasonably constant evaporating and condensing temperatures. Part-load operation of screw and centrifugal compressors should be avoided whenever possible.

Box 5.3: Compressor Capacity Control. Reproduced from: GPG280, Energy Efficient Refrigeration Technology – the Fundamentals, http://www.actionenergy.org.uk.

- **On/Off operation** is a simple and reliable way of ensuring that refrigeration is only undertaken when it is required. However, frequent on/off cycling of the compressor can cause maintenance problems.
- **Cylinder unloading (with reciprocating compressors)** is frequently used to provide capacity control of multi-cylinder reciprocating compressors. The power requirement is reduced as the capacity decreases, although not in exactly the same proportion as frictional and other parasitic losses that exist within the unload cylinder.
- **Variable speed drives** provide efficient capacity control. Positive displacement compressors (reciprocating and screw) can employ variable speed drives but careful consideration must be given to the reduction in compressor efficiency. Other practical issues, such as oil pump performance at low rotational speeds, should not be neglected. Similarly, the part-load performance of centrifugal compressors operating at off-design speed conditions requires careful analysis. For all compressors, a particularly important issue concerns variable speed drives and avoidance of speeds that create resonant frequencies and attendant vibration problems.

    Since no linear relationship exists between speed (rpm), capacity and system COP, reference should be made to the manufacturer's published data in order to establish the optimum solution. Variable speed drives must only be incorporated with the manufacturer's guidance.
- **Slide valves** are used to provide capacity control of screw compressors. The slide valves provide smooth steeples regulation down to 10 % of design capacity. The relationship between capacity and shaft power

requirements is, however, non-linear and a screw compressor utilizes the significant proportion of full-load power when operating at part-load.

- **Variable inlet guide vanes** can be employed on centrifugal compressors to provide capacity control. The vanes are installed in the compressor housing directly ahead of the impeller inlet. Capacity reduction is achieved by changing the direction of flow, prior to the refrigeration vapor entering the impeller. This is extremely inefficient way of providing capacity control since it reduces the isentropic efficiency of the compressor and hence the system COP.
- **Suction throttle** devices used to provide capacity control are particularly wasteful. The suction throttle achieves capacity control by lowering the compressor inlet pressure. This is equivalent to operating with evaporating temperature lower than designed, with all the inherent efficiency penalties. Any existing installations employing a suction throttle device should be investigated carefully so as to consider the adoption of an alternative method of providing capacity control.
- **Hot-gas bypass** involves capacity control by circulating hot high pressure gas from the compressor discharge through an expansion valve to the low pressure of the system. This is a very inefficient method of providing capacity control since no useful cooling is undertaken by the bypass gas and there is no attendant reduction in compressor power input. Also, unless correctly designed, the hot-gas bypass can result in excessive superheating of the suction vapor which can lead to compressor overheating.

  For energy efficient operation of the refrigeration plant, hot-gas bypass capacity control should be avoided. Existing installations using hot-gas bypass for capacity control should be examined so as to consider the use of alternative methods.
- **Compressor sequencing.** It is not uncommon to find large refrigeration systems, with a number of compressors all operating at part-load, and other related auxiliaries (fans, pumps, etc.), operating at full-load. Major savings can be made by ensuring that a sequence control system is adopted. Under low-load conditions, compressors should be sequenced so that one machine operates under part-load, with all other compressors either on full-load or isolated with all related auxiliary equipment automatically switched off. This is very effective way of increasing energy efficiency.

5.3 Performance Analysis

Operational performance analysis must be based on the actual measurements of parameters that determine refrigeration system performance. Table 5.4 lists those parameters and the recommended accuracy of measurements for a refrigeration plant.

A certain number of measuring points is pre determined at the designing stage, as well as corresponding measuring equipment for monitoring the operational parameters of the process. Manufacturers of equipment usually prescribe the manner of utilizing measuring equipment, time intervals for readouts, acceptable values, etc. The best practice is to adhere to the guidelines provided by equipment manufacturers.

We distinguish the following measurements or tests of refrigeration plants:

- *Functional measurements*: The minimum measurements prescribed by the equipment manufacturers enabling permanent functionality of the plant. The standard equipment of more recent refrigeration plants implies also the computation of performance indicators such as COP, load factor, temperature of condensation and evaporation, etc.
- *Special measurements*: In cases when some reconstructions of the refrigeration plant have been performed for the purpose of improving its efficiency, it is necessary to execute additional measurements intended to confirm achieved improvements. For example, building in a new evaporator with a larger surface for heat transfer requires specific measurements which will confirm the effect of proposed and executed energy conservation measures regarding increased energy efficiency. This type of measurement is used in the process of energy auditing.

Table 5.4 Relevant parameters for operational performance analysis and recommended accuracy of measurements ($\pm$)

Measuring Value	Accuracy of Measurement
Compressor suction temperature	0.3 °C
Compressor discharge temperature	0.3 °C
Vapor at the evaporator outlet	0.3 °C
Vapor at the condenser inlet	0.3 °C
Vapor in front of the flow meter, if any	0.3 °C
Liquid at the condenser outlet	0.3 °C
Liquid in front of the expansion device	0.1 °C
Water at the condenser inlet and outlet	0.05 °C
Secondary refrigerant (water or brine) at the evaporator inlet and outlet	0.05 °C
Temperature of vapor or vapor-liquid mixture at the heat exchanger inlet and outlet	0.05 °C
Temperature of compressor oil	0.5 °C
Suction pressure of compressor	1 %
Discharge pressure of compressor	2 %
Vapor pressure in front of the flow meter, if any	1 %
Evaporator pressure	1 %
Condenser pressure	2 %
Pressure of compressor oil	1 %
Pressure difference after and before oil pump	5 %
Vapor flow	2 %
Water or brine flow	2 %
Power of electrical motor of compressor	0.5 %
Sped of compressor	1.5 %

5.4 Performance Improvement Opportunities

Good operation and maintenance is one of the most important issues for operation of a refrigeration system with the lowest energy costs. It is common to find systems that in practice use up to 20 % more energy than they should. Mostly the 'faults' can be easily corrected and require little or no capital investment. Operation and maintenance issues need to be checked constantly. Regular monitoring programs should be established throughout an adequate energy management system so that problems are identified quickly and proper corrections are implemented promptly. Performance improvement opportunities can be classified in the following categories: (a) maintenance; (b) control system; (c) operational procedures; and (d) technological improvements.

5.4.1 Maintenance

As it has already been mentioned, maintenance is one of the most powerful and certainly the cheapest manner of providing high energy efficiency for the existing refrigeration system. Figure 5.11 shows a simple refrigeration plant and its regular checks. Failure to carry out these basic checks and effect timely removal of possible reasons for energy efficiency reduction will undoubtedly significantly decrease energy efficiency and increase operating costs.

A maintenance check list is presented in Box 5.4. This list should be considered only as an example of actions that are proposed for further elaboration and incorporation into the regular Maintenance Plan.

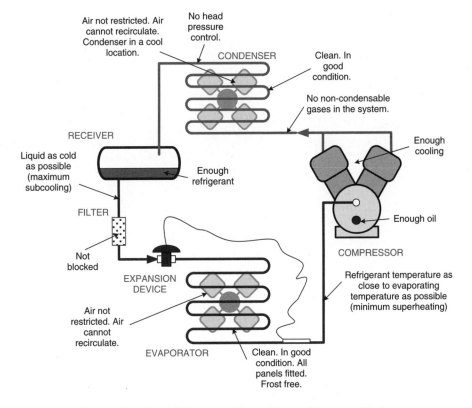

Figure 5.11 Simple Refrigeration Plant and Focus of Permanent Checks

Box 5.4: Maintenance Check List for Refrigeration System. Reproduced from: GPG 364, Service and Maintenance Technical Guide, http://www.actionenergy.org.uk.

- Clean evaporators and drains. Check drain line and pen heaters are working.
- Make sure the defrost system is working correctly and that defrost is frequent enough and for the correct duration.
- Check the condensing or standing pressure of the system – if it is higher than it should be, then there may be air or other non condensable gases in the system. These should be purged from the system.
- Check the system for leaks.
- Make sure all valve caps are fitted.
- Check the temperature difference across filter driers – if there is a difference of more than 1°C the filter is partially blocked and should be changed.
- Check condition of insulation of cold stores and cabinets and repair if they are damaged.
- Check doors, lids, seals, strip curtains and night blinds and repair or replace if damaged.
- Check suction line insulation (and liquid line insulation, if applicable) and replace or repair if they are damaged or missing.
- Check that product loading does not impede air flow from coolers.

Box 5.5: The Chiller's Good Maintenance Practice

- Ensure the air flow onto air-cooled condensers, evaporative condensers and cooling towers is not restricted, or warmer than necessary.
- Introduce maintenance regime that includes as a minimum:

    – Cleaning the condenser
    – Checking for refrigerant leaks
    – Checking the operation of fans and pumps
    – Checking control settings
    – Checking compressor operation
    – Ensuring components and pipe-work are free of vibration
    – Ensuring that manufacturers' schedules of compressor service are followed on large chillers.

- Do not set the chilled liquid temperature too low (this will make the overall power consumption higher than necessary).

It is of particular importance for the staff in the maintenance group to be well trained for routine checks and certainly for the removal of possible irregularities in the system's operations.

Box 5.5 contains a precise list which refers only to chillers. However, the list as such is not sufficient. Without proper organization and adequate training of staff, it will be difficult to expect good maintenance results.

5.4.2 Control System

Control of the refrigeration system, as of any other technical system, starts with measurements (Fig. 5.12). Measurements should cover both the supply and demand side of the system in order to enable adjustment of supply side operation to the demand side variations.

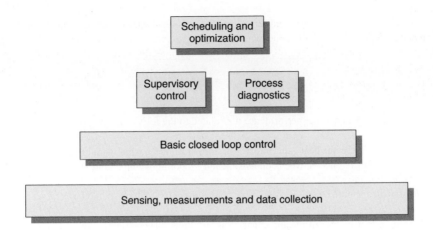

Figure 5.12   Measurement and Control Hierarchy

Coordinating the different levels of control and processing of measured data requires total familiarity with the process that requires refrigeration. One example will be used to explain the manner in which the above can be accomplished in the case of a simple refrigeration system.

Figure 5.13 illustrates what can happen when the variability in the system is reduced by improved control. The change of temperature of refrigerant vapor at the suction of the compressor is represented. In the case of poor control, the average value of this temperature is 10.2 °C, which is significantly higher than the limit value of ≈4 °C. The existence of this limit is extremely important as it protects the compressor from sucking in liquid drops from the outlet of the evaporator. By replacing the expansion device, it is possible to significantly reduce temperature variations at the outlet of the evaporator (improved control in Fig. 5.13). This enables the reduction of the average superheating temperature of the vapor to the desired 4.2 °C.

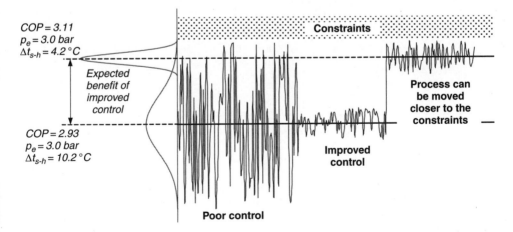

Figure 5.13   Benefits of Reduced Variability

The basic parameters of the considered refrigeration plant are: single stage: ammonia; discharge pressure 12 bar and suction pressure 3 bar. The increased vapor temperature at the compressor's suction point will reduce the isentropic efficiency of the compressor which directly decreases the COP.

5.4.3 Operational Procedures

The operational factors that affect performance are related to the following:

- *Refrigeration load* (e.g. the heat gains into a cold store or cabinet), which should be as low as possible;
- *Temperature lift* (difference between condensing and evaporating temperatures) which should also be as low as possible; or
- *Superheat*, which should be as low as possible, and *sub-cooling*, which should be as high as possible.

In Box 5.6, the effect of two influencing factors is presented. These two factors are temperature of superheating vapor at the compressor suction and sub-cooling of liquid before entering the expansion device. The control of these parameters within operational procedures has to be permanent. As there are usually several control loops, they should be adjusted in such a manner that simultaneous actions produce optimum results.

Box 5.6: Superheating and Sub-Cooling Temperatures of Refrigerant

Superheat and Sub-Cooling Factors	What to do?
Suction superheat should be as low as possible, while ensuring superheated gas returns to the compressor.	– Do not set the expansion valve superheat too high – usually 5°C; – Superheat (5°C temperature difference between evaporating and evaporator outlet) is sufficient to prevent liquid return to the compressor; – Insulate the suction line.
Liquid sub-cooling should be as high as possible.	– Insulate the liquid line if it passes through an area of high ambient temperature (e.g. interior of a supermarket), or if the liquid refrigerant is sub-cooled, e.g. by an economizer; – Where multiple evaporators are used in one system (e.g. a supermarket system) ensure the liquid lines are sized and routed so that one or more evaporators are not left without refrigerant; – If the liquid line has a high vertical lift check that the pressure drop from this has not caused flash gas in the liquid going into the expansion valve.

5.5 Performance Monitoring

5.5.1 Values to be Monitored

For a simple refrigeration cycle (Fig. 5.3), there are six values that can be used to assess condenser, evaporator and compressor performance. Permanent metering should be installed for this purpose, and should be calibrated regularly in order to achieve the accuracy specified in Table 5.4. The following values need to be monitored:

- *Compressor power consumption.*
- *Evaporating and condensing temperatures.*
- *Ambient temperature*: For air cooled condensers, the *dry* bulb temperature is required. For evaporative condensers or cooling towers, the *wet* bulb temperature is required.
- *Process temperature*: The inlet and outlet temperatures of the fluid being cooled have to be measured.
- *Evaporator cooling load*: This parameter can be measured directly or indirectly. Direct type assumes measurement of refrigerant vapor after leaving the evaporator and the enthalpy drop across the evaporator. Indirect method assumes measurement of secondary fluid flow rate and temperature drops across the evaporator or, if the flow rate cannot be measured, by using fan or pump characteristic curves or known heat exchanger characteristics. However, the accuracies of measurements specified in Table 5.4 have to be achieved.
- *Condenser heat load*: Either directly in a similar manner as for the evaporator, or more commonly, indirectly by adding the evaporator load and the performed compressor work.
- *COP* of the individual compressors and the whole refrigerant plant – can be performed by using Part III Toolbox 13 Software 11.

Monitoring these values continuously and comparing them with the nominal values provided by the manufacturers will enable the detection of potential performance problems when they appear, and if so, which particular component or group of components is causing it.

5.5.2 Software 11: Refrigeration Systems Analysis (Vapor-Compression Cycles)

The main purpose of this software is to determine refrigeration system performance indicators in changed exploitation circumstances and to analyze regimes that will result in greater efficiency. In the approach that will be presented here, the basis consists of data obtained in direct measurements. These data represent input variables for the program application guiding the user through the analysis of energy efficiency.

Software 11 (Part III Toolbox 13) offers seven types of refrigeration plants and five types of the most frequently used refrigeration fluids (Step 1 in Fig. 5.14). Therefore, it can be used for a number of different installations. Figure 5.14 shows the available options for refrigeration plants and accessible refrigeration fluids.

Each stage of analysis requires a corresponding group of input parameters, which is defined firmly and organized logically by interface. Input parameters are measured values or arbitrarily selected values limited by corresponding validation. The user oversees the calculation, input accuracy and quality of output parameters with the constant opportunity to change or correct inputs.

The software application has been devised in order to guide the user intuitively in two stages through the analysis of essential performance indicators, not only of the plant as a whole but also of plant components (condenser and evaporator).

In Step 3, the essential performance of the refrigeration system for working conditions on the basis of measured values which represent starting parameters is defined. After the performed calculation (Step 3), in addition to the numerical values of calculation, the software also offers a graphical presentation of the refrigeration cycle with all state variables in the characteristic points of the cycle (Fig. 5.15). In Step 4 the appropriate MS Excel report is generated.

The mathematical model, which describes the function of the plant, is very sensitive to input (measured) data. Thus, it is very important for these data to be entered carefully in compliance with recommendations that the software offers to the user in the form of approximate limits. The measured values that are necessary to enter in the second step of the analysis, as well as the values which represent the software output (Step 3), are as follows:

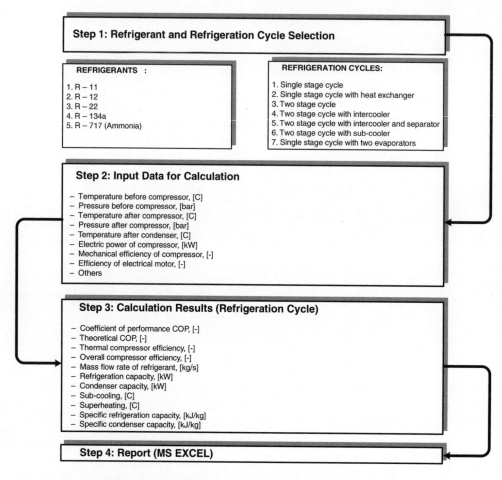

Figure 5.14   Flow Chart of Software 11: Refrigeration Systems Analysis

(a) *Input of Measured Data*
- Temperature before compressor, [°C]
- Pressure before compressor, [bar]
- Temperature after compressor, [°C]
- Pressure after compressor, [bar]
- Temperature after condenser, [°C]
- Electric power of compressor, [kW]
- Mechanical efficiency of compressor, [-]
- Efficiency of electrical motor, [-]

(b) *Results of Refrigeration Cycle Calculation*
- Coefficient of performance COP, [-]
- Theoretical COP, [-]
- Thermal compressor efficiency, [-]
- Overall compressor efficiency, [-]
- Mass flow rate of refrigerant, [kg/s]
- Refrigeration capacity, [kW]
- Condenser capacity, [kW]

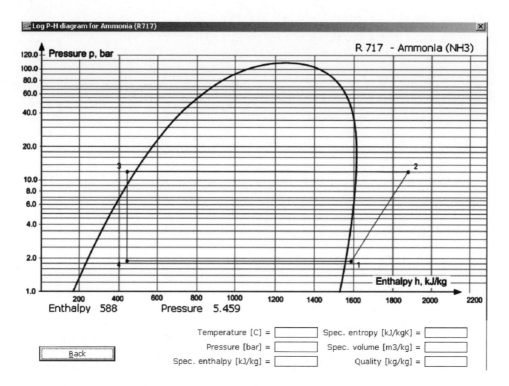

Figure 5.15   p–h Diagram of Selected Refrigeration Cycle

- Sub-cooling, [°C]
- Superheating, [°C]
- Specific refrigeration capacity, [kJ/kg]
- Specific condenser capacity, [kJ/kg]

In order to obtain accurately calculated values, the theoretical refrigeration cycle and single change of parameters (for upper and lower tolerance values) are used to observe the manner in which calculated values change. By varying one of parameters (the temperature tolerance is ± 0.3 °C and the pressure tolerance is ± 2 % of given values) with other parameters constant, the following accuracies are obtained:

- specific refrigeration capacity, ±0.111 %;
- specific work, ±0.53 %;
- refrigeration coefficient, ±0.534 %;
- mass refrigerant flow, ±0.535 %;
- refrigeration capacity, ±0.534 %.

The desired accuracies of measured values are given in Table 5.4.

The calculation of condenser and evaporator can be performed by using Part III Toolbox 13 Software 9 (Heat exchanger operating point determination).

5.5.3 Software 10: Cooling Towers Calculation

The cooling tower is a common component of refrigeration plants. Part III Toolbox 12 deals in detail with cooling towers and Software 12 (Part III Toolbox 13) enables the calculation of cooling tower performance based on measurements of the operational parameters of an existing cooling tower.

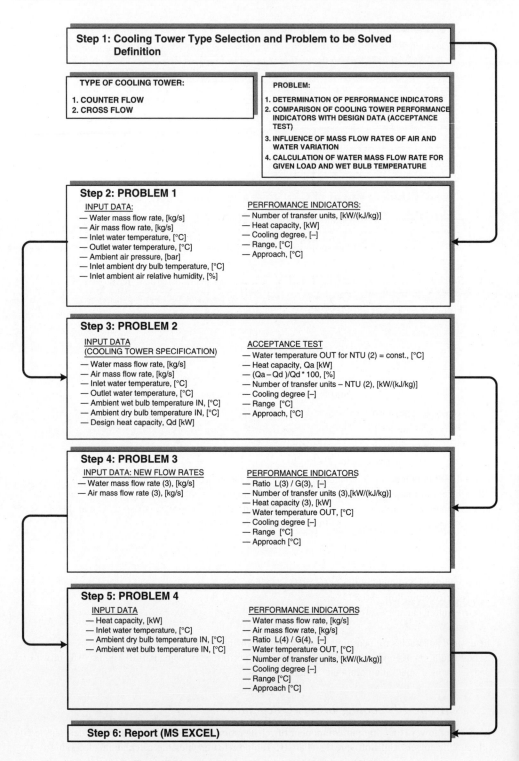

Figure 5.16   Flow Chart of Software IV-10: Cooling Towers

Cooling towers are found in numerous industrial energy systems. For example, relevant to the energy systems mentioned in this book, the cooling tower can be seen in compressed air and refrigeration systems and also in the cogeneration system that follows. In addition to that, the system for supplying a factory with industrial water is, as a rule, equipped with cooling towers. This is the reason why this software belongs to a group of software which links various energy systems horizontally.

A flow chart for this software is presented in Figure 5.16. This software is also characterized by the fact that based on executed measurements and input of available data related to the corresponding cooling tower, it first determines energy performance indicators and then, based on an analysis of the obtained results, it estimates the effects of improvement by employing available performance improvement measures.

5.6 Example: Improvement of Chilled Water System Operation

Let us consider the operation of one chiller which supplies the process with chilled water at a temperature of 14 °C. The designed return water temperature is 19 °C. Only the water temperature at the evaporator outlet is measured and it is the governing value for controlling the cooling capacity of the compressor.

The annual operating time of the factory (and the analyzed system) is 8448 hours. The relevant factory produces textile (yarn and cloths). The raw material is polyester and cotton. The majority of chilled water is used for supplying air washers which have, besides removing the dust from production departments, to provide the working ambient dry bulb temperature at 29 °C and relative humidity at 75 ± 5 %.

5.6.1 Basic Data on Chilled Water System

CHILLER:

- Centrifugal compressor (380 V, 50 Hz)
- Refrigeration capacity (100 % load)   = 2462 kW
- Power of electric motor   = 405 kW
- Refrigerant   = R11
- Evaporator (Shell & Tube):

    – Water flow rate   = 7067 l/min (117.8 kg/s)
    – Inlet water temperature   = 19 °C
    – Outlet water temperature   = 14 °C

- Condenser (Shell and Tube):

    – Water flow rate   = 7948 l/min (132.5 kg/s)
    – Inlet water temperature   = 33 °C
    – Outlet water temperature   = 38 °C

COLLING TOWER:

- Water flow rate   = 9100 l/min (151.7 kg/s)
- Air flow rate   = 3750 m³/min
- Power of the fan's motor   = 20 kW
- Inlet/Outlet water temperature   = 38/33 °C
- Design wet bulb temperature   = 28 °C (t_{db} = 36 °C, ϕ = 55 %)
  (Tropical climate)

The water flow in the cooling tower (9100 l/min) is obviously higher than necessary (7948 l/min). This is not a deficiency, but it should be noted down as it indicates that capacity is larger than required.

PUMPS:

	Evaporator pump	Condenser pump	Supply water pump
Water flow rate, [m³/h]	425	450	425
Head, [mWC]	45	35	45
Power, [kW]	75	75	75
Speed, [rpm]	1450	1450	1450

The pumps are identical; however, the condenser pump operates at a different working point then other two.

The design requirements have specified the maximum cooling capacity to be as follows:

$$Q_{Process} = \frac{6667}{60} \cdot 4.186 \cdot (19 - 14) = 2326 \,[kW]$$

The cooling capacity of the compressor is 2462 kW. This is not a significant difference.

5.6.2 As Built Scheme of Chilled Water System

Figure 5.17 illustrates the designed scheme of the cooling system. All of the motors which drive pumps, compressors and ventilators operate with a constant number of revolutions. The water flows from the side of the condenser and evaporator are constant with variable refrigeration loads. Regulation of refrigeration load from the side of end-users is achieved by three way valves which are governed by the signal provided by a temperature sensor, as shown in the scheme. The heat exchanger shown here symbolizes the load which consists mainly of air washers. If the consumers' load is decreased, the temperature sensor registers a lowering outlet temperature of working fluid from the exchanger and reduces water flow through the exchanger and increases it in the bypass main. At the outlet of the exchanger, the bypassed water is mixed with water which has passed through the exchanger and which flows into the tank from which it is pumped and sent to the evaporator of the chiller by a special pump. The automatic controls of the compressor regulate the outlet cold water of the evaporator (14 °C) by using variable inlet guide vanes.

5.6.3 Data Collection, Measurement and Analysis

A detailed energy audit was carried out in September and October. Data were collected for 12 months starting from October in the previous year. The consumption of the electrical energy of the investigated system is represented in Figure 5.18. Figure 5.19 shows the monthly refrigeration capacity of the system and also the temperature of ambient wet bulbs. The temperature of ambient wet bulbs is high during the entire year. In April, it reached the design temperature of 28 °C. As we have already mentioned, the basic governing parameter of the system is chilled water temperature at the evaporator outlet (14 °C). It can be ascertained that the refrigeration capacity is fairly stable, but also that the refrigeration capacity is lower than the maximum during the whole year. That is to say that the maximum refrigeration capacity (100 % load) for the design temperature is 2462 kW and is mainly below 2000 kW.

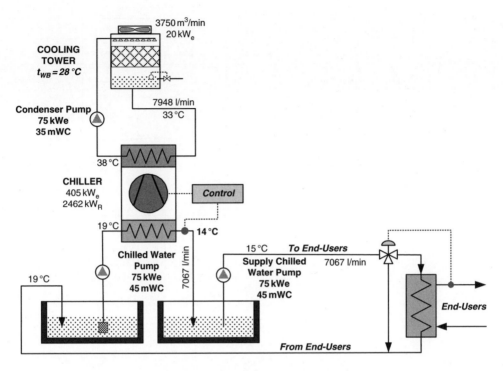

Figure 5.17 Current Scheme of Chilled Water System

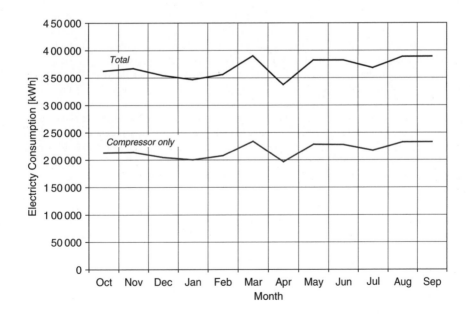

Figure 5.18 Monthly Electricity Consumption for Running the Chilled Water System

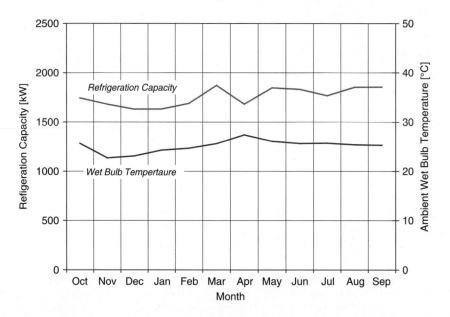

Figure 5.19   Monthly Refrigeration Capacity and Ambient Wet Bulb Temperature

The inspection of daily log sheets has determined that the supply and return water temperatures recorded manually are not reliable. The liquid in glass thermometers with very poor resolution was used. Practically, the recorded data were useless for ascertaining heat balance. The measurement of consumed electrical energy for the compressor, pumps and ventilators had been carried out by reading electric meters every two hours. For pumps and ventilators there is one measuring device and their individual share is estimated on the basis of their power.

On the basis of available data, it can be concluded that for 12 months the overall consumption of electrical energy of this chilled water system was 4277.90 MWh. The consumption of the compressor was 2639.89 MWh (62 %).

During four days (Sep 30–Oct 3), the relevant parameters of the chilled water system were measured. The objective of these measurements was to determine the actual operational regime which would then be used for determining more accurately the effects of potential performance improvement measures (such as, for example, *variable speed control* on some motor drives). The results of the measurements are presented in Figures 5.20 and 5.21 (Compressor Power and Refrigeration Capacity versus time and Condenser and Evaporator Water Temperature versus Time, respectively).

Table 5.5 shows the average values of the measured parameters, as well as their standard deviations. The monthly measured parameters of the observed process (those which are available – Figs 5.18 and 5.19) and their average values indicate a good coordination with those represented in Table 5.5 (average monthly refrigeration capacity was 1757 kW and compressor power was 308.8 kW). This means that the measurements performed during the detailed energy audit represent well the average work regime of the system during the whole year.

Figure 5.22 shows the refrigeration cycle a in p–h diagram with basic measured and calculated parameters.

Some relevant parameters of ambient air are shown in Table 5.6 in the manner in which they are recorded by the National Meteorological Institution.

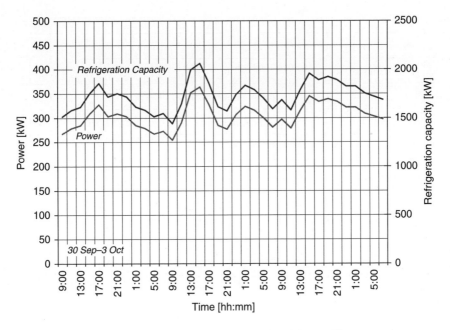

Figure 5.20 Power and Refrigeration Capacity versus Time

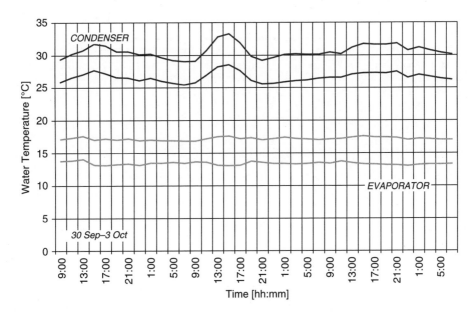

Figure 5.21 Water Temperatures versus Time

Table 5.5 Average Measured Parameters of Chilled Water System

	Compressor						Evaporator			Condenser		
	Power	Capacity	t_1	t_3	p_e	p_c	$t_{we,out}$	$t_{we,in}$	Q_e	$t_{wc,out}$	$t_{wc,in}$	Q_c
	kW	%	°C	°C	bar	bar	°C	°C	kW	°C	°C	kW
AVE	304.50	75%	18.10	32.00	0.668	1.360	13.39	17.20	1726.52	30.60	26.65	2031.1
SDEV	26.03	6%	0.43	1.04	0.01	0.06	0.24	0.21	147.55	1.01	0.73	173.59

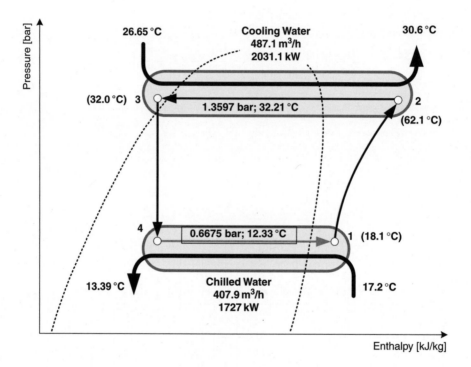

Figure 5.22 p–h Diagram

Table 5.6 Average Ambient Air Parameters

	t_{DB} [°C]	RH [%]	h [kJ/kg]	t_{WB} [°C]
MEAN Max	32.7	91.0	107.2	31.4
MEAN	27.8	76.6	74.4	24.6
MEAN Min	24.1	58.0	52.2	18.4
Ext Min	17.7	32.9	29.0	9.6

Table 5.7 illustrates the parameters of the refrigeration cycle for the average conditions of the regime during the detailed audit. Most of the parameters are calculated on the basis of the available data about the plant by applying Part III Toolbox 13 Software 10 and 11. Some missing parameters are estimated (e.g., mechanical compressor efficiency (0.88), electrical motor efficiency (0.95)).

Table 5.7 Calculation Results

			Design	Current	VSD (1)	VSD (2)	VSD (3)
Electrical motor power (installed)	$N_{el(in)}$	kW	405	405	405	405	405
Electrical motor power (for use)	$N_{c,el}$	kW	405.0	304.5	270.0	248.0	238.0
Refrigeration capacity	Q_e	kW	**2459**	**1727**	**1860**	**1750**	**1645**
EVAPORATOR							
Speed of pump (nominal)	n_e	rpm	1450	1450	1450	1450	1450
Evaporator water flow rate	m_e	m³/h	423.0	407.9	**313.7**	**301.0**	**282.9**
Evaporator water flow rate	m_e	kg/s	117.5	113.3	87.1	83.6	78.6
Power of pump	N_e	kW	66.0	65.2			
Network	$H_{e,net}$	mWC	43.6	41.3	28.5	27.0	25.1
Speed (VSD)	n_{VSD}	rpm			**1062**	**1027**	**978**
Power(VSD)	N_{VSD}	kW			25.9	23.4	20.3
Inlet water temperature of evaporator	$t_{we,i}$	°C	14.0	13.6	14	14	14
Outlet water temperature of evaporator	$t_{we,o}$	°C	19.0	17.2	19.1	19	19
Logarithmic temperature difference	$\Delta t_{ln,e}$	°C	3.75	2.65	3.60	3.50	3.46
Evaporation temperature	t_e	°C	12.21	12.33	12.56	12.56	12.56
Evaporation pressure	p_e	bar	0.6645	0.6675	0.6736	0.6736	0.6736
Temperature (suction)	t_1	°C	19.0	18.1	18.0	18.0	19.0
Enthalpy (suction)	h_1	kJ/kg	261.00	260.48	260.40	260.40	261.00
Entropy (suction)	s_1	kJ/kgK	0.96200	0.96000	0.95920	0.95920	0.96200
Specific volume	v_1	m3/kg	0.26082	0.25873	0.25625	0.25625	0.26082
Evaporator capacity	Q_e	kW	2459	1727	**1860**	**1750**	**1645**
Overall heat transfer conductance	$(UA)_e$	kW/K	656	652	516	500	475

(*continued overleaf*)

Table 5.7 (*continued*)

			Design	Current	VSD (1)	VSD (2)	VSD (3)
CONDENSER							
Speed of pump	n_c	rpm	1450	1450	**1281.5**	**1121.8**	**957.7**
Condenser water flow rate	m_c	m3/h	481.3	487.1	432.0	378.0	302.7
Condenser water flow rate	m_c	kg/s	133.7	135.3	**120.0**	**105.0**	**84.1**
Power of pump (nominal speed)	H_c	kW	68.7	68.9			
Network	$H_{c,net}$	mWC	38.1	38.7	33.4	28.9	23.5
Power(VSD)	N_{VSD}	kW			47.4	31.8	19.8
Inlet water temperature of condenser	$t_{wc,i}$	°C	33.00	27.05	29.35	28.74	27.83
Outlet water temperature of condenser	$t_{wc,o}$	°C	38.00	30.64	33.22	33.00	33.00
Logarithmic temperature difference	$\Delta t_{ln,c}$	°C	4.31	3.11	3.58	3.73	4.20
Condensing temperature	t_c	°C	40.28	32.29	35.12	35.12	35.12
Condensing pressure	p_c	bar	1.7619	1.3630	1.4954	1.4954	1.4954
Temperature (discharge)	t_2	°C	61.0	62.8	55.4	55.4	55.3
Enthalpy	h_2	kJ/kg	283.80	285.65	280.85	280.85	280.79
Entropy	s_2	kJ/kgK	0.9780	0.9986	0.9786	0.9786	0.9784
Specific volume	v_2	m³/kg	0.110729	0.145596	0.128569	0.128569	0.128524
Condenser capacity	Q_c	kW	2798	2031	2130	1998	1883
Overall heat transfer conductance	$(UA)_c$	kW/K	649	653	595.5	535.1	448.1
Temperature of liquid refrigerant	t_3	°C	38.5	32.0	34.5	34.5	34.5
Enthalpy of liquid refrigerant	h_3	kJ/kg	95.4	89.8	91.9	91.9	91.9
COOLING TOWER							
Wet bulb temperature	twb	°C	28.00	487.1	24.47	24.47	24.47
Number of transfer units	NTU		116.60	135.3	108.35	105.63	96.13
Cooling tower capacity	Q_{CT}	kW	2801	68.9	2130	1998	1883

Table 5.7 (*continued*)

			Design	Current	VSD (1)	VSD (2)	VSD (3)
COMPRESSOR							
Ratio of compression	p_c/p_e	–	2.65	2.04	2.22	2.22	2.22
Efficiency of compressor	η_i	–	0.644	0.868	0.868	0.868	0.868
Specific refrigeration capacity	q	kJ/kg	165.56	170.73	168.46	168.47	169.06
Indicator work	l_i	kJ/kg	22.80	25.17	20.45	20.45	19.79
Real work	l_{real}	kJ/kg	27.27	30.11	24.46	24.46	23.67
Coefficient of performance (cycle)	COP_c		6.071	5.671	6.887	6.888	7.142
Mass flow rate of refrigerant	m_{R11}	kg/s	14.85	10.11	11.04	10.39	9.73
Power of compressor		kW				270	254.07	230.33
SUPPLY SYSTEM							
Supply water temperature	t_{ws}	°C	14.0	13.6	14	14	14
Return water temperature	t_{wr}	°C	19.0	17.2	19.1	19	19
Speed of pump	n_s	rpm	1450	1450	**1016**	**975**	**920**
Supply water flow rate	m_s	m3/h	423.0	407.9	313.7	301.0	282.9
Head	H_s	mWC	43.9	45.4			
Power	N_s	kW	66.0	65.2			
Network	$H_{s,net}$	mWC	43.3	40.6	26.1	24.4	22.1
Power (VSD)	N_{VSD}	kW			22.7	20.1	16.9
POWER							
Compressor	N_{com}	kW	405.0	304.5	270.0	248.0	238.0
Power of chilled water pump	N_e	kW	66.0	65.2	25.9	23.4	20.3
Power of condenser pump	N_c	kW	68.7	68.9	47.4	31.8	19.8
Power of CT fan	N_{CTf}	kW	20.0	20.0	20.0	20.0	20.0
Power of supply pump	N_s	kW	66.0	65.2	22.7	20.1	16.9
Power of the system	N_{sys}	kW	625.7	523.7	386.0	343.3	315.0
COP system	**COPs**	–	**3.930**	**3.297**	**4.818**	**5.097**	**5.223**

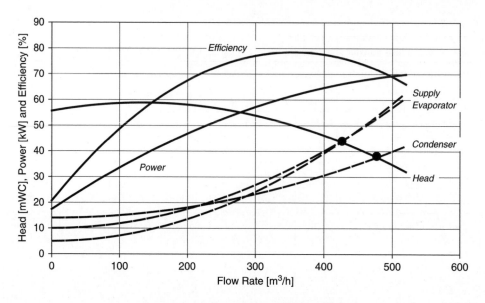

Figure 5.23   Pump and Network Characteristics

The pumps are identical. Their features are represented in Figure 5.23. The parameters of working points are measured during the detailed energy audit. The hydraulic characteristics of the network are calculated on the basis of these parameters (Part III Toolbox 7 and 8).

These results will be used as a reference for estimates related to performance improvement of the chilled water system leading to the reduction of electrical energy consumption.

5.6.4 Performance Improvement Measures (PIM)

In the presented regime, the overall power of the chilled system was 519.45 kW. Out of that the compressor power amounted to 58.6 %, and of the ventilator cooling tower to 3.8 %. The remaining power (37.6 %) was evenly distributed approximately between three pumps. The objective of this PIM is to optimize the operation of the pumps by variable speed regulation (VSD).

In the work conditions which were then present in the installation, the temperature differences of water at the inlet and outlet of the condenser and evaporator were 3.95 °C and 3.82 °C, respectively (Table 5.5).

The change of flow will change the conditions for heat transfer. Thus, if we reduce the water flow in the condenser, for example, this will cause the reduction of the water side heat transfer coefficient and certainly the overall heat transfer conductance (UA).The heat flow rate has to be constant and this is possible if the logarithmic temperature difference between fluids will increase. The heat flow rate is defined as follows:

$$Q = UA \cdot \Delta T_{ln} \qquad (5.10)$$

Where:

$$\Delta T_{ln} = \frac{(T_{con} - T_{c,out}) - (T_{con} - T_{c,in})}{\ln \frac{T_{con} - T_{c,out}}{T_{con} - T_{c,in}}}$$

is Logarithmic Temperature Difference, [K]

U   Overall coefficient of heat transfer, $[W/m^2K]$
A   Area of the heat transfer, $[m^2]$
UA   Overall heat transfer conductance

Generally speaking, changes in the water flow will cause the changes in the work of:

- cooling tower;
- condenser;
- evaporator; and
- end-users.

Indirectly, it will also cause a temperature change in condensing and evaporation, as well as compressor power. The consequences of these changes will be reflected in the work of the entire system. In the estimate presented in Table 5.7, Part III Toolbox 13 Software 9 and 10 are used. They offer the option of estimating the impact of water flow change on the individual work of the cooling tower and heat exchanger.

Figure 5.24 provides a schematic presentation of the observed chilled system and includes the changes by the proposed PIM. Speed regulators are anticipated for all three pumps. On the side of end-users it is necessary to install two way regulation valves instead of three way valves (or to adapt the existing ones). The intention is that each consumer is supplied with as much water as necessary and not, as before, to return the excess water into the tank (Fig. 5.17).

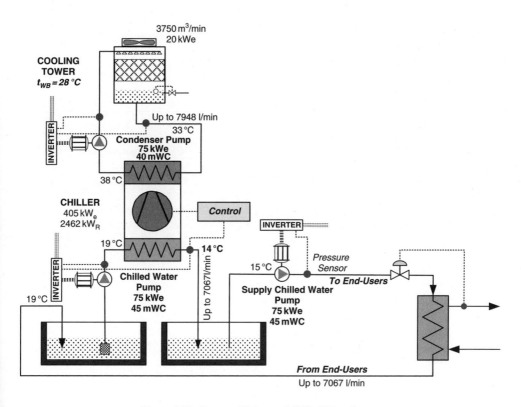

Figure 5.24   Improved Scheme of Chilled Water System

5.6.5 Cost Benefit Analysis

The next step is to estimate the duration of certain loads of the chilled water system during the year. This load depends on the intensity of the production process and on the parameters of the ambient air. For this estimate, we are going to use the *load duration curve* shown in Figure 5.25. Based on that Figure, we can conclude that during the year the mean maximum load (1860 kW) occurs at around 2,600 hours, mean load (1750 kW) at around 4400 hours and mean minimum load (1645 kW) at around 1448 hours per annum.

It is now necessary to calculate the consumption of electrical energy of the entire system for certain loads provided that only the water flow in the pumps is changed in order to meet the requirements of end-users in respect of refrigeration capacity and temperature.

With the load drop, the end-users' temperature sensors register the temperature drop at the outlet of the heat exchanger (consumer) and close the regulation valve in the supply pipeline, which is shown in the Figure. By closing the valve, the pressure sensor behind the supply pump registers pressure increase and actuates the speed regulator (VSD), i.e., reduces the frequency of electricity which causes a decrease of the number of revolutions and thus reduces the flow. Thus, the pressure behind this pump will be reduced to the predetermined value. The reduction in the pump's number of revolutions will in turn lead to the reduction in electrical power. The evaporator pump has its own speed regulator (VSD), which is set to $\Delta t_e = 5\,°C$ and with this value regulates the flow. The condenser pump is set to keep the water temperature difference at $\Delta t_c = 5\,°C$, also.

This calculation implies some assumptions and approximations. More complex measurements will certainly provide more reliable data but this would be achieved only at much greater expense.

The following calculations with the following assumptions or *scenarios* were performed:

(a) The chillers and pumps operate with full capacities. The ambient air is taken to be as designed (wet bulb air temperature 28 °C). The supplied chilled water temperature is 14 °C, and return is 19 °C (design column in Table 5.7).

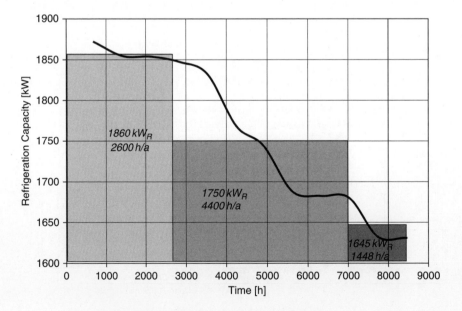

Figure 5.25   Load Duration Curve of Chilled Water System

(b) The chillers operate with capacity requirements intended to meet the necessary annual average load of end-users (air washers and heat exchangers). The supply chilled water temperature is 14 °C, and return is 16.7 °C. The pumps operate at full flow load. The calculation is performed for the average annual values of dry and wet bulb temperatures of ambient air (29.12/25.13 °C) (current column in Table 5.7).

(c) The water temperature differences are kept at 5 °C in the condensers, evaporators and on the side of end-users. This means that the flow rate of both water flows is reduced to meet the real load. The pump power is reduced also. The ambient air was the same as for case b. These calculations are performed for three different loads (1860 kW – VSD(1) column in Table 5.7, 1750 kW – VSD(2) column in Table 5.7 and 1645 kW – VSD(3) column in Table 5.7).

The results of the calculations for all cases are presented in Table 5.7. It is obvious that a significant reduction in electrical energy consumption can be achieved by the implementation of frequent regulators.

By introducing the variable speed drive (VSD) controls of the pumps (condenser, evaporator, and supply water), the system's COP will be increased from the current 3.297 to an average of 5.097. This means that electrical energy consumption will be reduced by:

$$\Delta E = [523.7 \cdot 8448 - (386.0 \cdot 2600 + 343.3 \cdot 4400 + 315.0 \cdot 1448)]$$
$$= 1454.0 \ [\text{MWh/year}] \tag{5.11}$$

or by 32.9 % compared with the consumption before the implementation of the proposed performance improvement measures. Taking into consideration the prices of energy, equipment and installation works at that time, the simple pay back period was 1.4 years.

5.7 Bibliography

GIL52, The Engine of the Refrigeration System: Selecting and Running Compressors for Maximum Efficiency, http://www.actionenergy.org.uk.

GIL79, Finding the Best Liquid Chiller Package, www.actionenergy.org.uk.

GPG 38, Commercial Refrigeration Plant: Energy Efficiency Installation, http://www.actionenergy.org.uk.

GPG215, Reducing Energy Costs in Industry with Modern Control Techniques, http://www.actionenergy.org.uk.

GPG280, Energy Efficient Refrigeration Technology – the Fundamentals, http://www.actionenergy.org.uk.

GPG347, Installation and Commissioning of Refrigeration Systems, www.actionenergy.org.uk .

GPG 364, Service and Maintenance Technical Guide, http://www.actionenergy.org.uk.

Gvozdenac, D., Kljajić, M. (2005) Energy Audit of Industrial Vapor Compression Refrigeration Systems, PSU-UNS International Conference on Engineering and Environment – ICEE-2005, Novi Sad 19–21 May.

Gvozdenac, D., Morvay, Z., Kljajić, M. (2003) Energy audit of cooling towers, PSU-UNS International Conference 2003: Energy and the Environment, Prince Songkla University, Hat Yai, Thailand, 11–12 December.

Stoecker, W. F., Jones, J. W. (1984) Refrigeration & Air Conditioning (Second Edition), McGraw-Hill.

Review of Energy Efficient Technologies in the Refrigeration Systems of the Agrofood Industry, Thermie Programme Action No I 185, http://edo.uniba.sk:8000.

6

Industrial Cogeneration

6.1 System Description

Cogeneration or **C**ombined **H**eat and **P**ower (CHP) refers to the recuperation of waste heat when electricity is generated and using it to create high temperature hot water or steam or hot gases. Steam, hot water or hot gases can then be used for some production processes, space heating, producing sanitary hot water, or powering absorption chillers for chilled water production.

CHP and *cogeneration* are basically the same thing. Cogeneration is generally identified with district heating and large utility owned power plants or industrial power production and plant operation, while CHP is generally associated with smaller scale energy units.

The conceptual scheme of a cogeneration system is presented in Figure 6.1.

Fuel (in a given case of natural gas) produces mechanical work which is used for the production of electrical energy in the generator. In addition to mechanical work, heat energy is produced which should also be used. There are several ways of utilizing heat energy from the prime mover, but in all cases a device for its utilization can be called a heat recovery unit. The output from this device is most often steam, hot water or hot gases. For each prime mover, the relation between the produced mechanical work (electrical energy) and heat energy (Table 6.1) is very clearly defined. This is one of the essential limitations of cogeneration plants as it is very difficult to find an industrial process in which the actual need for electrical and heat energy corresponds completely to the output values of the cogeneration plant. This means that there must be a standard source of either electrical or heat energy or both. The other possibility is that the surplus of one of two products of the cogeneration plant is simply 'thrown away'. This will drastically reduce the energy efficiency of the plant.

Parameters of outlet fluids from the heat recovery unit will depend on the quantity of available waste heat of prime mover. Sometimes it is necessary to additionally adjust parameters of these fluids for the needs of the process with additional consumption of fuel. It should be pointed out that in some industries, when allowed by the process, it is possible to use prime mover's products of combustion directly without further processing.

It can be concluded that basic products of the cogeneration plant are certain quantities of electrical and heat energy. Of course, the process for which this energy is intended dictates absolutely independently quantities and the dynamics of utilizing these energies. The process is always primary and the cogeneration plant should be subordinated in relation to the process. The Fig. 6.1 shows one possibility for using produced electrical and heat energy. In addition to direct utilization of steam and hot water, part of hot water is used for running the absorption chiller and production of chilled water for running the air conditioning units. Optionally, it is possible to use produced electrical energy for running the mechanical type of compressors for the production of the same temperature chilled water. What scheme of utilization

Applied Industrial Energy and Environmental Management   Zoran K. Morvay and Dušan D. Gvozdenac
© 2008 John Wiley & Sons, Ltd

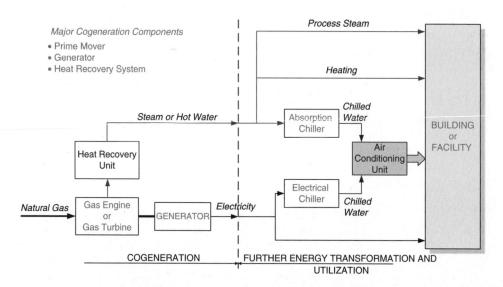

Figure 6.1   Conceptual Scheme of Cogeneration System

Table 6.1   Heat to Power Ratios[1] and others Parameters of Cogeneration Systems. Reproduced from: A Guide to Cogeneration, The European Association for the Promotion of Cogeneration, Brussels

Cogeneration System	Heat to Power Ratio* [kW$_h$/kW$_e$]	Power Output [% of fuel input]	Overall Efficiency [%]
Back-pressure steam turbine	4.0–14.0	14–28	84–92
Extraction-condensing steam turbine	2.0–10	22–40	60–80
Gas turbine	1.3–2.0	24–35	70–85
Combined cycle (gas plus steam turbine)	1.0–1.7	34–40	69–83
Reciprocating engine	1.1–2.5	33–50	75–85

* Higher heat to power ratio assumes additional firing.

of generated heat and electrical energy will be used depends strongly on the actual process and on the energy supply design.

It should be pointed out that the cogeneration concept of energy supply for industrial processes has been very much neglected although it has been traditionally used by some of industries for a long time (the petrochemical and chemical industries, for example). The reason for the insufficient use of cogeneration schemes in industrial practice used to be the relatively high investment costs and the relatively low prices of fuels. Today, the situation has changed significantly and energy costs weigh heavily on the final price of a product. Therefore, every increase in efficiency of energy transformations can pay off well.

---

[1] The heat to power ratio is the ratio between useful heat and electricity from cogeneration.

We are of the opinion that each factory should consider the options for using cogeneration plants, due in particular to the fact that a large number of countries have already adopted or are preparing to adopt corresponding legal and regulatory support for a widespread use of cogeneration schemes.

This chapter presents analyses based on estimates of technical and particularly economic indicators of the benefits concerning the introduction of these plants in industry. This will present reliable grounds for decision making in a particular company or in several companies which are situated within close proximity. For positive or negative conclusions, the prices of certain fuels or their mutual relations will be decisive.

The potential for using cogeneration as a measure for saving energy is underused in industry at present. The promotion of cogeneration based on useful heat demand has to be an industry priority given the potential benefits of cogeneration with regard to saving primary energy, avoiding network losses and reducing emissions, in particular of greenhouse gases.

6.2 Principles of Operation

A Typical cogeneration scheme is presented in Figure 6.2. For the production of 35 units of electrical energy and at the same time 50 same units of heat energy, it is necessary to introduce some fuel with a chemical energy value of 100 units. In this case the useful energies total 35 + 50 = 85 units.

The conventional manner of providing the necessary fuels for a process entails buying electrical energy from a public grid and the production of heat energy in the plant by using the chemical energy of the same or some other fuel in relation to the fuel used in the cogeneration plant. This case is illustrated in Figure 6.3. For the production of 35 units in a certain thermal power plant, it is necessary to invest around 106 units of energy. In this case, as much as 71 units of energy are lost into the surroundings. In order to produce for a process the required 50 units of heat energy in the plant, we must consume additional 59 units of fuel energy. This is a total of 165 units. It can be concluded that in the case of the conventional means of supplying the process with energy it is necessary to invest in considerably more fuel energy than in case of the cogeneration plant.

However, both figures show only the principal options and do not raise questions such as: Is it possible during the whole life span of cogeneration plant exploitation to produce transformed heat and electrical energy also? If this is not the case, are there any significantly changed relations and even the conclusion that a complex cogeneration plant cannot economically withstand competition from two simple plants for special production? Therefore, these issues will be given more attention and will be dealt with more detail in this chapter.

So far we can conclude that cogeneration is:

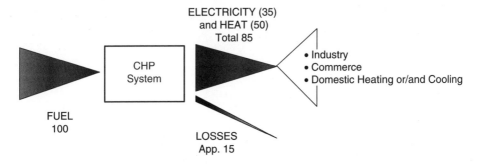

Figure 6.2   Cogeneration Scheme

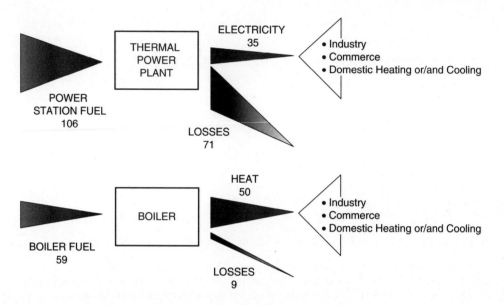

Figure 6.3   Conventional Power Generation Scheme

- *integrated* system;
  located at or near a building or facility;
- satisfying at *least a portion* of the facility's *electrical demand*; and
- *utilizing* waste heat to provide steam, hot water or hot gases.

Generated electrical energy can usually be used directly in the production process or it can be sold to a local distribution company at corresponding voltage. Heat energy can also be used directly but very often it is further transformed in order to provide more efficient utilization. The most common use of heat energy is for steam generation, space heating or for running absorption chillers which generate chilled water.

6.3 Types of Industrial Cogeneration Plants

Cogeneration technologies which are nowadays used in commercial and yet not fully commercial ways are as follows:

(a) combined cycle gas turbine with heat recovery;
(b) steam backpressure turbine;
(c) steam condensing extraction turbine;
(d) gas turbine with heat recovery;
(e) internal combustion engine;
(f) micro-turbines;
(g) stirling engines;
(h) fuel cells;
(i) steam engines;
(j) organic Rankine cycles;
(k) any other types of technologies or combination thereof falling under the definition of *cogeneration*.

Table 6.2 Technical Characteristics of Cogeneration Systems. Reproduced from: A Guide to Cogeneration, The European Association for the Promotion of Cogeneration, Brussels

System	Electric Power	Annual Average Availability	Electric Efficiency		Total Efficiency	Power to Heat Ratio
	MW	%	Load 100 %	Load 50 %	%	–
Steam turbine	0.5–100*	90–95	14–35	12–28	60–85	0.1-0.5
Open cycle gas turbine	0.1–100	90–95	25–40	18–30	60–80	0.5–0.8
Closed cycle gas turbine	0.5–100	90–95	30–35	30–35	60–80	0.5–0.8
Joule-Rankine combined cycle	4–100*	77–85	35–45	25–35	70–88	0.6–2.0
Diesel engine	0.07–50	80–90	35–45	32–40	60–85	0.8–2.4
Reciprocating internal combustion engine package	0.015–2	80–85	27–40	25–35	60–80	0.5–0.7
Fuel cells	0.04–50	90–92	37–45	37–45	85–90	0.8–1.0
Stirling engines	0.003–1.5	85–90 (expected)	35–50	34–49	60–80	1.2–1.7

* The value 100 MW is the usual upper limit for industrial applications. Systems of this type can have higher capacities, also.

General information on each of above-mentioned cogeneration technologies is presented in Table 6.2. Although their electrical efficiencies are very different, the total efficiencies are fairly equal for all of the presented technologies. The power to heat ratio varies significantly from technology to technology. In most industrial branches, the consumption of heat energy is either higher or considerably higher than the consumption of electrical energy (see Table 2 in Introductory Chapter), which means that the majority of available technologies can provide electricity for export. This is strategically a very important issue for the government's energy policy since various stimulation mechanisms can provide for the significant production of electrical energy in industry. The annual average availability is for most technologies very high but it should be pointed out that annual overhaul requires very specialized service.

A large majority of small and medium size factories (50–400 employees) has an installed electrical energy transformer station with a capacity in the range from 0.5 to 10 MW and which is connected to the public electricity supply grid with a voltage of 22 kV or similar. We may raise the question as to which cogeneration technological solutions should be taken into consideration.

If we decide primarily on fully commercial technologies, then with regard to the above-mentioned cogeneration technologies, it is possible to adopt steam turbines, gas turbines or gas engines. When the available capacity is concerned generally, these technologies meet practical requirements. However, within these technologies, the choice of a corresponding one requires additional conditions on which the decision making process will be based. For example, one of the major parameters is the plant's operations time scheme. For continuous operations in the course of several months during a year, all three technologies are suitable, but if the plant operates in daily cycles in an ON/OFF mode, then there is only gas engines technology, and for smaller plants there is also gas turbine technology, which is considered acceptable. In such conditions, the steam turbine is not acceptable as it has a slow start-up.

Many technologies and configurations that can generate power much more efficiently than the current industry average are available today in a wide range of sizes. Natural gas-fired combustion turbines have had their performance improved, with efficiencies that have tripled since the mid 1970s. New gas turbine units are typically in the hundreds of megawatt size, one-half to one-quarter the size of a conventional coal-fired steam turbine electricity generating plant and economical operation is possible in units as small as a few megawatts.

Big industrial cogeneration plants (over 100MWe) have a large number of annual operating hours and, as a rule, deliver a substantial part of electrical energy production to the distribution grid under special contacts. These plants will not be considered here.

Generally, it can be said that cogeneration technologies are developing in the direction of meeting industrial requirements and available packages are adapted to the specific features of certain industrial branches to a large extent.

In most cases, the choice of the prime mover will be determined by site requirements. This in turn will dictate the other items of the plant. The main indicators for the selection of a cogeneration plant are as follows:

Steam turbines may be the appropriate choice for sites where:

- electrical base load is over $250\,kW_e$;
- there is a high process steam requirement. Heat to power ratio is greater than 3 :1;
- cheap, low-premium fuel is available;
- adequate plot space is available;
- high grade process waste heat is available (e.g. from furnaces or incinerators);
- existing boiler plant needs to be replaced;
- heat to power ratio is to be minimized, using a gas turbine combined cycle.

Gas turbines may be suitable if:

- power demand is continuous and over $1\text{--}2\,MW_e$;
- natural gas is available (although this is not a limiting factor);
- there is high demand for medium/high pressure steam or hot water, particularly at temperature higher than $140\,°C$;
- demand exists for hot gases at $450\,°C$ or above – the exhaust gas can be diluted with ambient air to cool it or put through an air heat exchanger;
- there is a possibility to use a combined cycle with a steam turbine.

Reciprocating engines may be suitable for sites where:

- power or processes are cyclical or not continuous;
- low pressure steam or medium or low temperature hot water are required;
- there is a low heat to power demand ratio;
- natural gas is available, gas powered engines are preferred;
- natural gas is not available, fuel oil or LPG powered diesel engines may be suitable;
- electrical load is less than $1\,MW_e$ – spark ignition (units available from $3\,kW_e$ to $10\,MW_e$)
- electrical load greater than $1\,MW_e$ – compression ignition (units available from $100\,kW_e$ to $20\,MW_e$)

Although the above-mentioned indicators will enable a broad choice of prime movers, the final selection will be made on the basis of the particular site requirements.

Generating technologies are becoming smaller and more reliable, largely because of innovations, including automated control systems, turbine improvements, advanced materials (e.g. ceramics) and flexible manufacturing. These more recent technologies reduce emissions and improve portability, reliability and efficiency.

The vast majority of newly planned capacities are gas-fired with the greatest movement in countries that have proceeded furthest towards utility restructuring.

Considerations also include the long-term availability and cost of fuel, the cost of electricity purchased, including charges associated with the provision of a back-up supply, and the credit earned for any exported electricity. In addition, the service and technical support available from the equipment suppliers, and the proven reliability of particular machines, may have a significant impact on the outcome of the selection procedure.

6.3.1 Advantages and Disadvantages of Each System

This section lists the main advantages and disadvantages of each of the prime mover options applicable in industry (Table 6.3).

6.3.2 Thermally-Activated Technologies

In the cogeneration process, besides the production of mechanical work, heat energy is also generated. This heat energy usually consists of hot gasses and hot water. This process can be called cogeneration only if the large majority of generated energy is usefully deployed. The further use of generated heat energy is not part of cogeneration and it is only its useful utilization that allows the process which has generated this energy to be called cogeneration.

The technologies that use *thermal energy* as the *primary energy* for their operation are called *thermally activated technologies*. The two most common thermally activated technologies applicable to cogeneration systems are as follows:

(a) space and process heaters;
(b) absorption chillers.

These systems are not part of the cogeneration plant. They are added for the purpose of achieving more effective and more efficient utilization of generated heat energy.

(a) Space and Process Heating

The design of the heat recovery unit involves the consideration of many related factors, such as the thermal capacity of exhaust gases, the exhaust gas flow rate, the sizing and type of heat exchanger and the desired parameters over a varied range of operating conditions for the cogeneration system – all of which need to be considered for proper and economical operation.

The following cases can appear in practice:

- *Process heating*: Exhaust gases may be used either directly in the process or they may be used indirectly to heat water or generate steam.
- *Space heating*: Exhaust gases from the prime mover normally indirectly (via some form of heat exchanger) heat the water or air that is used for space heating.
- *Supplemental heating*: In some cases the exhausts may not be hot enough to provide the necessary thermal energy, so it may be used to preheat water or air in a secondary system or duct firing may be added to raise the temperature of exhaust gases.

Table 6.3   Cogeneration Advantages and Disadvantages versus Type of Prime Mover

	Advantages	**Disadvantages**
Steam Turbines	High overall efficiency; Any type of fuel may be used; Heat to power ratios can be varied through flexible operations; Ability to meet more than one site heat grade requirements; Wide range of sizes available; Long working life.	Generally high heat to power ratios; High cost; Slow start-up.
Gas Turbines	High reliability which permits long-term unattended operation; High grade heat available; Constant high speed enabling close frequency control of electrical output; High power to weight ratio; No cooling water required; Relatively low investment cost per kWe electrical output; Wide fuel range capability (diesel, LPG, naphtha, associated gas, landfill sewage); Multi fuel capability; Low emissions.	Limited number of unit sizes within the output range; Lower mechanical efficiency than reciprocating engines; If gas fired, requires high-pressure supply or in-house boosters; High noise levels (of high frequency can be easily alternated); Poor efficiency at low loading (but they can operate continuously at low loads); Can operate on premium fuels but need to be cleaned off dry; Output falls as ambient temperature rises due to thermal constraints within the turbine; May need long overhaul periods.
Reciprocating Engines	High power efficiency, achievable over a wide load range; Relatively low investment cost per kWe electrical output; Part-load operation flexibility from 30% to 100 % with high efficiency; Can be used in island mode; Good load following capability; Fast start-up time of 15 second to full load (gas turbine needs 0.5–2 hours); Real multi-fuel capability, can also use HFO as fuel; Can be overhauled on site with normal operators; Low investment cost in small sizes; Can operate with low-pressure gas (down to 1 bar).	Must be cooled even if the heat recovered is not reusable; Low power to weight ratio Out of balance forces require substantial foundation; High levels of low frequency noise; High maintenance costs.

(b) Absorption Chillers

Absorption chillers use heat and a binary mixture in order to produce chilled water. A gas burner is usually used to heat the mixture of lithium bromide and water (binary mixture). The recovered waste heat in the form of hot water or steam can be used to power indirectly the fired absorption chiller (electricity is used for binary mixture pumps, but only a small fraction of the electricity required by electric motor driven chillers).

Table 6.4 Absorption Chillers. Reproduced from: CHP Resource Guide, Oak Ridge National laboratory, 2003

<table>
<tr><td rowspan="20" style="writing-mode: vertical">ABSORPTION CHILLERS</td><td>**Capacity Range***</td><td>**Single-Effect**</td><td>**Double-Effect**</td></tr>
<tr><td>**COP (Coefficient of Performance)**</td><td>0.65–0.70</td><td>0.9–1.2</td></tr>
<tr><td>**Heat Source:**</td><td></td><td></td></tr>
<tr><td>Minimum Temperature, °C</td><td>82</td><td>175</td></tr>
<tr><td>Steam Flow Rate, kg/h per kW_R</td><td>2.3</td><td>1.3–1.4</td></tr>
<tr><td>Steam Pressure, barg</td><td>1</td><td>8–9</td></tr>
<tr><td>**Integration with Waste Heat from:**</td><td></td><td></td></tr>
<tr><td>– Reciprocating engines, kW_R/kW_e</td><td>0.8–1.0</td><td>1. –1.4</td></tr>
<tr><td>– Combustion turbines, kW_R/kW_e</td><td>1.0–1.3</td><td>1.4–1.8</td></tr>
<tr><td>**Average Electric Power Offset**</td><td>$0.17\,kW_e/kW_R$</td><td>$0.17\,kW_e/kW_R$</td></tr>
<tr><td>**Installed Cost [$US/$kW_R$]**</td><td></td><td></td></tr>
<tr><td>350 kW_R</td><td>280</td><td>340</td></tr>
<tr><td>1 750 kW_R</td><td>200</td><td>255</td></tr>
<tr><td>3 500 kW_R</td><td>185</td><td>240</td></tr>
<tr><td>7 000 kW_R</td><td>140</td><td>200</td></tr>
<tr><td>**Operation and Maintenance Costs** ($US/$kW_R$/year)</td><td></td><td></td></tr>
<tr><td>350 kW_R</td><td>8.5</td><td>8.5</td></tr>
<tr><td>1 750–2 000 kW_R</td><td>4.5–8.0</td><td>4.8–7.5</td></tr>
</table>

* Subscripts **R** and **e** denotes refrigeration and electrical power, respectively.

Some rough figures for absorption chillers are giving in Table 6.4. These figures can be used for quick calculation and estimation (the rated capacities of the absorption chillers are based on producing chilled water at 7 °C). Of course, the best way is to contact manufacturers and ask for the true figures that you need.

Commercially available absorption chillers that can be utilized within a cogeneration system use:

- steam;
- hot water;
- hot exhaust gases; or
- direct firing of natural gas or similar fuel (in this case, the absorption chiller is not a part of cogeneration plant).

An example of possible scheme of absorption chiller and gas engine cogeneration plant is presented in Figure 6.4. The electrical output of the generator is 2000 kW_e and refrigeration capacity is 1256 kW_R (7/12 °C chilled water).

The represented plant is very common in industrial practice. It is very often used both for heating and cooling, as the parameters of hot water are also suitable for heating. Such a solution is often found in commercial buildings.

A somewhat more recent plant for industrial use is presented in Figure 6.5. This is a compact combined cogeneration plant comprised of a gas turbine, steam turbine and double-end drive generator. Additional firing primarily regulates the necessary flow rate and the parameters of steam for the optimum operation of the steam turbine. The steam turbine can be either condensing or back pressure. The production of steam for the process may vary depending on the process needs, which makes such a plant very suitable for industrial utilization. The commercial power of the plant is up to 20–25 MWe, which is very acceptable for numerous industrial applications. Their overall efficiency can be up to 85 %. In case the plant is used

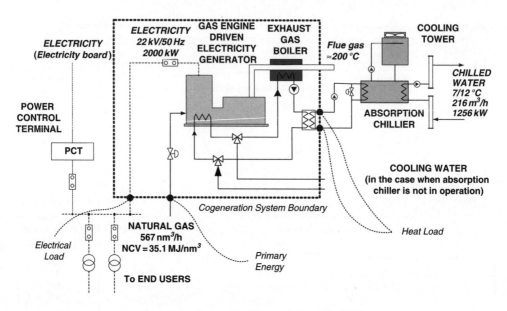

Figure 6.4   Cogeneration Plant and Absorption Chiller

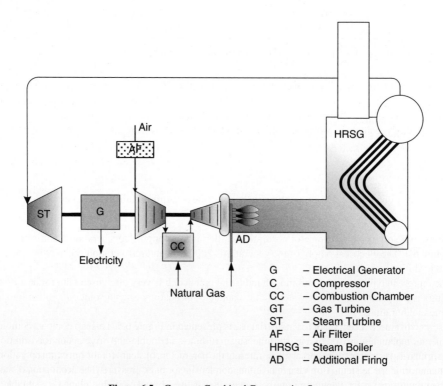

Figure 6.5   Compact Combined Cogeneration System

within the condensing regime of a steam turbine and serves only for the production of electricity, efficiency is up to 59 %. In this last case, the plant is not cogeneration as it produces useful heat for the process.

6.4 Operational Modes of Cogeneration Systems

The mode of operation is characterized by a criterion on which the adjustment of the electrical and useful thermal output of a cogeneration system is based. There are various modes of possible operations. The most distinctive are the following:

(a) *Heat-match mode*. The useful thermal output of a cogeneration system at any moment in time is equal to the thermal load (without exceeding the capacity of the cogeneration system). If the generated electricity is higher than the load, excess electricity is sold to the grid; if it is lower, supplementary electricity is purchased from the grid.
(b) *Electricity-match mode*. The generated electricity at any moment in time is equal to the electrical load (without exceeding the capacity of the cogeneration system). If the co-generated heat is lower than the thermal load, an auxiliary boiler supplements the need; if it is higher, excess heat is rejected to the environment through coolers or exhaust gases.
(c) *Mixed-match mode*. At certain periods of time, the heat-match mode is followed, while in other periods the electricity-match mode is followed. The decision is based on considerations such as load levels, fuel price and electricity tariffs on a particular day and at a particular time.
(d) *Stand-alone mode*. There is a complete coverage of electrical and thermal loads at any moment in time with no connection to the grid. This mode requires the system to have reserve electrical and thermal capacities, so that in the case where a unit is out of service for some reason, the remaining units are capable of covering the electrical and thermal loads. This is the most expensive strategy, at least from the standpoint of the initial costs of the system.

In general, the heat-match mode results in the highest fuel utilization rate (fuel energy savings ratio) and perhaps in the best economical performance for cogeneration in the industrial and building sectors.

In the utility sector, the mode of operation depends on the total network load, the availability of power plants and the commitments of the utility towards customers regarding the supply of electricity and heat. However, applying general rules is not the most prudent approach in cogeneration. Every application has its own distinctive characteristics.

For the operation of cogeneration systems, microprocessor-based control systems are very welcome. They can provide the capability to operate in a base load mode, to track either electrical or thermal loads, or to operate in an economical dispatch mode (mixed-match mode). In the latter mode, the microprocessor can be used to monitor cogeneration system performance, including:

- system efficiency and the amount of useful heat available;
- the electrical and thermal requirements of a user, the amount of excess electricity which has to be exported to the grid and the amount of heat that must be rejected to the environment;
- the cost of purchased electricity and the value of electricity sales, as they may vary with the time of the day, the day of the week, or the season.

The microprocessor can determine which operating mode is the most economical or even whether the unit should be shut down. Moreover, by monitoring operational parameters such as efficiency, operating hours, exhaust gas temperature and coolant water temperatures, the microprocessor can help in scheduling maintenance. If the system is unattended, the microprocessor can be linked by a telephone line with a remote monitoring centre, where the computer analysis of data may notify skilled staff about an impending need for scheduled or unscheduled maintenance. Furthermore, as part of a data acquisition system, the microprocessor can produce reports on the system's technical and economical performance.

6.5 Performance Definition

The overall efficiency of cogeneration plant can be defined as follows:

$$\eta_{Cogen} = \left(\frac{\text{Net Electricity Output} + \text{Heat Recovered}}{\text{Energy Input}} \right) \cdot 100\,\% \tag{6.1}$$

The basic disadvantage of overall efficiency defined in this manner is in the fact that electrical and thermal energy are merely summed up. It is clear that these two 'values' are very different in the thermodynamic sense and that electrical energy is much more valuable. In the economies of developed countries distinction in values is well balanced and reflected in their prices. A much more real estimate of efficiency is given by exergy efficiency definition. We are not going to pursue this analysis any further as industrial practice still uses exergy analysis insufficiently.

For the example given in Figure 6.2, we have:

$$\eta_{Cogen} = \frac{35 + 50}{100} \cdot 100\,\% = 85\,\% \tag{6.2}$$

For the conventional scheme shown in Figure 6.3, the efficiencies of both processes are as follows:
Electricity:

$$\eta_E = \frac{35}{106} \cdot 100\,\% = 33.0\,\% \tag{6.3}$$

Heat:

$$\eta_H = \frac{50}{59} \cdot 100\,\% = 84.7\,\% \tag{6.4}$$

and overall efficiency is:

$$\eta_{Conventional} = \frac{35 + 50}{106 + 59} \cdot 100\,\% = 51.5\,\% \tag{6.5}$$

There is a really significant difference in the overall efficiency of two analyzed arrangements for supplying the process with absolutely the same amount and type of energy.

The overall efficiency and sustainability of cogeneration is dependent on many factors, such as the technology used, fuel types, load curves, the size of the unit and also the properties of heat. The most reliable overall efficiency can be calculated if the *annual* sum of electricity or mechanical energy production and useful heat output is divided by the fuel input used.

The amount of *primary energy savings*[2] provided by cogeneration production can be calculated on the basis of the following formula:

$$PES = \left(1 - \frac{1}{\dfrac{H\eta_{Cogen}}{H\eta_{Ref}} + \dfrac{E\eta_{Cogen}}{E\eta_{Ref}}} \right) \cdot 100\,\% \tag{6.6}$$

---

[2] Directive 2004/8/EC of the European Parliament and of the Council of 11 February 2004 on the promotion of cogeneration based on a useful heat and demand in the internal energy market and amending Directive 92/42/EEC [2004] OJ L52.

where:

PES = Primary energy savings.

$H\eta_{Cogen}$ = Heat efficiency of the cogeneration production defined as annual heat output divided by the fuel input used to produce the sum of useful heat output and electricity from cogeneration.

$H\eta_{Ref}$ = Referent efficiency value for separate heat production.

$E\eta_{Cogen}$ = Electrical efficiency of the cogeneration production defined as annual electricity from cogeneration divided by the fuel input used to produce the sum of useful heat output and electricity from cogeneration.

$E\eta_{Ref}$ = Referent efficiency value for separate electricity production.

At a first glance, Equation (6.6) is very simple. Unfortunately, it is very complicated to use for determination of cogeneration process efficiency and in particular the referent efficiency values for separate heat and electricity production are very complex. In addition to the basic definition, some other reference documents are used as well.

This performance indicator is used primarily for defining the efficiency of the observed cogeneration process in relation to some other prescribed reference processes. The results of this indicator are used mostly outside the factory for the purpose of qualifying cogeneration in front of some national regulatory body in order to obtain subventions or stimulations.

For example, as presented in Figures 6.2 and 6.3, primary energy saving (PES) is equal to:

$$PES = \left(1 - \frac{1}{\frac{50/100}{50/59} + \frac{35/100}{35/106}}\right) \cdot 100\,\% = 39.4\,\% \tag{6.7}$$

It is obvious that the factory's management does not have to pay too much attention to primary energy consumption in a thermal power plant which can be somewhere far away from the factory itself. The concern grows with the electricity price and with a well balanced national energy policy. Namely, many countries have recognized the national benefits from implementing the distributed production of electrical and heat energy and the importance of industry for this utilization. Therefore, they have prepared comprehensive programs for the widespread use of cogeneration.

An analysis of the possible implementation of this advanced technology in a factory has to:

- Identify all the opportunities for useful heating and cooling demand, suitable for the application of cogeneration, as well as the availability of fuels and other energy resources to be utilized in cogeneration.
- Recognize the barriers which may prevent the implementation of cogeneration in the factory. The most common barriers are related to prices and costs of and access to fuels, barriers in relation to grid system issues and barriers in relation to administrative procedures.

6.6 Factors Influencing Performance

A cogeneration plant has a very apparent technical advantage in a large number of cases in industrial practice. However, this is not enough to decide immediately on its application without a substantial analysis of the economic indicators. These plants always have technical alternatives in the form of conventional boilers and electricity supply from the public grid. Invariably, the factory already has the required standard infrastructure for the regular energy supply of the process. A cogeneration plant as a performance improvement measure is an expensive investment and only lower operational costs can provide an advantage in relation to conventional plant. A somewhat more favorable situation arises in the case of building a new factory. Then, in the designing process, both options are considered at the same time. The economics of the cogeneration plant's operations are influenced primarily by the price of fuels. For the most part, this is the price of fuel which is used in the prime mover and the price of electricity. As the structure for calculating the price of electrical energy is very complex, it is necessary to examine the current tariff system carefully (active energy price, power price, back-up electricity, top-up electricity, etc.).

6.6.1 Load of Cogeneration Plant

Efficiency drops if the cogeneration plant operates with a load lower than the nominal one. Such an operation always occurs in a real situation and has to be analyzed carefully. Gas turbine and gas engine efficiency versus load is presented in Figure 6.6. Changes of efficiency are represented by changes of loads for three types of gas turbine and one gas engine. For loads of less than 60 % of nominal power, the efficiency of both turbines and engine drops significantly.

The prime mover used for electricity generation must operate at a constant speed. To maintain constant speed at varying loads, the steam supply to a steam turbine is throttled using a valve or series of valves in the supply. In common with gas turbines with constant speed, expansion efficiency decreases when power output is reduced. Depending upon the design of the steam turbine and the inlet and outlet steam conditions, the reduction in efficiency with a falling load may be modest or significant, but definitely below that of the gas engine.

The other factor having an influence on energy efficiency is load duration. It is connected directly with the previous factor, i.e., with the prime mover efficiency change versus load. It is very important to consider in detail the duration of certain loads even at an early stage in design when the nominal capacity of the plant is defined and in the case of several units when their individual power is determined.

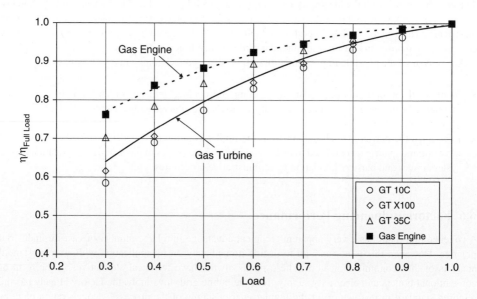

Figure 6.6   Gas Engine and Gas Turbine Efficiency versus Load

6.6.2 Load Duration Curve

Load duration curve represents the dependence of power (electrical and heat) on time for a certain factory. Figure 6.7 shows the principal appearance of this curve for heat power in one factory. Why is this diagram important? Let us assume that we have chosen a cogeneration plant which has output heat power of 1000 kW. In this case, the plant will operate annually for as much as 6000 hours under full load, which also means with the highest efficiency. If, however, we install a larger cogeneration plant (of, for example, 1500 kW), in addition to significantly increased investments, the plant will operate under full capacity for only 1500 hours per year. For the remaining time it will operate with lower efficiency and thus with increased operational costs. It is true that a smaller plant will operate partially with reduced load but the time of such work is considerably shorter than in the case of a large plant.

It is, of course, assumed in both cases that insufficient heat energy will be provided from the conventional plant.

Let us now explain in more detail the *load duration curve* using a real life example. Energy consumption (electrical, heat and refrigeration for air conditioning) in different months is presented in Figure 6.8. It is possible to distinguish the typical seasonal variations of heat and refrigeration energy.

Average electrical demand versus time of appearance (load duration curve) is presented in Figure 6.9. As the project assumes the replacement of classical (vapor-compression) chillers for chilled water generation by an absorption one, the average demand will be lover during the summer period. That is the reason why the new load duration curve is below the current one. Of course, the new one will be relevant for further calculations.

Figure 6.10 illustrates the heat energy versus time of appearance. The main idea of the project is to use cogeneration heat for producing hot water, which will be used for heating during winter and for producing chilled water by absorption chiller during summer. The upper curve in Figure 6.10 shows the cumulative heat energy consumption for both heating and cooling. The heat energy for running the absorption chiller is calculated by using the estimated cooling load divided by 0.7 (COP of single-effect absorption chiller).

In Figures 6.9 and 6.10, it can be seen that average electrical power over 1200 kW appears for approximately 5000 hours per year. At the same time, the necessary heat energy for more than 5000 hours per

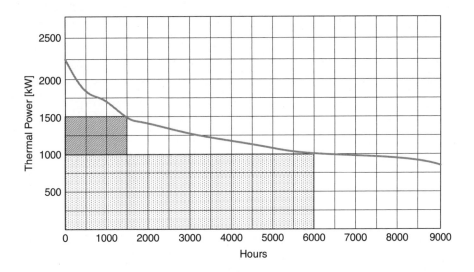

Figure 6.7   Example of Load Duration Curves

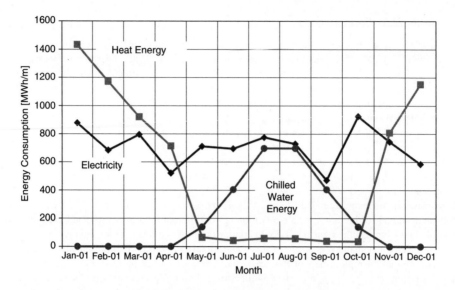

Figure 6.8   Energy Consumption versus Time

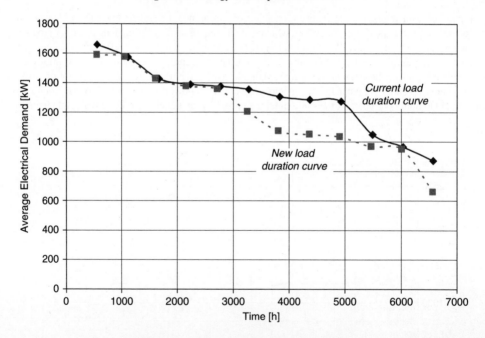

Figure 6.9   Load Duration Curve – Electrical Power

year is over 600 MWh a month, or the heating power is approximately 1100 kW. The heat to power ratio is approximately 0.91. This figure is one of those that are decisive for the selection of the prime mover.

Taking into consideration the heat-to-power ratio (Table 6.1), the required size of the prime mover (Fig. 6.4) and the advantages and disadvantages of different prime movers (Table 6.2), the selection of a gas engine is imposed as the best solution.

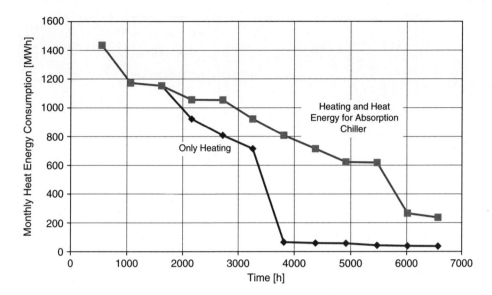

Figure 6.10   Load Duration Curve – Heat Energy

The shape of both valid curves for calculation are very similar for more than 5000 hours which is very beneficial for the operation of a future cogeneration system and which assumes that both electricity and heat energy will be used simultaneously.

6.7 Economic Aspects of Cogeneration as a Performance Improvement Measure

If the technical assessment shows that several alternative cogeneration schemes may be acceptable (as is frequently the case), an economic assessment needs to be prepared for each one before the final choice is made. During this evaluation, there will be areas of interface with technical assessments which may themselves be modified as a result. Some technical and economical reiteration is to be expected in order to develop the optimum package.

6.7.1 Capital Cost

This is the expenditure required for the establishment of an operational cogeneration on the site and is comprised of the following:

- cost of cogeneration unit(s) and associated plant, installation, testing and commissioning;
- connection charges including reinforcement of the local or national electrical network;
- cost of associated mechanical and electrical services, installation and commissioning;
- civil engineering works (new building or reconstruction of existing buildings) and foundations;
- training of staff, sets of spare parts and special tools needed;
- engineering design;
- compliance with all relevant regulations (environmental requirements, fire protection, etc.).

Prices are obtained from the appropriate manufacturers, suppliers, contractors and engineering consultants and are added together to reach the 'first cut' capital cost.

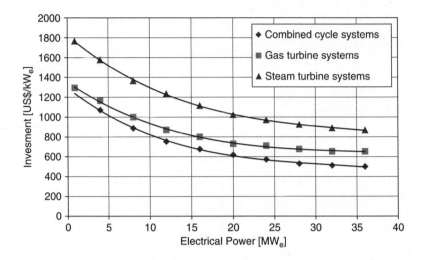

Figure 6.11   Investment Costs for Different Types of Cogeneration Plants versus Electrical Power

Capital costs typically vary from 600 US/kW_e$ for large schemes to more than 2000 US/kW_e$ for the very small ones and also depend on the choice of cogeneration plant and auxiliaries.

For gas turbine and large reciprocating engine cogeneration plants, the prime mover and generator package and associated equipment frequently represent 40 %–60 % of total cost. The heat recovery equipment and associated equipment can account for a further 15 %–30 % of the cost. Electrical switchgear and protection equipment amount to 5 %–15 %. The remaining percentage up to 100 % is for design costs, project management and installation (piping, civil engineering works, etc.).

If heat energy is further used for chilled water production in an absorption chiller, additional capital cost has to be included.

The specific investment per kW of electrical output is presented in Figure 6.11. It is obvious that steam turbine systems are the most expensive and the combined systems are the cheapest for the same electrical power. But, as we have already mentioned, there are many other criteria which determine which possible scheme is the best for a particular company.

Specific investment in a gas engine based cogeneration plant is illustrated in Figure 6.12. It has to be noted that the investment for the whole plant can be and is much higher than for the basic elements of the plant itself.

6.7.2 Operating and Maintenance Costs

The operating costs of cogeneration plant are as follows:

- fuel for the prime mover and for supplementary and auxiliary firing, if applicable;
- labor for operating and servicing the plant;
- maintenance materials and labor;
- consumables (lubricating oil, water treatment chemicals, etc.);
- back up electricity[3] prices and top up[4] and export electricity prices.

---

[3] 'Back-up electricity' means the electricity supplied through the electricity grid whenever the cogeneration process is disrupted, including maintenance periods, or is out of order.

[4] 'Top-up electricity' means the electricity supplied through the electricity grid in cases where the electricity demand is greater than the electrical output of the cogeneration process.

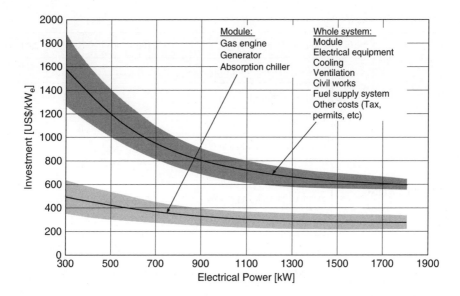

Figure 6.12 Specific Investment in Cogeneration Plant with Gas Engines versus Electrical Power

The typical operating and maintenance costs are as follows:

- reciprocating engines   0.0090–0.0180 $US/kWh;
- gas turbine engines   0.0055–0.0125 $US/kWh;
- steam turbines   ≈ 0.0035 $US/kWh.

In many cases, the heat cost will be the same or somewhat higher than before and the critical cost in cogeneration economics is the cost per generated electricity in kWh.

6.7.3 Overall Economics of Cogeneration Projects

Under favorable circumstances, cogeneration projects can result in simple pay-back periods of three to five years. The economics of cogeneration projects are more sensitive to changes in electricity price than to changes in fuel prices. Because of that, a sensitivity analysis should be a part of the feasibility study irrespective of the economic analysis employed.

In the case of cogeneration projects, the factors favoring a short pay-back period are as follows:

- low investment cost;
- low fuel price;
- high electricity price;
- minimum cogeneration fuel price in comparison to boiler fuel price;
- high operating hours;
- high overall efficiency.

The draft parameters of prime movers are presented in Table 6.5. At first glance, it can be seen that the capacity of reciprocating prime movers is much lower than for turbines. There are bigger reciprocating engines and smaller turbines than are mentioned in Table 6.4, but they are still not in common use.

Table 6.5   Draft Parameters of Prime Movers

	Capacity Range [kWe]	100 – 500	500 – 2,000
RECIPROCATING IC ENGINES	**Electrical Generation Efficiency**		
	% of NCV of Fuel	24–28	2–38+
	Heat Rate, kW_{Fuel}/kW_e	4.10–3.50	3.50–2.60
	Recoverable Useful Heat		
	Hot Water (90/70 °C), kW_h/kW_e	1.17–1.45	1.17–1.45
	Steam (1 barg), kg/h per kW_e	1.8–2.3	1.8–2.3
	Installed Cost, $US/$kW_e$		
	(with Heat Recovery)	1800–1400	1400–1000
	O & M Costs, $US/$kWh_e$	0.015–0.012	0.012–0.010
	NO_x Emission Levels, g/kWh_e		
	Rich Burn with 3-way catalyst	$\approx$ 13.5–18.0	$\approx$ 13.5–18.0
	Lean Burn with SCR treatment	$\approx$ 0.9–2.8	$\approx$ 0.9–2.8
GAS TURBINES	**Capacity Range, [kW_e]**	1000–10 000	10 000–50 000
	Electrical Generation Efficiency		
	% of NCV of Fuel	24–28	31–36
	Heat Rate, kW_{Fuel}/kW_e	4.10–3.50	3.20–2.80
	Recoverable Useful Heat		
	Hot Water (90/70 °C), kW_h/kW_e	1.45–1.75	1.45–1.75
	Steam (1 barg), kg/h per kW_e	2.3–2.8	2.3–2.8
	Installed Cost, $US/$kW_e$		
	(with Heat Recovery)	1500–1000	1000–800
	O & M Costs, US/kWh_e	0.015– 0.012	0.012– 0.010
	NO_x Emission Levels, ppm		
	With Dry Low NO_x Burner	< 25	< 25
	With SCR	< 10	< 10

6.8 Performance Assessment

As opposed to other energy systems when we estimate the plant's operation in relation to designed parameters or in relation to the best achieved parameters of a given plant over a certain period of time, with cogeneration plants the situation is completely different. Almost invariably a cogeneration plant has been built some time after the completion of a factory. Even new factories have, besides the cogeneration plant, supplementary plants which provide for process implementation at a time when the cogeneration plant is not operational. This certainly increases the reliability of the process but also significantly raises overall investment.

Therefore, it is necessary to estimate the operation of the cogeneration plant not only on the basis of its designed parameters but also on the basis of comparison with the plant which can replace it at a time when it does not operate for of some reason. Then, economic factors become much more important than technical ones. Unit prices of power supplying the process such as: steam ($US/t), electricity ($US/kWh_e), hot water ($US/kg), chilled water ($US/m^3), etc. should be presented depending primarily on the current prices of electrical energy and used fuels. These prices must be calculated for both existing systems. Only then it will be possible to make a decision concerning the manner in which the process will be managed. In doing so, it is important to estimate local and global emissions if there is an opportunity for their valuation in terms of money.

From the technical standpoint, it is sufficient to use the performance indicators Overall Efficiency (Eq. 6.1) and Primary Energy Savings (PES) (Eq. 6.6).

6.9 Performance Monitoring and Improvement

At the prime mover itself, there is nothing much to change in relation to what has been delivered by the manufacturer or in relation to what has been bought. All interventions are risk burdened unless initiated by the manufacturer and are very expensive. The same applies for an absorption chiller if it is an integral part of the cogeneration plant scheme. Improvements within the factory can be made only on the side of waste heat by reducing the temperature of gases and increasing the efficiency of utilization. These improvements are achieved principally in the same manner as explained for boilers (Part II Chapter 2).

As will be explained somewhat later in a concrete example, it is very important to monitor the operating philosophy of the cogeneration plant continually and find out whether it is really possible to accomplish significant savings there. The present control system of cogeneration plant enables process management that is very flexible and the preparation of specific software for optimum process guidance.

Modern control systems enable the monitoring of all the relevant parameters of the plant's operations. However, as the plant's products are intended for the process, it is very useful and technically easily accomplished to monitor performance indicators permanently, as mentioned in the previous paragraph.

6.10 Environmental Impacts

On the basis of the plant's energy balance, the environmental impact of a considered cogeneration plant can be estimated. This analysis should be executed on both local and global levels.

Table 6.6 presents the average values of emissions for various cogeneration plants. The degree of usefulness but also the quality of fuel may vary, thus the consequences of presented emissions may vary, also.

Table 6.7 shows the same emissions caused by the production of electrical energy in large central power plants. The existence of these plants in national economies is very diverse and it is difficult to estimate emissions at a global (national) level without extensive analysis. Certainly, losses of electrical energy caused by transmission should be taken into consideration as they are around 10 %. In cogeneration plants these losses are practically eliminated.

Table 6.6 Typical Values of Emissions from Cogeneration Systems. Reproduced from: Cogen 3: Proven, Clean & Efficient Biomass, Coal and Gas Cogeneration, Cal Bro Intelligent Solution, 2004.

System	Fuel	Electrical Efficiency [%]	Specific Emissions [g/kWh_e]					
			CO_2	CO	NO_x	HC	SO_x	Particulates
Diesel	Diesel 0.2% S	35	738.15	4.08	15.56[2]	0.46	0.91	0.32
	Dual[1]		593.35	3.81	11.30[3]	3.95	0.09	0.04
Gas Engine	Natural Gas	35	577.26	2.80	1.90	1.00	≈ 0	≈ 0
Gas Turbine	Natural Gas	25	808.16	0.13	2.14	0.10	≈ 0	0.07
	Diesel 0.2 % S		1033.41	0.05	4.35	0.10	0.91	0.18
Gas Turbine (low NO_x)	Natural Gas	35	577.26	0.30	0.50	0.05	≈ 0	0.05
Steam Turbine (new)	Coal	25	1406.40	0.26	4.53	0.07	7.75	0.65
	Heavy Fuel Oil		1100.00	≈ 0	1.94	0.07	5.18	0.65
	Natural Gas		808.16	≈ 0	1.29	0.26	0.46	0.07

(1) 90 % of energy supplied by natural gas and 10 % by diesel oil.
(2) Modern design engines emit 11–12 g NO_x/kWh_e
(3) Modern design engines emit 7–8 g NO_x/kWh_e

Table 6.7 Typical Values of Emissions from Central Power Plants

System	Fuel	Efficiency [%]	Specific Emissions [g/kWh_e]					
			CO_2	CO	NO_x	HC	SO_x	Particulates
Steam Turbine (old)	Coal 3 % S	34	1034.12	0.18	3.13	0.05	19.87	1.41
	Heavy Fuel Oil 1 % S	31	887.06	0.18	3.18	0.05	4.76	0.23
	Natural Gas	31	651.74	0.09	3.04	0.18	≈ 0	0.05
Steam Turbine (new)	Coal 3 % S	31[1]	1134.20	0.18	2.50	0.05	6.00	0.14
	Heavy Fuel Oil 1 % S	31	887.06	0.18	1.36	0.05	3.63	0.14
Gas Turbine	Diesel 0.2 % S	34	759.86	0.55	2.40	0.18	0.14	0.18
	Natural Gas		594.24	0.55	1.95	≈ 0	≈ 0	0.05
Gas Turbine (low NO_x)	Natural Gas	38	531.68	0.30	0.50	≈ 0	≈ 0	0.04

(1) Lower efficiencies of new steam turbine systems are due to NO_x and SO_2 abatement equipment.

Table 6.8 Typical Values of Emissions from Water and Steam Boilers (An efficiency of 80 % has been used for all boilers)

System	Fuel	Specific Emissions [g/kWh$_e$]					
		CO_2	CO	NO$_x$	HC	SO$_x$	Particulates
Hot Water Boiler	Natural Gas	252.55	0.03	0.19	0.02	≈ 0	0.02
	Diesel 0.2 % S	322.94	0.06	0.25	0.02	0.37	0.03
Steam Boiler	Coal 2 % S	439.50	0.08	1.36	0.02	2.32	0.20
	Heavy Fuel Oil 1 % S	343.73	0.06	0.57	0.02	1.55	0.20
	Natural Gas	252.55	0.03	0.39	≈ 0	≈ 0	0.02
Industrial Steam Boiler	Coal 2 % S	439.50	0.16	1.12	0.08	5.65	0.98
	Heavy Fuel Oil 1 % S	343.73	0.06	0.78	0.02	2.03	0.30
	Natural Gas	252.55	0.03	0.33	≈ 0	≈ 0	0.03

Cogeneration plants undoubtedly involve the recuperation of waste heat of exhaust gases and its utilization for the production of steam, hot water or hot gases which are used in the process. Without the cogeneration plant, this quantity of heat will have to be produced in the plant by using again some of the available fuels. These additional emissions of some boilers are given in Table 6.8. Taking into consideration this additional combustion, it can be said that cogeneration can but does not necessarily have to increase emissions at a local level, but it definitely reduces emissions at a global level.

6.11 Case Study: Drying Kiln (Gas Turbine Operation Philosophy Improvement)

This real life example refers to a factory for the production of gypsum boards (35 664 000 m^2 per year). The factory employs around 200 workers and it operates for 7488 hours per year. The overall annual expenses of the company for energy are 2.45 million $US.

The Detailed Energy Audit (DEA) itself will not be discussed here; however, some of the data will be used for analyzing one of the recognized energy conservation opportunities. This is the case of a drying KILN which is used for drying gypsum boards. This is the largest single Energy Cost Center (ECC) in the factory. Specifically, the DEA defined the system's boundaries for several ECCs which have been analyzed separately and which have been submitted to the surveying and controlling of energy performances.

Figure 6.13 presents the gas consumption (*without consumption for running the gas turbine*) and overall consumption of electrical energy depending on the monthly production in the factory. The gas turbine, which is mentioned here, is an integral part of the drying kiln and is used for improving the energy efficiency of the drying process. Namely, the drying kiln is driven by natural gas. This natural gas is used partly for driving the turbine and partly for the production of heat energy directly. Exhaust gases from the turbine are mixed directly with gases from independent combustion chambers. Gases prepared in this way are introduced into the drying kiln. The maximum turbine output power at the generator terminal is 7000 kW$_e$ (with inlet air temperature of $-40\,°C$ at sea level). However, ever since the factory commenced work, the gas turbine has been operating with considerably lower power than possible under the given circumstances. The normal operating philosophy of this cogeneration plant has been to cover factory electrical energy consumption by about 90 %–95 %.

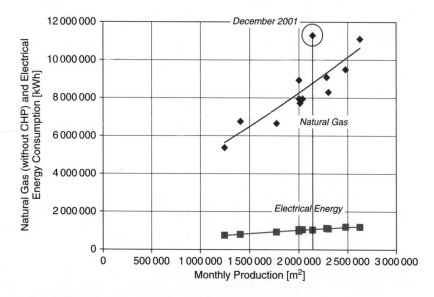

Figure 6.13   Natural Gas and Electrical Energy Consumption versus Production

Let us now go to Figure 6.13. It can easily be noticed that one point, which indicates natural gas consumption in the process, considerably deviates from the other eleven points. In that month and each December in a year, the general overhaul of the turbine is carried out. It is obvious that in that period a considerable increase of gas consumption occurs in the drying kiln, and that there is no substantial increase of electricity consumption. During the turbine's overhaul, the required electricity is taken over completely from the local electric grid and the process in the drying kiln is carried out undisturbed by the utilization of two existing combustion chambers. This and some other facts have indicated the need to optimize the operating philosophy with the aim of reducing electricity costs and performing a detailed check of the current manner of utilization of the cogeneration plant, as well as examining the opportunities and financial conditions for the permanent and extensive production of electricity for the local distribution company. During the DEA (Detail Energy Audit) and related to this ECC (Energy Cost Center), the above-mentioned has been the primary plan and this is what the example is concerned with.

6.11.1 Facts

Here are some basic data related to the analyzed ECC:

- Total factory energy cost per year (TEC):   2 445 738 $US/a
- Number of operating hours (NOH):   7448 h/a
- Analyzing period: Apr 1 – Mar 31 next year
- Total 12 months electricity consumption (EC):   12 292 700 kWh/a
  (Generated by cogeneration plant (ECC):   *10 378 700 kWh/a*
  (Purchased from local grid (ECG):   *1 914 000 kWh/a*
  (Purchased from local grid during shut-down of gas turbine)   *949 520 kWh/a*
- Average purchased electricity cost: 0.0825 $US/kWh
- Average unit price of electricity produced by cogeneration
  plant (only cost of natural gas is calculated)   0.0781 $US/kWh
- Average price of electricity paid by local electricity authority
  to the company   0.0363 $US/kWh

- Total annual NG consumption (NGC):   161 114 988 kWh/a
- Annual NG consumption for cogeneration only
  (NGCG):   57 080 000 kWh/a
- Average unit NG cost (NGP):   0.0142 $US/kWh
- Annual average ambient temperature (T_a):   30 °C

6.11.2 Description of Kiln

The simplified scheme of the drying kiln is presented in Figure 6.14. As we are going to analyze only the energy aspects of the drying process, we will describe only that part. There are two consumers of natural gas: (1) gas turbine; and (2) two additional combustion chambers for hot gas generation and direct supply of the kiln. The proper temperatures of the drying fluids are controlled by mixing the flue gases from the turbine exhaust, the flue gases from the combustion chambers and fresh air.

In the scheme, the circles designate measuring points which have been used for measuring the parameters of the whole plant within the detailed energy audit. The majority of these measuring points have also been used in the regular monitoring and governance of the process. Measurements have been mostly executed with the existing (built in) measuring equipment. It is important to point out that daily log sheets have been used and kept in an orderly manner.

The main design data for the relevant gas turbine are:

Power
(ISO: 15 °C, sea level No inlet or exhaust losses;
  Relative humidity 60 %; NCV = 31.5 to 43.3 MJ/nm³)   5200 kW$_e$
  Fuel Rate   17 200 kW$_{NG}$
  Exhaust Flow   78 800 kg/h
  Temperature (turbine inlet):   1010 °C
  Temperature (turbine outlet)   485 °C
  Electrical Efficiency   30.2 %
  Compressor Discharge Pressure   10.8 barg
  Heat-to-Power Ratio
  (Exhaust temperature 156.3 °C)   1.45

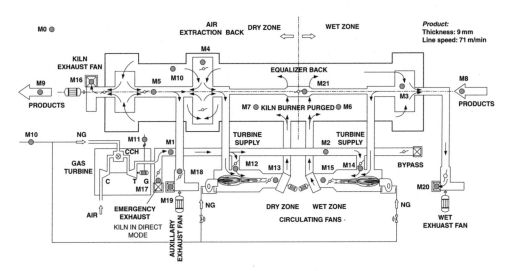

Figure 6.14 Drying Kiln and Waste Heat Recovery Unit (grey circles denote measuring points. Measurements are given in Table 6.8)

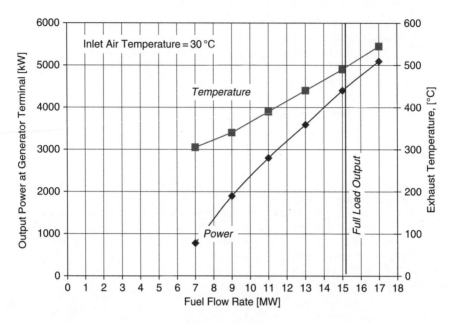

Figure 6.15   Output Power and Exhaust Temperature versus Fuel Flow Rate

Figure 6.15 shows the dependence between the output power at generator terminal and exhaust temperature and the fuel flow rate of natural gas for the average annual ambient temperature of 30 °C. This diagram and other technical data for the given turbine have been used for further calculations.

6.11.3 Objectives of Analysis

The preliminary analysis of the gathered information, the processed data referring to the entire factory (overall production, specific consumptions, etc.) and the preliminary data for the defined Energy Cost Center – Drying Kiln have enabled the identification of objectives and the development of a working plan within the detailed energy audit.

The relation between the cost of electricity and gas is as much as 5.81 (0.0825/0.0142). Such a relation has existed for several years. This is extremely favorable for the maximum production of electrical energy. However, in negotiations with the local electricity distribution company, a very low price for buying up the produced electrical energy has been offered. This price has been only 0.0363 $US/kWh. However, there is an advantage in that the delivery of excess electrical energy to the local grid has not been conditioned by delivery periods (peak on, partial peak or peak off). The medium price which the factory pays to the local electricity distribution company was during the analyzed period equal to 0.0825 $US/kWh (this unit price is calculated according to monthly invoices and consumed active energy). Also, the price of produced electrical energy in the cogeneration plant was calculated and amounted to 0.0781 $US/kWh (the calculation was performed only on the basis of gas consumed for its production). The review of the daily and monthly log sheets of the plant's operations resulted in the conclusion that the average power of the turbine is chiefly around 1500 kW.

As shown in Figure 6.15, the existing turbine can achieve a much higher power at a working temperature of intake air of around 30 °C.

The above-mentioned facts have induced us to propose to the factory's management that it abandon the previous operation philosophy of meeting only its own consumption needs and to take up an operation philosophy which entails the maximum production of electrical energy and the sale of surplus electricity to the local distribution company. Considering the consumption of heat energy in the drying kiln, there can been no doubt that all available waste heat from the turbine will be absorbed in it.

The inspection of fans, exhaust gases ducts, dumpers and controlling equipment has indicated that such a change of operation philosophy can be implemented without any need for modifications.

We will deal now with the measurement results and technical estimations which have been used to assess the expected effects of the suggested energy conservation measures.

6.11.4 Measurement, Technical Calculation and Analysis

The results of measurement are presented in Table 6.9. Some measured values have been recalculated in previously used units. On the basis of measured values, it is possible to calculate the degree of the turbine's efficiency. It is equal to 26.0 % (2112/8118). The turbine's exhaust gas temperature was 348.3 °C during

Table 6.9   Measurements Report: Product Dimensions: 9 mm SC x 1200 mm: Line Speed: 71 m/min, October 14th

Point	Description	Value			
		Name	Symbol	Unit	Result
M0	Ambient	Temperature	t_o	[°C]	32.7
		Relative humidity	R.H.$_o$	[%]	57.9
M1	Turbine supply to kiln	Temperature	t_1	[°C]	348.3
M2	Turbine supply to wet zone burner	Pressure	p_2	[WC]	−0.82
M3	Wet zone – returned gas	Temperature	t_3	[°C]	177.1
		Pressure	p_3	[WC]	0.02
M4	Air extraction	Temperature	t_4	[°C]	121.5
		Dumper position	o/p_4	[%]	59.1
M5	Dry zone – waste gas	Temperature	t_5	[°C]	95.4
		Pressure	p_5	[WC]	0.05
		Dumper position	o/p_5	[%]	39.8
M6	Wet zone	Temperature	t_6	[°C]	237.3
M7	Dry zone	Temperature	t_7	[°C]	147.8
M8	Product – inlet	Moisture	w_8	[%]	40.82
M9	Product – outlet	Moisture	w_9	[%]	13.04
M10	NG consumption	Flow rate	m_{10}	[MW]	8.12

(*continued overleaf*)

Table 6.9 (*continued*)

| Point | Description | Value | | | |
		Name	Symbol	Unit	Result
M11	Electricity production	Energy	E_{11}	[kWh]	**2112**
M12	Turbine supply to dry zone combustion	Dumper position	o/p_{12}	[%]	**96**
M13	Combustion chamber supply to dry zone	Dumper position	o/p_{13}	[%]	**0**
M14	Turbine supply to wet zone combustion	Dumper position	o/p_{14}	[%]	**100**
M15	Combustion chamber supply to wet zone	Dumper position	o/p_{15}	[%]	**54.3**
M16	Kiln exhaust fan	Temperature	t_{16}	[°C]	**103**
		Dumper position	o/p_{16}	[%]	**47.5**
		Flow rate	m_{16}	[m³/h]	**56 520**
M17	Emergency exhaust	Temperature	t_{17}	[°C]	−
		Dumper position	o/p_{17}	[%]	**0**
		Flow rate	m_{17}	[m³/h]	**0**
M18	Auxiliary exhaust fan	Temperature	t_{18}	[°C]	**87**
		Dumper position	o/p_{18}	[%]	−
		Flow rate	m_{18}	[m³/h]	**33 840**
M19	Wet exhaust fan	Temperature	t_{19}	[°C]	**155**
		Dumper position	o/p_{19}	[%]	**34.6**
		Flow rate	m_{19}	[m³/h]	**43 920**
M20	Equalizer back	Dumper position	o/p_{18}	[%]	**100**

the measurement. By comparing other measured values with those established for the technical procedure for making the of gypsum boards, we have concluded that they are all within the allowed ranges.

The cogeneration plant worked during the entire analyzed period (except during the annual maintenance period for the gas turbine in December (it lasted 21 days or 504 hours)). The average generating power (P_e) for NOH = 7,448 (number of operating hours per year) is:

$$P_e = \frac{E_e}{NOH} = \frac{10,378,000}{7,448 - 504} = 1,495 \, kW \tag{6.8}$$

This value has been obtained as an arithmetic mean of monthly electrical power.

The average consumption of natural gas for the generation of average electrical power of 1495 kW is equal to $8.22 \times (7448 - 504) = 57,080 \, MWh$ (Fig. 6.15).

The useful heat power, which is used in the kiln, at average output temperature of gases from the kiln of 117 °C, is equal to:

$$Q_{Cogen} = m_{eg} \cdot \bar{c}_{p,eg} \cdot \left(T_{eg} - \overline{T}_{kiln}\right) = \frac{72,000}{3600} \cdot 1.052 \cdot (325 - 117) = 4,376 \, kJ/s \text{ (or kW)} \qquad (6.9)$$

The average output gases temperature from the turbine is 325 °C for average annual power of the turbine of 1495 kW (Fig. 6.15). The exhaust gases mass flow rate is 72,000 kg/h and is practically constant for all operational loads.

The useful heat energy of the turbine's exhaust gases is as follows:

$$\dot{Q}_{Cogen} = 4376 \cdot (7448 - 504) = 30,386,944 \, kWh \, (30,387 \, MWh) \qquad (6.10)$$

The overall efficiency of the cogeneration plant can be found as follows:

$$\eta = \frac{1,495 + 4,376}{8,220} \cdot 100 = 71.4\% \qquad (6.11)$$

For driving the cogeneration plant during the analyzed period of twelve months, the following natural gas energy was used:

$$\Sigma NG_{Cogen} = 8,220 \cdot (7,448 - 504) = 57,079,680 \, kWh \, (57,080 \, MWh) \qquad (6.12)$$

This, at prices valid at that time, was worth 810,531 \$US.

During that period, 10 378 MWh of electrical energy was produced, as well as 30 387 MWh of useful heat energy.

The turbine's waste energy was used together with the hot gases made by two combustion chambers (see Fig. 6.14). The total average annual natural gas power of both the turbine and combustion chambers was 14.7 MW. This value is calculated as mean monthly consumption which is measured regularly. The total annual natural gas consumption for the drying kiln was as follows:

$$\Sigma NG_{KILN} = 14,700 \cdot (7,448 - 504) = 102,076,800 \, kWh \, (102,077 \, MWh) \qquad (6.13)$$

This was worth 1 449 491 \$US.

It is now possible to calculate the quantity of consumed gas only for the combustion chambers. It is equal to:

$$\Sigma NG_{CC} = \Sigma NG_{KILN} - \Sigma NG_{Cogen} = 102,076,800 - 57,079,680 = 44,997,120 \, kWh \\ (44,997 \, MWh) \qquad (6.14)$$

The average power of the combustion chamber burners is:

$$NG_{CC} = \frac{44,997,120}{7448 - 504} = 6,480 \, kW \qquad (6.15)$$

If the efficiency of combustion in chambers is equal to 0.90, then the useful heat introduced into the kiln, but only from the combustion chambers, equals $0.9 \times 44\,997\,120 = 40\,497\,408 \, kWh$. It is now possible to estimate the total delivered heat energy into the drying kiln. It equals $40\,497\,408 + 30\,386\,944 = 70\,884\,352 \, kWh \, (70\,884 \, MWh)$. The participation of heat from the cogeneration plant is 42.3 %.

During the turbine's overhaul, the production process does not stop, however the consumption of gas in the combustion chambers is increased. The power of natural gas during this period can be estimated on the basis of the total required heat energy, which has just been calculated. Thus:

$$NG_{CC-SD} = \frac{70,884,352}{7448 - 504} \times \frac{1}{0.9} = 11,342 \, kW \qquad (6.16)$$

The increased consumption of gas for driving the kiln during the overhaul of the turbine is equal to:

$$\Sigma NG_{CC-SD} = (11,324 - 6,480) \times 504 = 2,441,376 \, kWh \qquad (6.17)$$

This increase can easily be seen in Figure 6.13 in the month of December when the overhaul was performed. However, it should be pointed out that the turbine's overhaul is a very complex operation and cannot be performed during the factory's short shut-downs.

6.11.5 Energy Performance Improvements

The suggested measures entail the following:

(a) The drive of the cogeneration plant with maximum capacity (with consequential increase of the natural gas consumption);
(b) complete supply of the factory with electrical energy;
(c) the sale of surplus electrical energy to the local electricity distribution company at the previously contracted price without preconditions regarding time and quantity of deliveries.

This measure does not require additional investment as has already been mentioned because the existing control system enables the transfer to the new operation philosophy without any limitations.

The turbine's operational parameters at the maximum load can be read from Figure 6.15. The flow of the turbine's exhaust gases practically does not change with varied loads. The new operational parameters of the turbine at the full load are as follows:

Generated output power: $P_e = 4485\,kW$
NG consumption: $M_{NG} = 15.2\,MW$
Exhaust gas temperature: $T_{eg} = 493\,°C$
Exhaust gas flow rate: $m_{eg} = 72\,000\,kg/h$

The available waste heat energy from the cogeneration plant for the kiln (for average temperature of the exhaust gases from the kiln $\overline{T}_{kiln} = 117\,°C$ is as follows (this is the value that has been measured and confirmed by reviewing the daily log sheets) :

$$Q_{Cogen,new} = m_{eg} \cdot \bar{c}_{p,eg} \cdot (T_{eg} - \overline{T}_{kiln}) = \frac{72,000}{3600} \cdot 1.052 \cdot (493 - 117) = 7,911\,kJ/s\,(kW) \qquad (6.18)$$

As opposed to the previous case, the temperature of the output gases from the turbine is now 493 °C. The annual useful heat of the cogeneration plant is:

$$\dot{Q}_{Cogen,new} = 7,911 \cdot (7448 - 504) = 55,498,504\,kWh\,(55,499\,MWh) \qquad (6.19)$$

The power of the natural gas required for producing 5485 kW of electrical power and useful heat power of 7911 kW is equal to 15.22 MW (Fig. 6.15).
The efficiency of this cogeneration process is:

$$\eta_{new} = \frac{4,485 + 7,911}{15,220} \cdot 100 = 81.4\,\% \qquad (6.20)$$

It can be noticed that the efficiency of the cogeneration plant has been improved significantly. It used to be 71.4 % and now expectations can be as high as 81.4 %.

The load duration curves for the current and proposed operating philosophy of this cogeneration plant are presented in Figure 6.16. It has already been pointed out and can easily be seen that the plant operated under a significantly lower load than the load which is possible under given circumstances. This caused the overall efficiency to be 10 % lower than the optimum.

The gas consumption for driving the turbine in the new regime is as follows:

$$\Sigma NG_{Cogen,new} = 15,200 \cdot (7448 - 504) = 105,548,800\,kWh\,(105,549\,MWh) \qquad (6.21)$$

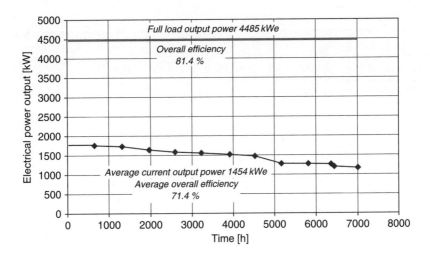

Figure 6.16   Load Duration Curve

The required heat energy for driving the drying kiln from the combustion chambers can be determined as we know the total required heat energy for running the process and the heat energy released under the turbine's full load. In this manner, the additional heat energy that is required can be obtained, which is equal to:

$$\dot{Q}_{CC,new} = 70, 884, 064 - 55, 489, 504 = 15, 394, 560 \, kWh \, (15, 395 \, MWh) \tag{6.22}$$

and the gas consumption for driving the combustion chambers is:

$$\Sigma NG_{CC,new} = \frac{15, 394, 560}{0.9} = 17, 105, 067 \, kWh \, (17, 105 \, MWh) \tag{6.23}$$

It is now possible to estimate the overall consumption of natural gas under the full load of the gas turbine which produced the surplus electrical energy. Thus:

$$\Sigma NG_{KILN,new} = 105, 548, 800 + 17, 105, 067 = 122, 653, 867 \, kWh \, (122, 654 \, MWh) \tag{6.24}$$

This quantity of gas is worth 1,741,685 $US (for the current regime it is 1,449,491 $US). The increased expense for purchasing gas is 292,194 $US. However, the produced electrical energy is equal to:

$$E_{Cogen,new} = 4, 485 \cdot (7, 448 - 504) = 31, 143, 840 \, kWh \, (31, 144 \, MWh) \tag{6.25}$$

The overall consumption of electrical energy in the whole factory is 12 292 700 kWh. Energy purchased from the local grid during the shut-down of the gas turbine is 949 520 kWh. This means that it is possible to deliver 31 143 840 − (12 292 700 − 949 520) = 19 800 660 kWh (19 801 MWh). At the price of 0.0363 $US/kWh the value of the electrical energy is 718 764 $US.

Finally, it is possible to calculate the ultimate benefit achieved by introducing this simple and no cost energy conservation measure. The following is obtained:

$$\text{Expected Benefit} = 718, 764 - 292, 194 = 426, 570 \, US\$ \, (0.43 \, million \, US\$) \tag{6.26}$$

This saving amounts to as much as 17.4 % of the overall expense for energy existing at that time.

Table 6.10 CO_2 Balance

	Current	New
Cogeneration Plant	$10\,378\,700 \times 808.16 = 8388\,t/a$	$31\,143\,840 \times 808.16 = 25\,169\,t/a$
Combustion Chamber	$44\,997\,120 \times 252.55 = 11\,364\,t/a$	$15\,394\,560 \times 252.55 = 3888\,t/a$
Electrical Energy Generated somewhere else	$31143\,840 - 10\,378\,700 = 20\,765\,140$ Coal: $20\,765\,140 \times 1134.20 = 23\,552\,t/a$ HFO: $20\,765\,140 \times 887.06 = 18\,420\,t/a$ NG: $20\,765\,140 \times 594.24 = 12\,335\,t/a$	
Total:	**Coal: 43 303 t/a** **HFO: 38 172 t/a** **NG: 32 087 t/a**	**29 057 t/a**

6.11.6 Environmental Impact

As it has already been mentioned, it is very difficult to estimate the environmental impact of any energy conservation measure as precise standards do not exist. One of the most serious issues involves the lack of reliable data concerning emissions of large plants for the production of electrical energy which are usually part of national energy systems.

In this given case the consumption of natural gas has been increased from the current 102 076 800 kWh to 122 653 867 kWh (20.2 %), but the production of electrical energy in the same plant has been increased by almost 200 % (from 10 378 700 to 31 143 840 kWh), and the production of heat energy has been increased by 82.6 % (from 30 386 944 to 55 498 504 kWh). At the same time, the consumption of natural gas in combustion chambers has been reduced by 62.0 % (from 44 997 120 to 15 394 560 kWh).

By using data concerning emissions of certain plants (Tables 6.6, 6.7, and 6.8), it is possible to establish the balance of CO_2 emissions for the observed plants (Table 6.10). We have not made estimates of the real emissions of certain exhaust gases as this requires the proper chemical composition of fuels (natural gas) and the measured data for exhaust gases. We have used the estimates given in the above-mentioned tables. The increased production of electrical energy under the new operation philosophy has been expressed by the production of electrical energy in central power plants fired by coal, heavy fuel oil (HFO) and natural gas (NG). It is obvious that the results are different in these three cases. However, it is evident that there is a significantly reduced emission of CO_2 in the case of the suggested solution. Valuation of this positive effect varies in different countries, but it is possible to say that presently a generally accepted evaluation of this effect does not exist. It is the same is with the emission of other gases and particles.

The general conclusion is that by applying these measures, it is possible to achieve a significant reduction in harmful emissions.

6.12 Bibliography

A Guide to Cogeneration, The European Association for the Promotion of Cogeneration, Brussels, http://www.cogen.org.

CHP Resource Guide (2003) Oak Ridge National laboratory.

Cogen 3 (2004) Proven, Clean & Efficient Biomass, Coal and Gas Cogeneration, Cal Bro Intelligent Solution.

Combined Heat and Power Project Model, RETScreen International (Clean Energy Decision Support Center), Canada, http://www.retscreen.net.

Directive 2004/8/EC of the European Parliament and of the Council of 11 February 2004 on the promotion of cogeneration based on a useful heat and demand in the internal energy market and amending Directive 92/42/EEC, Official Journal of the European Union, L 52, 21.02.2004, 'Official Journal of the European Union'.

Gvozdenac, D., Menke, C., Vallikul, P. (2006) Potentials of Natural Gas Based Cogeneration in Thailand, 2nd Joint International Conference on: Sustainable Energy and the Environment (SEE 2006), 21–23 November, Bangkok, Thailand.

Gvozdenac, D., Marié, M., Kljajié, M. (2003) Industrial CHP, PSU-UNS International Conference 2003: Energy and the Environment, Prince Songkla University, Hat Yai, Thailand, 11-12 December.

Griffiths, R. T. (1995) Combined Heat and Power, Energy Publications.

Roadmapping of the Paths for the Introduction of Distributed Generation in Europe, http://www.dgfer.org.

The European Educational Tool on Cogeneration (2001) Second Edition, December (http://www.cogen.org/projects/educogen.htm).

Part III

Toolbox – Fundamentals for Analysis and Calculation of Energy and Environmental Performance

Part III – Toolbox represents a special asset of the book. It includes a set of software on the accompanying website and fundamental tools for the analysis of energy and environmental performance of industrial plants. These 'tools' are: data on material and fuel properties, analytical methods, definitions, procedures, questionnaires and software to support the practical application of the methods elaborated in the first two parts of the book. The software was developed and used during our consulting practice. It facilitates calculations and problem solving for specific solutions for energy and environmental performance improvement, financial evaluation and performance monitoring. It will help readers to understand the relevant topics and to assess their own industrial energy systems. The context on why, when, and how to use the tools and software is provided consistently throughout the book but, again, the knowledgeable reader should be able to use the tools directly without referring to the book. The Toolbox also contains definitions of terms or basic explanations of subjects that are dealt with in the textbook but which could divert the reader's attention if explained in the main text.

Toolboxes 1–12 contain explanations of physical processes and provide formulae for the computation of process parameters, conversion factors, properties of commonly used fluids in industry, etc. Toolbox 13 contains eleven Software Packages that were used to prepare concrete examples in Parts I and II of the Book, but can also be used by readers for their own problem solving.

This book and the related Toolboxes embody our philosophy and consulting experience in developing and applying energy and environmental management solutions for the performance improvement of industrial plants. As with our endeavors in training and implementing specific solutions, where we have also learned from the very people that we have trained or assisted, we welcome comments and suggestions, in this case from our readers.

The following material is available on the accompanying website www.wiley.com/go/morvay industrial.

1 Auditing of Energy and Environmental Management Practice
2 Energy Auditing in Industry
3 Basis of Economic and Financial Analysis
4 Energy Units, Conversions, Thermo-Physical Properties and Other Engineering Data
5 Fuels, Combustion and Environmental Impacts

6 Thermodynamic and Transport Properties of Moist Air
7 Applied Hydraulics
8 Pumps and Fans
9 Heat Transfer in a Thermal Apparatus
10 Industrial Insulation
11 Heat Exchangers Operating Point Determination
12 Cooling Towers
13 Set of Software for Performance Analysis of Industrial Energy Systems:

    Software 1:   ***Performance Target Setting***
    Software 2:   ***Performance Monitoring and Control***
    Software 3:   ***Economic and Financial Analysis (EFA)***
    Software 4:   ***Fuels, Combustion and Environmental Impacts***
    Software 5:   ***Thermodynamic Properties of Water and Steam***
    Software 6:   ***Thermodynamic Properties of Moist Air***
    Software 7:   ***Efficiency of Steam Boilers***
    Software 8:   ***Steam System Insulation***
    Software 9:   ***Heat Exchangers Operating Point Determination***
    Software 10: ***Cooling Towers Calculation***
    Software 11: ***Refrigeration Systems Analysis (Vapor-Compression Cycles)***

Index

Applied Industrial Energy and Environmental Management   Zoran K. Morvay and Dušan D. Gvozdenac
© 2008 John Wiley & Sons, Ltd